Manual of Nearctic Diptera

Willi Hennig, 1913–1976

Manual of Nearctic Diptera *Volume 3*

Coordinated by

J. F. McAlpine (Editor)
D. M. Wood

Biosystematics Research Centre (formerly Institute)
Ottawa, Ontario

Research Branch
Agriculture Canada

Monograph No. 32
1989

Canadian Cataloguing in Publication Data

Manual of Nearctic Diptera

(Monograph/Agriculture Canada; 27-28, 32)

Includes bibliographical references and index.
Cat. no. A54-3/27E (v. 1)
Cat. no. A54-3/28E (v. 2)
Cat. no. A54-3/32E (v. 3)
ISBN 0-660-10731-7 (v. 1)
ISBN 0-660-12125-5 (v. 2)
ISBN 0-660-12961-2 (v. 3)

1. Diptera—Handbooks, manuals, etc. 2. Insects—Handbooks, manuals, etc. I. McAlpine, J. F. II. Canada. Agriculture Canada. Research Branch. III. Series: Monograph (Canada. Agriculture Canada); 27-28, 32

QL534.M36 1981 595.77 C82-071621-9

Available in Canada through

Authorized Bookstore Agents
and other Bookstores

or by mail from

Canadian Government Publishing Centre
Supply and Services Canada
Hull, Quebec, Canada K1A 0S9

Catalogue No. A54-3/32E
ISBN 0-660-12961-2

Price subject to change without notice

Cover and index figure: Diagrammatic cladogram of the Diptera.
Frontispiece: Willi Hennig, 1913–1976. From photograph taken at Biosystematics Research Centre, Ottawa, November 1967

staff editors
Sharon M. Rudnitski and Jane T. Buckley
Research Program Service

AUTHORS

A. BORKENT, B.Sc., M.Sc., Ph.D.
Biosystematics Research Centre, Agriculture Canada, Ottawa, Ont. K1A 0C6, Canada

J. F. MCALPINE, B.S.A., M.Sc., Ph.D.
Biosystematics Research Centre, Agriculture Canada, Ottawa, Ont. K1A 0C6, Canada

D. M. WOOD, B.A., M.A., Ph.D.
Biosystematics Research Centre, Agriculture Canada, Ottawa, Ont. K1A 0C6, Canada

N. E. WOODLEY, B.Sc., Ph.D.
Systematic Entomology Laboratory, B.B. II, Agricultural Research Service, U.S. Department of Agriculture, c/o National Museum of Natural History, Washington, D.C., 20560, U.S.A.

CONTENTS

PREFACE

D. M. Wood

Volumes 1 and 2 of the *Manual of Nearctic Diptera* were concerned with the anatomy, terminology, and identification of all the known families and genera of two-winged flies from America north of Mexico. This third volume, dedicated to the late Willi Hennig (1913–1976) (for obituary see Schlee 1978), explains the rationale underlying the classification that was adopted in Volumes 1 and 2, using the methodology of phylogenetic systematics developed by Hennig (1950, 1965, 1966) and based largely on his most recent phylogenetic treatment of the order (Hennig 1973). The subject matter of Volume 3 was originally intended to appear as a single chapter on the phylogeny of Diptera, to be written by Hennig himself. However, his manuscript was only in draft at the time of his death and the editors were reluctant to make the required changes and additions without his permission. Since then, many new facts and ideas have accumulated, which have necessitated an entire volume devoted to this subject. A few of these ideas do not entirely correspond to the most recent opinions of Hennig, and some of them are not even in complete accordance with the classification that we, ourselves, utilized in Volumes 1 and 2. We make no apologies for these changes because classification is not static but is subject to revisions that result from new information or interpretation.

Volume 3 comprises three chapters, each to a considerable degree independent of the other two. In all three, however, cladistic methodology is used; monophyletic taxa are identified by their shared, derived character states, or evolutionary novelties (synapomorphies), and these derived states are recognized by outgroup comparison, i.e. by their absence in other taxa. Different synapomorphies are used to group two monophyletic taxa (sister groups) into larger monophyletic taxa of higher rank (Hennig 1966, Wiley 1981). Chapter 114 presents a phylogenetic interpretation of the infraorders in the suborder Nematocera as well as a discussion of the sister group of the Diptera. In Chapter 115, the groups of the suborder Brachycera formerly called "Orthorrhapha" (Tabanomorpha and Asilomorpha of Volume 1) are treated. The remainder of the Brachycera (the cyclorrhaphous Brachycera, formerly called "Cyclorrhapha"), comprising the Muscomorpha of Volumes 1 and 2, which includes those Diptera that pupate entirely enclosed by the physically and chemically modified, last larval cuticle, are dealt with in Chapter 116.

Dipterists are in general agreement that the order Diptera, the suborder Brachycera (Chs. 115, 116), and the Muscomorpha (Ch. 116) are each a demonstrably monophyletic group, i.e. a taxon containing all the descendants of a common ancestor recognized as such by shared derived traits or synapomorphies. The same is probably not true, however, of the Nematocera (Ch. 114) or of the taxa covered in Chapter 115. Each of these taxa is probably paraphyletic, i.e. it does not contain all the descendants of its common ancestor. The Nematocera is considered to be paraphyletic because the ancestor of the Brachycera is presumed to have, as its sister group, some part of the Nematocera, not the Nematocera as a whole. Unfortunately, general agreement, as to which part is that sister group, has not been reached, although this topic is discussed in Chapter 115. The sister group of the Muscomorpha is even more contentious; the authors of Chapters 115 and 116 present different hypotheses of this difficult problem. Both of these hypotheses differ from others advanced elsewhere (Griffiths 1972, Hackman and Väisänen 1982). These discrepancies arise in part from differing interpretations of sister-group relationships between the Muscomorpha and the Asiloidea and Empidoidea (the Asilomorpha of Vol. 1). Such is the nature of phylogenetic interpretation, which depends entirely on, and can be no better than, a correct interpretation of homology.

> *... there is no simple and absolutely dependable criterion for deciding whether corresponding characters in different species are based in synapomorphy. Rather it is a very complex process of conclusions by which, in each individual case, "synapomorphy" is shown to be the most probable assumption.* [Hennig 1966: 128]

The term Muscomorpha is also used in this volume in two different senses because of different hypotheses of its origins. In Chapter 116, McAlpine uses Muscomorpha, as in Volumes 1 and 2, to refer to the cyclorrhaphous Brachycera. Woodley, in Chapter 115, has expanded the concept of Muscomorpha to include the cyclorrhaphous Brachycera plus the Asilomorpha *sensu* Volume 1. For the equivalent of the infraorder Muscomorpha *sensu* McAlpine, i.e. the cyclorrhaphous Brachycera alone, Woodley uses the superfamily Muscoidea.

Although the views expressed in Volume 3 are not entirely consistent with those of Hennig, they are based largely on his last classification of the order (Hennig 1973), and they owe much to discussions between Hennig and the editors of the Manual during a series of workshops on the subject held in Ottawa in 1967 (Hennig 1969). It is safe to say that no one has contributed more to an understanding of the phylogenetic relationships of Diptera than has Hennig; yet he, himself, was the first to recognize the preliminary nature of some aspects of his phylogenetic proposals and to modify his views as new information became available. He decried the insufficiency of reliable data on many aspects of dipteran morphology, which forced him, at times, to place much faith on details of wing venation. We hope that the wealth of new data introduced in Volume 3 and in Volumes 1 and 2 will partially alleviate this problem.

Acknowledgment is made here to G. W. Byers, D. A. Craig, J. F. Cumming, and N. P. Kristensen for their helpful comments on Chapter 114; to Ralph Idema, who continued the tradition of Volumes 1 and 2 by illustrating larval labral structures (Figs. 114.3–11); to Bruce Cooper and Barbara Bissett, who attended to the labeling; to Barry Flahey, who prepared the cover illustration, the frontispiece, and all the cladograms; and to Sharon M. Rudnitski and Jane T. Buckley, Research Program Service, Agriculture Canada, for editing Volume 3; and to all other members of the Centre who assisted in the preparation of Volume 3.

References

Griffiths, G. C. D. 1972. The phylogenetic classification of Diptera Cyclorrhapha, with special reference to the structure of the male postabdomen. Series entomologica, 8. Dr. W. Junk, N.V., The Hague.

Hackman, W., and R. Väisänen. 1982. Different classification systems in the Diptera. Anns zool. Fenn. 19 (3): 209–220.

Hennig, W. 1950. Grundzüge einer Theorie der phylogenetischen Systematik. Deutscher Zentralverlag, Berlin.

Hennig, W. 1965. Phylogenetic systematics. A. Rev. Ent. 10: 97–116.

Hennig, W. 1966. Phylogenetic systematics. *Translated by* D. D. Davis and R. Zangerl. University of Illinois Press, Urbana, Ill.

Hennig, W. 1969. Neue Gattungen und Arten der Acalypteratae. Can. Ent. 101: 589–633.

Hennig, W. 1973. Ordnung Diptera (Zweiflügler). Handb. Zool. 4 (2) 2/31 (Lfg. 20): 1–337.

Schlee, D. 1978. In Memoriam Willi Hennig 1913–1976—Eine biographische Skizze. Ent. Germ. 4 (3/4): 377–391.

Wiley, E. O. 1981. Phylogenetics. The theory and practice of phylogenetic systematics. John Wiley & Sons, New York, N.Y.

PHYLOGENY AND CLASSIFICATION OF THE NEMATOCERA 114

D. M. WOOD AND A. BORKENT

It is inconceivable that, rightly understood, the evidence as to ancestry provided by the larvae should be in conflict with that given by the adult forms. Where such appears to be the case, it can only be due to a mis-reading of the facts.
[Edwards 1926: 112]

INTRODUCTION

One of the primary goals of systematics is to provide a phylogenetic interpretation of the group under study. Our consideration of the features of Nematocera has provided a basis for such an interpretation.

In this chapter we discuss the origins (sister groups) of the Nematocera and the evolutionary relationships between infraorders and families. From our hypotheses we go on to suggest a classification for these families.

Most attempts at understanding relationships between various groups of Nematocera have been based on a subjective assessment of the degree of similarity or difference shown by their constituent members. Not only has this approach generated diverse schemes and futile, untestable arguments throughout the literature on Nematocera, but the assumption itself, namely that similarities or differences may be trusted as reliable indicators of relationship, has been shown to be false.

Edwards noted, as early as 1926, that phylogenetic interpretation of Diptera was often confused because similarities between taxa could be due to different causes and authors often failed to distinguish between them. He recognized the following sources of similarity: convergence of an acquired character; convergence through character loss; and homologous characters. Further, he distinguished, among homologous characters, between "archaic" character states, which are similarities that arose far back in the ancestry of the organism, and "new" character states, or evolutionary novelties. These concepts and, most important, the methodology used in distinguishing between archaic and new character states were not clearly understood until the works of Hennig (1950, 1965, 1966). In our analysis, we have adopted Hennig's methodology of cladistics, grouping taxa only on the basis of shared evolutionary novelties (synapomorphies) as identified by outgroup comparison, i.e. these novelties do not occur in other taxa. We use the term synapomorphy, with reference to the members of a taxon, as "a homologous character (state) found in two or more taxa that is hypothesized to have arisen in the ancestral species of these taxa and no earlier" (Wiley 1981). It is, thus, an indicator of a monophyletic group. A monophyletic group is one we "believe contains all known descendents of a single stem-species" (Hennig 1981). Although some authors use the term autapomorphy for the same concept, and Wiley (1981) defines it as a character evolved in a single species, we use autapomorphy in a special sense to refer to a synapomorphy, which, although important in recognizing a monophyletic group, is of no value in determining its sister group because it is lacking in all other taxa.

Similarities owing to ancient ancestry (symplesiomorphies), which are now almost universally regarded as of no value in establishing relationships, are rejected. Differences between taxa (resulting from autapomorphies), no matter how great, are also discounted as evidence of lack of relationship. For example, two families as dissimilar as Blephariceridae and Deuterophlebiidae are considered, despite their differences, as sister taxa because they share some apomorphies and, thus, appear to be more closely related to one another than to any other extant taxon. Eldredge and Cracraft (1980), Wiley (1981), and other authors discuss cladistic methodology in further detail.

In practice, phylogenetic analysis of the Nematocera still rests to a discouraging degree on misinterpreted or inadequately established homologies and on an insufficient understanding of most structures and their function. The extant Nematocera are remarkable for their diversity of structure, particularly regarding larval mouthparts and adult terminalia. Coupled with an astonishing degree of convergence exhibited by many taxa, opportunities for misinterpretation have been considerable. So many careful, detailed observations remain to be made that many hypotheses of relationship must be regarded only as a provocation for further investigation.

The most recent comprehensive phylogenetic analysis of the Nematocera, within an analysis of the Diptera as a whole, is that of Hennig (1973). Supplemented by our own observations, particularly on larvae, and with those of others published since 1973, Hennig's foundation has provided us with the basis for our analysis. To a certain degree we have substantiated Hennig's conclusions, which he based primarily on evidence from adults, especially on features of wing venation. Where our conclusions differ, it is mainly because of our different interpretation of larval character states.

In each section we summarize Hennig's (1973) classification, the data upon which it is based, and any subsequent contributions, followed by our own classification. The analysis of character states provides outgroup comparisons and concludes with a discussion. The data are summarized in the cladogram of all families.

Our phylogenetic discussion generally follows a phyletic sequence (Nelson 1974, Patterson and Rosen 1977). Phylogenetically older groups are discussed before younger ones, so that the reader is led from a discussion of the sister group of Diptera to what we believe are the youngest infraorders within the Nematocera. Unfortunately, our understanding of the relationships between infraorders is still highly speculative and is based on only a few tenuous synapomorphies.

In a few instances (Anisopodidae, Thaumaleidae, Trichoceridae, and Tanyderidae), we were not sure which character states ought to be regarded as synapomorphies indicating monophyly of those families, although we have no reason to doubt that they are indeed monophyletic. This condition can be recognized by lack of a black square in the uppermost row in our cladogram (see Fig. 114.2). In addition, in one instance, the Mycetophilidae, evidence indicates that this family is probably a paraphyletic grouping.

Alternatively, we have not included all the known synapomorphies for some families (e.g. Culicidae) that we consider to be undoubtedly monophyletic. Both these situations reflect the emphasis in this chapter on understanding the relationships between families (as recognized by most workers and in Volume 1 of this Manual) rather than the monophyly of each family.

The classification provided in this chapter differs in some aspects from that outlined in Volume 1 of this Manual. This divergence is a consequence of our continued investigations, since the original organization of Volume 1, of data collected during and since that compilation.

HIGHLIGHTS IN THE DEVELOPMENT OF CLASSIFICATION OF THE NEMATOCERA

Historical background. An historical account of the development of classification of Diptera was presented by Lindner (1949: 82), preceded by bibliographic sketches of the early entomologists who contributed works on Diptera.

Latreille (1802) was the first to recognize that the Diptera could be arranged in two groups based on the structure of the antenna. In 1817 he formally proposed the Nemocera (later emended to Nematocera) for those taxa with antennae composed of many freely articulating flagellomeres. The remaining group, consisting of taxa with fewer, more compacted flagellomeres, was named the Brachycera by Macquart (1834, as Brachocères). Brauer (1863) proposed dividing the Diptera in a quite different way, into two groups, Orthorrhapha and Cyclorrhapha, based on whether or not the pupa was enclosed in a puparium. The two schemes were combined by Schiner (1864) who further divided the Nematocera, as follows:

- Diptera Orthorrhapha
  - Nematocera
    - Polyneura
    - Oligoneura
  - Brachycera
- Diptera Cyclorrhapha

Polyneura was proposed for those Nematocera with a more complete wing venation, and included the Chironomidae, Psychodidae, Culicidae, Tipulidae, and Anisopodidae. Oligoneura, with a more reduced venation, comprised the Cecidomyiidae, Mycetophilidae, Simuliidae, and Bibionidae. On the basis of the type of larval head capsule, Brauer (1869) divided the Nematocera into three tribes: Oligoneura, for the Cecidomyiidae alone; Polyneura, for the Tipulidae, which at that time also included the Trichoceridae; and Eucephala, for the remaining families. Believing that this classification was one-sided in favor of larvae, Osten Sacken (1891) proposed to divide the Nematocera into two taxa: *Nemocera vera*, containing the Cecidomyiidae, Mycetophilidae, Culicidae, Chironomidae, Psychodidae, Tipulidae, and Dixidae; and *Nemocera anomala*, containing the remaining families, the Bibionidae, Simuliidae, Blephariceridae, Anisopodidae, and Thaumaleidae. Based almost entirely on the degree of separation of the eyes in adult males, a character now known to differ even among congeneric members of the same family, particularly among the Simuliidae, Blephariceridae, and Anisopodidae, Osten Sacken's classification was hardly less one-sided than was that of Brauer. Coquillett (1901) proposed a division of families rather similar to that of Osten Sacken, calling the first Tipuloidea and the second Bibionoidea.

Lameere (1906) maintained the categories Polyneura and Oligoneura, but with rather different constituents than that of Brauer or than the one subsequently adopted by Hennig. Polyneura contained two families, the Culicidae and Tipulidae. The Culicidae consisted of two subfamilies; Ptychopterinae, containing the present families Ptychopteridae and Psychodidae; and Culicinae, equivalent to the modern Culicomorpha minus Simuliidae and Thaumaleidae. Oligoneura contained the Mycetophilidae, including Sciaridae, and the Cecidomyiidae. The Anisopodidae and Bibionidae, including the present Simuliidae, Thaumaleidae, and Blephariceridae, were transferred to the Brachycera. Thus, Lameere was as strongly influenced by the holoptic or dichoptic condition as was Osten Sacken, although his association of Anisopodidae with Brachycera foreshadowed recognition of a possible relationship between these two groups.

Knab (1915) proposed discarding the taxon Nematocera altogether, as being "unnatural." Believing that the presence or absence in the larva of abdominal spiracles 1–7 was of fundamental importance, he placed those families with amphipneustic and metapneustic larvae, including the Psychodidae, Culicidae, Ptychopteridae, Tipulidae, Anisopodidae, and Thaumaleidae, in the Polyneura. The remaining peripneustic families were placed in the Oligoneura. His observations were unfortunately superficial or erroneous; for example, the Simuliidae and Chironomidae, in which all spiracles are vestigial (*Archaeochlus* Brundin, with well-developed posterior spiracles, was then unknown), were included in the Oligoneura, whereas the Culicidae, which also have vestigial remnants of spiracles 1–7, were placed in the Polyneura.

A family tree presented by de Meijere (1916: 307) was one of the first attempts to express a phylogenetic arrangement of the families of Diptera. This tree is noteworthy for its two major lineages, one grouping the families that now constitute the Bibionomorpha (with addition of Mycetobiinae, now regarded as belonging to Anisopodidae), and the other grouping all the families of the modern Culicomorpha, including the Simuliidae and Thaumaleidae. His treatment was in stark contrast to the schemes of all his predecessors. This latter lineage also included, at its base, the Psychodidae, Blephariceridae, and Tipulidae. The Ptychopteridae and Anisopodidae were the only families left on branches of their own. This classification of de Meijere probably had an important influence on the subsequent schemes proposed by Edwards and Hennig.

Malloch (1917) also presented a phylogenetic tree and a classification to accompany it, which unfortunately embodied some of the disadvantages, and few of the advantages, of previous classifications. He maintained Brauer's three tribes, but in assigning the Ptychopteridae and Anisopodidae to the Polyneura, he ignored Keilin's (1912) work, which showed that these two families had more in common with, and ought to belong to, the Eucephala. He also, inexplicably, isolated the Bibionidae, Scatopsidae, and Simuliidae from the Eucephala and transferred them to the Oligoneura with the Cecidomyiidae, while maintaining the Mycetophilidae and Sciaridae in the Eucephala.

Crampton's (1924) phylogenetic proposals and particularly his morphological studies of thoracic structures of Nematocera (Crampton 1925, 1926) exerted a profound influence on subsequent phylogenetic studies, notably those of Edwards and Hennig. Unfortunately, he made no distinction between similarities based on symplesiomorphy and those based on synapomorphy; for although recognizing affinities between the Trichoceridae and Anisopodidae, which we regard as synapomorphies, he continued to group the Trichoceridae with the Tipulidae on what we now regard as symplesiomorphy. Thus he separated arbitrarily tipuloid and nontipuloid Nematocera. Furthermore, despite evidence of the anisopodid nature of *Mycetobia* Meigen ("a remarkable resemblance in all of [its] thoracic features"), he continued to regard it as a mycetophilid. He correctly postulated that *Hesperinus* Walker should be associated with Bibionidae and placed the Simuliidae with the Chironomidae rather than the Bibionidae, with which they were formerly associated. Interestingly, he also regarded *Axymyia* McAtee as representing a separate monotypic subfamily in the Anisopodidae. Thus, his treatment of the taxon now called Culicomorpha, subsequently adopted by Edwards and Hennig, has remained essentially unchanged to the present.

In a comprehensive summary of existing knowledge of nematoceran structure, Edwards (1926) proposed three lineages arising early in the Jurassic. Although well aware of the "striking similarity" between larval mouthparts of the Trichoceridae, Anisopodidae, Psychodidae, and Scatopsidae, particularly in the form of the premandibles, resemblance that had been described only the previous year by Goetghebuer (1925), he concluded that the lack of concordance in adult structure indicated that such character states in larval mouthparts were primitive features, retained in these four families but lost in the others. He retained the traditional association between Trichoceridae and Tipulidae as being the most primitive lineage. He also united in a single lineage the Mycetophiloidea and Bibionoidea, which Crampton had implied belonged together, thus foreshadowing the Bibionomorpha as later developed by Hennig. Furthermore, he adopted Crampton's association that we now call the Culicomorpha, although he was reticent to include the Simuliidae and Thaumaleidae in the group. The Ptychopteridae and Psychodidae were included in the Culicomorpha lineage. Edwards' ideas, like those of Crampton, were not formalized as a classification, although they profoundly influenced subsequent phylogenetic interpretation of Nematocera and were surely an important early influence for Hennig.

In his classification of the Palaearctic Nematocera, Lindner (1949) divided the Nematocera into the Polyneura, containing the Tipulidae and Trichoceridae, and the Oligoneura, comprising the remaining families.

White (1949) suggested that loss of male chiasmata is a synapomorphy, grouping the Anisopodidae, Bibionidae, Scatopsidae, Thaumaleidae, Blephariceridae, Mycetophilidae, Sciaridae, Cecidomyiidae, and Brachycera. Excluded from this supposed monophyletic group were the Tipulidae, Psychodidae, Simuliidae, Culicidae, and Chironomidae. However, subsequent work on the Simuliidae (Rothfels and Mason 1975) has shown that species with achiasmate males have arisen independently in several lineages, suggesting that the character state is susceptible to parallelism. White (1973: 487) discussed further examples of parallelisms and reversals in other families.

Hennig's classification versus that of Rohdendorf. Hennig (1948, 1954, 1968, 1969, 1973, 1981: 422) and Rohdendorf (1964, 1974, 1977) have each developed the most comprehensive recent classifications of Diptera. Each author modified his own scheme with the passage of time, while expending many paragraphs analyzing the perceived shortcomings of the other's schemes. Whereas Hennig adhered strictly to cladistic analysis, Rohdendorf (1977) denigrated this approach, attacking the problem of evolution from the bias of dialectic views (see review by Griffiths 1975). The two resultant classifications were compared diagrammatically by Hennig (1969, 1981), showing clearly the incompatibilities between their two systems.

Rohdendorf (1964, 1974) regarded the Nymphomyiidae as the most primitive of living Diptera, creating for them a new suborder, the Archidiptera, subsequently variously spelled as Archodiptera (Rohdendorf 1977), Archaediptera (Cutten and Kevan 1970), and Archaeodiptera (Hackman and Väisänen 1982). All other Diptera, forming the suborder Eudiptera, were considered the sister group of the Archidiptera. Rohdendorf also included in the Archidiptera two fossils infraorders, the Dictyodipteromorpha and the Diplopolyneuromorpha, known only from the upper Triassic as portions of wings. Further discussion of Rohdendorf's ideas concerning the placement of Nymphomyiidae is given in the sections "Phylogeny of the Infraorders of the Nematocera" and "Phylogeny of the Infraorder Blephariceromorpha." Rohdendorf (1964, 1974) divided the Eudiptera into four infraorders, the Deuterophlebiomorpha, Blephariceromorpha, Tipulomorpha, and Bibionomorpha. The first two each contained the single nominal family. The Tipulomorpha contained seven superfamilies most ending in *idea*: the Pachyneuridea, with Pachyneuridae; the Tipulidea, with Trichoceridae, Ptychopteridae, and Tanyderidae, as well as Tipulidae *sensu lato*; the Psychodidea, for Psychodidae *sensu lato*; the Culicoidea and the Chironomoidea, each constituted as described in this Manual, except for Thaumaleidae; the Orphnephilidea, for Thaumaleidae; and the Rhaetomyiidea, for Perissommatidae. The Bibionomorpha contained six superfamilies: the Bolitophilidea, for Bolitophilidae; the Fungivoridea, for remaining Mycetophilidae *sensu lato*; the Cecidomyiidea, for Cecidomyiidae *sensu lato*; the Scatopsidea, for Scatopsidae and Synneuronidae; the Bibionidea, for Bibionidae; and the Rhyphidea, for Anisopodidae *sensu lato* and Cramptonomyiidae. The Pachyneuridae and Cramptonomyiidae were thus widely separated. In Rohdendorf's later classification (1977), all the superfamily endings were changed to the *oidea* form and the Orphnephilidea, Rhaetomyiidea, and Rhyphidea were changed to their present form. The Pachyneuridae and Cramptonomyiidae were brought together in the Bibionomorpha, but no other changes were made.

Hennig's classifications were established using the principles of phylogenetic (cladistic) analyses, which he, himself, developed. They are easily understood and, furthermore, are testable. His latest concepts (Hennig 1969, 1973, 1981) have, therefore, formed the basis for our present hypotheses, presented in the remainder of this chapter. An outline of his 1973 classification is presented under "Phylogeny of the Infraorders of the Nematocera," and his ideas concerning the phylogeny within each infraorder precede our own.

MONOPHYLY AND PHYLOGENETIC ORIGINS OF THE DIPTERA

Despite marked structural differences and diversification within the order, the Diptera are universally accepted as a monophyletic taxon. The group's most conspicuous synapomorphy is the modification of the hind pair of wings into halteres and related modifications in thoracic structure.

Hennig (1973, 1981) also listed other synapomorphies, some of which are as follows: CuP reduced, developed basally as a vein but usually occurring as a fold; abdominal spiracle 8 absent in adult males; and apical segment of labial palpus modified into a labellum, with pseudotracheae on its ventral surface for absorbing liquids. He also suggested that loss of larval legs was a synapomorphy. However, larvae of Siphonaptera, the order to which Diptera may be most closely related, also exhibit this feature.

Although Hennig (1973) was convinced that the Brachycera were a monophyletic group, he was noncommital about the monophyly of the Nematocera, because he did not know whether the sister group of the Brachycera was the Nematocera as a whole, one of the infraorders of the Nematocera, or perhaps even a lineage within one of these infraorders. The Bibionomorpha were suggested as one possibility.

Until a suitable sister-group relationship can be postulated, there is no point in addressing further the problem of the monophyly of the Nematocera. Even if the Nematocera could eventually be shown to be a paraphyletic group, as we suspect it will, the phylogenetic arguments presented here would not be affected. The Brachycera would merely become the sister group of one of the lineages dealt with later. The matter is discussed further in the following chapter, dealing with the orthorrhaphous Brachycera.

There has been some debate as to whether the sister group of the Diptera is the Siphonaptera (Boudreaux 1979), the Mecoptera (Mickoleit 1981, Hennig 1981), only the Nannochoristidae from within the Mecoptera (Tillyard 1935, Imms 1944), or the Mecoptera plus the Siphonaptera (Kristensen 1975). Another possibility, which we entertain here, may not as yet have been expressly proposed in the literature. This new hypothesis regards the Siphonaptera and the Diptera as sister groups whose sister group in turn is the Nannochoristidae, not the Mecoptera as a whole.

Character states. Analysis of six of the character states (1–6) that have been presented in support of these views follows:

Larva

1. *First instar without egg burster on dorsum of head capsule (plesiomorphic) / first instar with egg burster (apomorphic)*

The apomorphic feature is unique in the Insecta. Within the Mecoptera, an egg burster has been recorded by Gassner (1963) for *Panorpa nuptialis* Gerstaecker and by D. A. Craig (personal commun., University of Alberta) for *Microchorista philpotti* (Tillyard) (Nannochoristidae). At least for *M. philpotti* the egg burster is very small and easily overlooked so that it may be present in other Mecoptera as well. However, Cooper (1974) could find no trace of it in *Hesperoboreus notoperates* (Cooper), a state that we consider here to be a loss.

2. *Eye covered with convex, lens-like cuticle (plesiomorphic) / eye composed of a group of pigmented cells located beneath a region of transparent, unmodified cuticle (apomorphic)*

The plesiomorphic condition is present in some Coleoptera, some Hymenoptera (at least the Symphyta), the Megaloptera, Neuroptera, Trichoptera, and Lepidoptera, and Mecoptera other than Nannochoristidae. The apomorphic condition is shown by the Nannochoristidae and all Nematocera. Larvae of the Siphonaptera apparently lack eyes altogether.

3. *Thoracic legs present (plesiomorphic) / thoracic legs absent or at most represented by small bristles (apomorphic)*

The Siphonaptera and Diptera are the only panorpoid orders that exhibit the apomorphic condition, with the exception of a few Lepidoptera in which absence of legs is clearly a secondary loss. Some larvae of Hymenoptera and Coleoptera have also lost their thoracic legs, a condition suggesting that this character state may be susceptible to parallelism.

Pupa

4. *Movable mandibles present (decticous) (plesiomorphic) / mandibles not movable (adecticous) (apomorphic)*

All Megaloptera, Neuroptera, Trichoptera (a few have become adecticous secondarily), primitive Lepidoptera (Zeugloptera), and Mecoptera have functional mandibles (decticous) (Hinton 1971). However, Coleoptera and Hymenoptera (except Xyelidae which may be the sister group of all other Hymenoptera) are adecticous as pupae, which suggests that the apomorphic condition is susceptible to parallelism. A possible exception in Diptera (Nymphomyiidae) is discussed under "Phylogeny of the Infraorder Blephariceromorpha."

Adult

5. *Lacinia and galea present, both blunt apically (plesiomorphic) / galea absent and lacinia elongate, blade-like, usually with a row of recurved teeth (apomorphic)*

The apomorphic condition is unique to the Nannochoristidae, Siphonaptera, and Diptera (Imms 1944), although the recurved teeth have been secondarily lost and the lacinia has become vestigial in many lineages of Diptera, including all Bibionomorpha, Asilomorpha, and Muscomorpha.

6. *Sperm pump absent (plesiomorphic) / sperm pump present (apomorphic)*

The apomorphic condition is unique to all Mecoptera (except Boreidae, in which this absence is regarded by Mickoleit (1971, 1974) as a secondary loss), Siphonaptera, and to many, but not all, Nematocera. Hennig (1969, 1981) and Kristensen (1975) regarded this character state as a synapomorphy of the Mecoptera, Siphonaptera, and Diptera, a group Hennig named the Antliophora (pump bearers). A detailed comparison between sperm pumps in the three orders has evidently never been made and, thus, it is uncertain whether this character state is homologous in the taxa possessing it.

Discussion. The hypotheses presented here result in the cladogram of Fig. 114.1. However, we are aware of two character states that may provide evidence that the Diptera and Nannochoristidae are sister taxa. These are the shared unique presence of a sensory pit on the third maxillary palpal segment of adults (lost in some Diptera) and the presence of finger-like, membranous protrusions of the larval anus called anal papillae. Within the Holometabola, only Trichoptera (probably primitive taxa) (Wiggins 1977), Nannochoristidae (Pilgrim 1972), and some Nematocera possess anal papillae. In all these taxa, the papillae are eversible, exerted by hemolymph pressure and retracted by muscular action (see also character 75 under Culicomorpha). The anal papillae function as osmoregulatory organs for these aquatic groups.

We suspect that the maxillary palpal sensory pit and anal papillae may have been lost secondarily in Siphonaptera. The maxillary palpal sensory pit has been lost secondarily in many Diptera and anal papillae are never evident in terrestrial larvae (e.g. those of Siphonaptera).

Mickoleit (1971, 1981) has suggested that the Mecoptera form a monophyletic group, contrary to the paraphyly suggested here. The characters he used to support this idea were fusion of the clypeus and labrum in the adult and loss of some muscles of the labrum, mandible, and hypopharynx (possibly correlated with fusion of the labrum to the clypeus). A fused clypeus and labrum is also present in some primitive Lepidoptera, for example in the Mnesarchaeidae (Kristensen 1968), in some primitive Trichoptera (Klemm 1966), and also in at least three lineages of Coleoptera (Crowson 1981). The Siphonaptera have an extremely reduced labrum hidden behind the anterior margin of the clypeus (Snodgrass 1946). Apparently, therefore, outgroup comparisons do not allow us to conclude decisively that this character state is synapomorphic for Mecoptera. We are not sure how to interpret the second character, reduced muscles of the labrum, mandible, and hypopharnyx, for lack of further outgroup comparisons. Until these two characters are more rigorously investigated (at least in other Holometabola) we do not recognize their absence as a synapomorphy of the Mecoptera.

Whether or not the fused condition of the clypeus and labrum is a synapomorphy of the Mecoptera or whether it

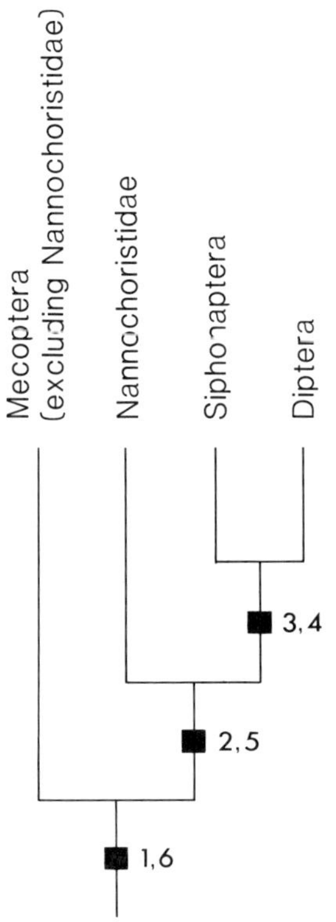

Fig. 114.1. Cladogram showing relationships of the Diptera and its sister taxa. Numbers refer to character states described in the text.

is synapomorphic of possibly the panorpoid orders, in our scenario we would still have to postulate a reversion in the Diptera to the plesiomorphic, unfused condition.

Willmann (1981, 1987) also supported the idea that Mecoptera form a monophyletic taxon. He suggested two additional synapomorphies: fusion of tergite and sternite 9 of the adult male abdomen and presence of a membranous organ on the distal portion of the gonostylus. The first is also present in nearly all Nematocera and is probably plesiomorphic within the Diptera. The second appears to be unique to the Mecoptera. Kristensen (1981) noted two additional character states, which he suggested indicate monophyly of the Mecoptera: metathoracic spiracle of the larva nonfunctional and bases of male adult gonocoxites fused dorsally and ventrally. The metathoracic spiracle is lacking in larvae of some Lepidoptera (but present in Micropterigidae), all Trichoptera, and most Diptera, indicating its susceptibility to homoplasy. Dorsal and ventral fusion of the gonocoxites appears, on the basis of outgroup comparisons, to be restricted to the Mecoptera.

Boudreaux (1979: 254) listed other apparent synapomorphies shared by the Siphonaptera and Diptera, which might support our conclusions (hind wings reduced; metanotum smaller than mesonotum; short adult antenna). However, these character states are either not restricted to the two orders (discussed above) or they seem superficial or susceptible to parallelism. He also discussed the lobes that are present on the mesothorax of ceratophylloid flea pupae (no lobes are present on the metathorax). If these are homologous to the wing buds of other insects, the missing metathoracic lobes of these flea pupae may be synapomorphic with the reduced wing bud in which the halteres of Diptera develop. This structure warrants further comparison, but as discussed by Rothschild (1975), the homology of these lobes with wing buds is questionable.

Kristensen (1975) suggested that the Mecoptera and Siphonaptera were sister groups and that these together formed the sister group of the Diptera. He listed four character states as synapomorphies of the Mecoptera and Siphonaptera: no extrinsic labral muscles; unique proventricular processes called acanthae; specialized spermatozoa; fused nervi corpora cardiaca. Boudreaux (1979) discussed each of these (except for the fused nervi corpora cardiaca) and showed most of them to be suspect as apomorphies. Fused nervi corpora cardiaca are also known in the Hymenoptera (Lafon-Cazal 1983). We consider this character state to be too poorly investigated to accept as good evidence of a relationship between the Mecoptera and Siphonaptera.

Based on sperm ultrastructure, Kristensen (1981) further supported the concept that Mecoptera and Siphonaptera are sister groups. We agree that the 9 + 2 configuration of the axoneme and the coiled flagellum must be considered apomorphic, which, therefore, contradicts our conclusions.

Several character states may contradict our hypothesis of a sister-group relationship between the Diptera plus Siphonaptera and the Nannochoristidae; namely the presence of a gonostylar organ and the dorsal and ventral fusion of the gonocoxites in all Mecoptera and the shared 9 + 2 configuration of the sperm of Mecoptera and Siphonaptera. Fusion of the clypeus and labrum of Mecoptera warrants further research in the light of outgroup comparisons.

Our conclusion that the Nannochoristidae is the sister group of the Diptera plus Siphonaptera rests on correct interpretation of homology between the blade-like lacinia, loss of galea, presence of a sensory vesicle on the third palpomere (not present in fleas), and loss of the lens-like cuticle covering the larval eye.

Should our conclusion that the Mecoptera are a paraphyletic group be confirmed by future studies, the family Nannochoristidae should be recognized as a separate order. Hinton (1981) has already provided the name: Nannomecoptera.

PHYLOGENY OF THE INFRAORDERS OF THE NEMATOCERA

Hennig's classification

Hennig (1973) divided the Nematocera into four infraorders as follows: the Tipulomorpha, Psychodomorpha, Culicomorpha, and Bibionomorpha.

Infraorder Tipulomorpha
 Superfamily Trichoceroidea
 Superfamily Tipuloidea
Infraorder Psychodomorpha
 Superfamily Blephariceroidea
 Superfamily Nymphomyioidea
 Superfamily Ptychopteroidea
 Superfamily Psychodoidea
Infraorder Culicomorpha
 Superfamily Culicoidea
 Superfamily Chironomoidea
Infraorder Bibionomorpha
 Section Pachyneuriformia
 Section Anisopodiformia
 Section Bibioniformia
 Section Mycetophiliformia
  Superfamily Scatopsoidea
  Superfamily Cecidomyioidea
  Superfamily Mycetophiloidea

Hennig (1954, 1968, 1969, 1973, 1981) regarded the Tipulomorpha as the most primitive infraorder. He included in this infraorder only the families Trichoceridae and Tipulidae *sensu lato* (as used in this Manual, including the Cylindrotomidae and Limoniidae *sensu* Hennig and most other European authors). This group, which he called the Polyneura, was considered to be the sister group of the rest of the Diptera, the Oligoneura. Hennig (1968) regarded the Oligoneura as a monophyletic group because the wing in all of its members was differentiated into a stalk and a blade, a derived condition compared to the Mecoptera and Polyneura. He also postulated (Hennig 1968) that the Polyneura formed a monophyletic group, based on the following character states, which he believed were synapomorphies: wing elongated, with junctions of all longitudinal veins and the discal cell shifted to distal half of wing; R_2 ending in R_1, never alone in the wing margin; scutal suture V-shaped, not interrupted medially; adult without mandibles; adult lacking a tergal depressor muscle of the trochanter; larva amphipneustic or metapneustic.

Hennig based his Psychodomorpha on the fusion of the meron of the mid coxa with the metepimeron, a character suggested earlier by Crampton. Hennig, however, expressed reservations about the value of this character as a synapomorphy. Within the Psychodomorpha, he grouped Tanyderidae and the Ptychopteridae (superfamily Ptychopteroidea) on the basis of the shared capability of adult males to grasp the female by folding the last tarsomere of each leg against the fourth tarsomere. The Blephariceridae and Deuterophlebiidae (superfamily Blephariceroidea) were grouped by the manner in which the wing was folded during its development within the pupa (Hennig 1973). He tentatively grouped the Culicomorpha, Bibionomorpha, and Brachycera on the shared presence of lobe-like pulvilli and of an alula. Finally, he placed the Brachycera as the sister group of the Bibionomorpha.

Proposed classification of the Nematocera

We have recognized seven infraorders:

Infraorder Tipulomorpha
Infraorder Blephariceromorpha
Infraorder Axymyiomorpha
Infraorder Bibionomorpha
Infraorder Psychodomorpha
Infraorder Ptychopteromorpha
Infraorder Culicomorpha

Character states. The six character states that were used in interpreting the relationship between the infraorders are analyzed below. All character states we use to advance our interpretation of the phylogeny and classification of the Nematocera are numbered sequentially and correspond to those appearing on Fig. 114.2. Other character states discussed herein are left unnumbered.

Larva

1. *Mandible with prostheca on articulated lobe ("lacinia mobilis" of Pilgrim 1972) (plesiomorphic) / prostheca arising directly from median surface of mandible (apomorphic)*

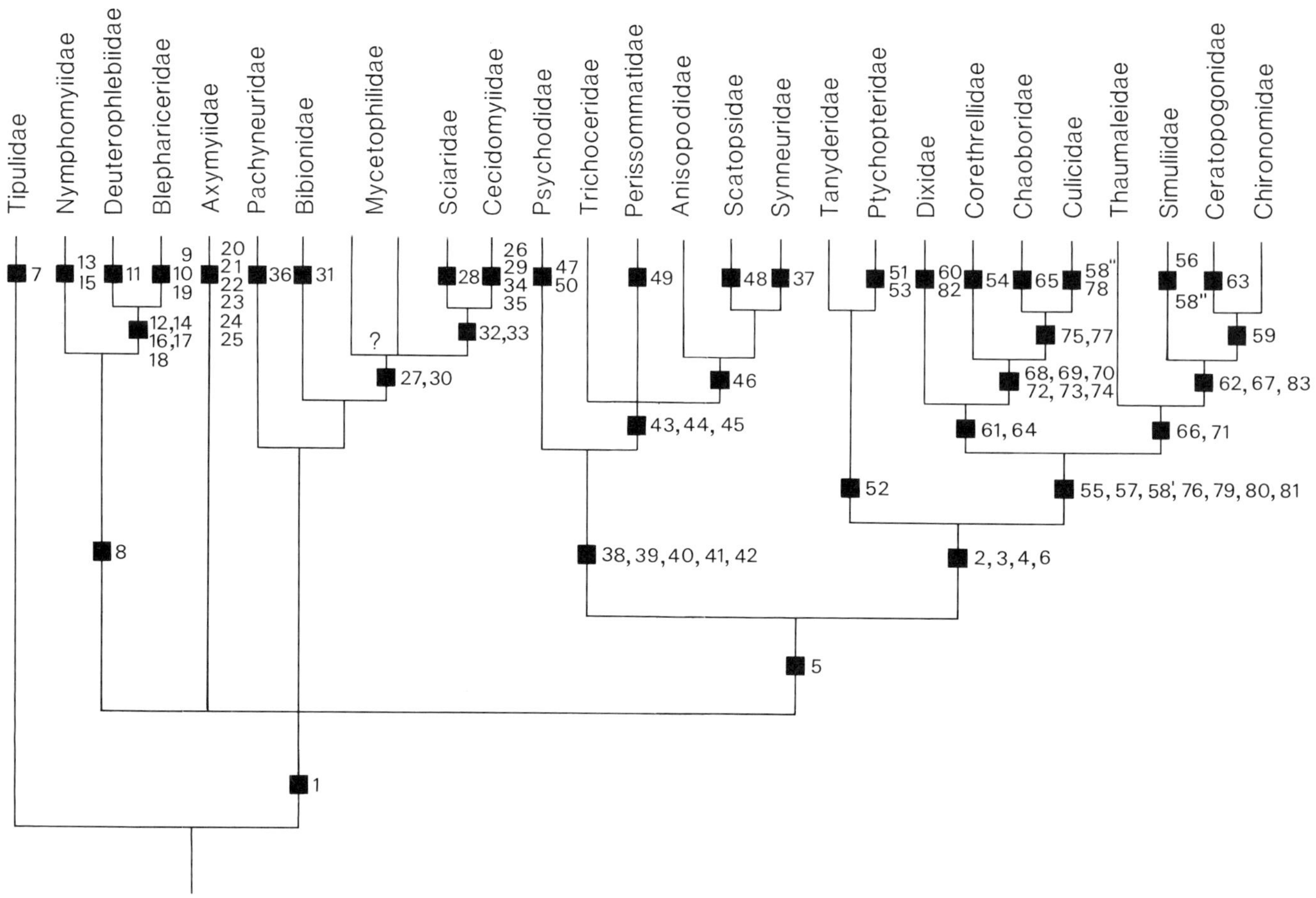

Fig. 114.2. Cladogram showing relationships between the families of the Nematocera. Numbers refer to character states described in the text.

In the Nannochoristidae, the lacinia mobilis is large and distinct, with a toothed apex, whereas in the Tipulidae it is much smaller and poorly defined (Chiswell 1955). In all other Diptera, the tuft of setae that constitutes the prostheca seems to arise directly from the mandible itself, rather than from a lobe. Hence, we tentatively regard the lack of this lobe in all Diptera except the Tipulidae as a synapomorphy, leaving the Tipulidae as the sister group of the rest of the Diptera.

2. *Ventral surface of the labrum without labral brushes (plesiomorphic) / labrum with a pair of labral brushes, each in the form of a convex, cushion-like area of cuticle covered with parallel, transverse rows of long setae (apomorphic)*

The labral brush is a complex feeding structure, unique to the Tanyderidae, Ptychopteridae, Dixidae, Culicidae, and Simuliidae. Setae are erected apparently by internal pressure and are closed by contraction of exceptionally well-developed labral retractor muscles. Wood (in preparation) argues that such a complex structure, with associated morphological and behavioral modifications for ingesting particulate food, is probably homologous in all five of these families, and we present it here as evidence of a sister-group relationship between Ptychopteromorpha and Culicomorpha. We assume it has been secondarily lost in some Culicomorpha.

3. *Premandible consisting of a simple sclerite lacking invaginated portion (plesiomorphic) / premandible with invaginated apodeme for insertion of labral retractor muscle (apomorphic)*

The apomorphic condition is found in the Ptychopteridae, Dixidae, Culicidae, Simuliidae, Ceratopogonidae, and Chironomidae. Corethrellidae and Chaoboridae appear to lack premandibles altogether, whereas the Thaumaleidae have a reduced premandible.

The plesiomorphic character state is found in the Psychodomorpha and Tanyderidae. If the Tanyderidae are the sister group of the Ptychopteridae, as proposed by Hennig (1973) and supported by us here, the reduced premandible of the Tanyderidae may possibly be a secondary loss, perhaps occurring as a result of their burrowing habits and a change of diet. On the other hand, the occurrence of a more complex premandible, with invaginated apodeme, in the Ptychopteridae may be more derived than in Tanyderidae. In this event, the Ptychopteridae alone could be the sister group of the Culicoidea. We wonder, however, why larvae of Tanyderidae should possess a fully developed labral brush, to all intents and purposes identical to that in Ptychopteridae, yet lack all the other devices that are present in the Ptychopteridae for collecting food particles from it, i.e. premandibular comb, epipharynx, and mandibular comb (see character 4). Faced with this conflicting evidence, we have chosen to place the Tanyderidae and Ptychopteridae together, as constituting the sister group of the Culicomorpha.

4. *Mandible without mandibular comb (plesiomorphic) / mandible with comb-like or brush-like row of long, curved setae along dorsal surface (apomorphic)*

The apomorphic condition is found in the Ptychopteridae, Dixidae, Corethrellidae, Chaoboridae, Culicidae, Simuliidae, and at least in the primitive *Archaeochlus* Brundin of the Chironomidae, but not in the Tanyderidae (and is thus correlated with the previous character). Cranston *et al.* (1987) recently revised *Archaeochlus*. This genus is probably the sister group of the remaining Chironomidae and, although not fully shown in their Figure 37, the larva of *Archaeochlus bicirratus* Brundin possesses a well-developed mandibular comb.

The argument applied in the previous case may also be applied here: either the Ptychopteridae alone is the sister group of the Culicomorpha, based on the shared presence of a mandibular comb, or the Tanyderidae plus the Ptychopteridae together are the sister group of the Culicomorpha, assuming that the Tanyderidae have lost their mandibular comb in acquiring stout chewing mandibles.

The mandibular comb is used for combing food particles from the labral brush as the mandibles and the labrum are adducted.

5. *Mandibular articulations located more or less dorsoventral to each other, with the mandibles operating in a horizontal plane (Fig. 3.11) (plesiomorphic) / epicondyle displaced medially and hypocondyle displaced laterally, with the mandibles operating obliquely (Fig. 3.11) or vertically (apomorphic)*

The plesiomorphic condition is characteristic of all outgroups of Diptera, e.g. the Siphonaptera, the Mecoptera (including Nannochoristidae), and the other panorpoid orders. It is also found in the Tipulomorpha, Bibionomorpha, and Axymyiomorpha. Movement of the mandibles to an oblique or vertical position is apparently always associated with feeding on particles in a liquid or semiliquid medium. Because the position of articulation points has probably shifted more than once, no clear transformation series can be established. Conversely, reversals have undoubtedly occurred, e.g. in wood-boring groups such as the Trichomyiinae (Psychodidae) and *Stenochironomus* Kieffer (Chironomidae) (Borkent 1984). The Axymyiidae may also represent such a reversal, which may account for our difficulty in placing this family phylogenetically. The vertical shift of the mandible is shown by most Psychodomorpha, whereas the Ptychopteromorpha and Culicomorpha exhibit various degrees of mandibular rotation.

Adult

6. *Ocelli present (plesiomorphic) / ocelli vestigial or absent (apomorphic)*

Loss of ocelli was given by Hennig (1973) as a synapomorphy of the Culicomorpha. In our arrangement we consider it a synapomorphy of the Culicomorpha and Ptychopteromorpha. Ocelli are present in nearly all Mecoptera (absent in Meropeidae, Boreidae, and *Apteropanorpa* Carpenter) but are lacking in Siphonaptera. Within the Nematocera, there are additional examples of convergence: ocelli are absent in the Tipulidae, Deuterophlebiidae, Psychodidae, and in some Cecidomyiidae. We suspect that this character state, by itself, is only a weak indication of relationship.

Discussion. We have segregated the Trichoceridae from the Tipulidae, the two families included by Hennig (1954, 1968, 1973) in his Tipulomorpha. Our Tipulomorpha is thus left with only the Tipulidae *sensu lato* as used in this Manual, including Cylindrotomidae and Limoniidae *sensu* Hennig and most other European authors.

Because Hennig (1973) dealt extensively with monophyly of the Tipulidae and Trichoceridae (his Tipulomorpha), a more detailed analysis of his arguments is given in the following section "Phylogeny of the Infraorder Tipulomorpha."

Similarly, the arguments so far presented in support either of a primitive position for the Nymphomyiidae, or of its being considered the sister group of the rest of the Diptera, are all based on *a priori* assumptions of supposed primitiveness. No previous author has presented rigorous arguments, based on synapomorphies, to suggest that the Diptera, exclusive of the Nymphomyiidae, form a monophyletic group. We deal with the specific arguments in the section "Phylogeny of the Infraorder Blephariceromorpha."

Recognizing the contradiction shown between character states of adults of the Trichoceridae and Tipulidae *sensu lato* and those of the larvae, Keilin (1912) suggested that the Trichoceridae might be allied to the Anisopodidae rather than to the Tipulidae. In his earliest phylogenetic interpretation, Hennig (1948, 1950, 1952) included the Trichoceridae, along with the Tanyderidae and Ptychopteridae, in his Psychodiformia, then as part of his Culicomorpha. Later, he concluded (Hennig 1953, 1954), in agreement with Edwards (1926), that these larval character states, found in the Trichoceridae, Anisopodidae, Scatopsidae, and Psychodidae, must represent the most primitive larval states among dipterous larvae. Judging them to be symplesiomorphies, he discounted their significance as an indication of a close relationship between these families. From this position, he concluded that the larval character states exhibited by the Tipulidae were derived from a trichocerid-like ancestor.

We have decided to regard similarities in the structure of larval mouthparts in the Trichoceridae, Anisopodidae, Scatopsidae, and Psychodidae, as described by Keilin (1912), Goetghebuer (1925), and Anthon (1943), as being synapomorphies. Accordingly, we have placed the four families together in the Psychodomorpha. Finally, details of these challenges to the grouping of the Psychodomorpha as supported here are discussed in the section dealing with that infraorder. Regardless, it should be noted that, because of the rather profound differences between larvae of Nannochoristidae and those of all living Diptera, recognition of primitive character states in larvae of Diptera by outgroup comparison has not yet been attained in many instances. However, superficial comparison of the larvae of Siphonaptera and Diptera indicates that study of these may be fruitful.

In Chapter 1 of this Manual, the Tanyderidae were included with the Tipulidae in the Tipulomorpha, but we have abandoned that arrangement for lack of justification. Admittedly, the Tanyderidae show such primitive features as a five-branched radius (Hennig 1954), but we cannot propose any synapomorphies between this family and the Tipulidae. We now refer the Tanyderidae, along with the Ptychopteridae, to the Ptychopteromorpha, leaving the Tipulidae *sensu lato* as the sole representative of the Tipulomorpha.

We have removed the Blephariceroidea and Nymphomyioidea from Hennig's Psychodomorpha, an arrangement he, himself, was not convinced of, and have combined them to form the Blephariceromorpha. The relationship of this infraorder to others is also enigmatic. The Axymyiomorpha, containing only the Axymyiidae, was isolated from Hennig's Bibionomorpha because resemblances seemed to be superficial rather than based on synapomorphies. However, the relationship of the Axymyiomorpha to the other infraorders has proven as elusive as have those of the Blephariceromorpha. Our most radical departure from Hennig's scheme is a reorganization of the infraorder Psychodomorpha, based on larval character states. It contains families taken from Hennig's Tipulomorpha (i.e. Trichoceridae), Bibionomorpha (Anisopodidae, Scatopsidae, and Synneuridae), and Psychodomorpha (Psychodidae). We isolated the Tanyderidae and Ptychopteridae from Hennig's Psychodomorpha and placed them in a separate infraorder because we propose a sister-group relationship between them and the Culicomorpha. Our concept of Culicomorpha remains essentially the same as that of Hennig's, except for recognition of a new family, the Corethrellidae, which we propose as the sister group of the remaining Chaoboridae plus the Culicidae.

PHYLOGENY OF THE INFRAORDER TIPULOMORPHA

Hennig's classification

Hennig (1973) included two groups in this infraorder, the families Tipulidae and Trichoceridae, as follows:

Infraorder Tipulomorpha (Polyneura)
  Superfamily Trichoceroidea
    Family Trichoceridae
  Superfamily Tipuloidea
    Family Tipulidae
    Family Cylindrotomidae
    Family Limoniidae

Hennig dealt extensively with the monophyly of the Trichoceridae and Tipuloidea, which comprised his infraorder Tipulomorpha. Because we regard the character states he used to establish this concept as being plesiomorphic or misinterpreted, we present the following detailed analysis of his arguments, character by character.

Larva

- *Larva peripneustic (plesiomorphic) / amphipneustic or metapneustic (apomorphic)*

The apomorphic condition was one of the characters given by Hennig (1973) as a synapomorphy of the Tipulidae plus Trichoceridae. The plesiomorphic condition is found in larvae of the Nannochoristidae, which are aquatic, as well as in other members of the Mecoptera and other panorpoid orders, which are terrestrial. Larvae of the Bibionomorpha and Scatopsidae, which are also essentially terrestrial, except when inhabiting semiliquid, rotting media, are also peripneustic. Most other nematoceran larvae are either amphipneustic or metapneustic. However, all dipteran larvae, even fully aquatic forms, retain at least vestiges of their first seven abdominal spiracles (Keilin 1944); vestiges of all nine pairs are found in such groups as the Simuliidae, which otherwise have no functional spiracles. Many aquatic forms, or those living regularly in liquid or semiliquid media, have lost their first seven pairs of abdominal spiracles, regardless of phylogenetic affiliation, whereas terrestrial forms still retain all or most of their spiracles. These first seven abdominal spiracles, once vestigial, could conceivably still be regained if the genetic mechanism of spiracle formation that affects those of segment 8 affects the first seven. In summary, we are reluctant to attach much phylogenetic importance to an interpretation of the presence or absence of spiracles.

Adult

- *Some wing veins ending before middle of wing, with discal cell centrally located (plesiomorphic) / wing elongated, with all veins (except A_2) ending beyond middle of wing and discal cell positioned in distal half of wing (apomorphic)*

Henning (1968), noting that members of the Trichoceridae and Tipulidae had rather long, narrow wings, postulated that this feature was a synapomorphy of the two families. However, he noted a general correlation between slenderness of body and elongation of wing. We can see no essential difference between wings of the Tipulidae, Trichoceridae, and other slender Nematocera such as *Cramptonomyia* Alexander (Figs. 4.6–8, 22.2–4). Therefore, we reject this character state as a possible synapomorphy of the Trichoceridae and Tipulidae.

- *Vein R_2* ending separately in wing margin (plesiomorphic) / R_2 *ending in R_1* (apomorphic)

A small, transverse vein extends between the first and second branches of the radius in the Tipulidae and Trichoceridae (Figs. 4.6, 4.8). First called the radial crossvein, it was interpreted by Alexander (1927) as R_2. Hennig (1954) regarded its presence as a synapomorphy of the two families and, hence, of his Tipulomorpha. We are uncertain whether it is homologous in these two families. In the Trichoceridae it lacks setae, which are found on adjacent veins, and appears to be a crossvein. In the Tipulidae, on the other hand, it is quite variable in position and direction and usually bears setae. Alexander (1927) made a case for regarding this vein in the Tipulidae as R_2; but in the Trichoceridae we feel that it could be a crossvein, perhaps homologous to the radial crossvein found in the Nannochoristidae and other Mecoptera. If so, its occurrence would be plesiomorphic. We are unable to decide whether this vein is indeed R_2 or a crossvein in either family or even whether it is homologous in both families.

- *Venation with base of M separate from base of R (plesiomorphic) / base of M connected to base of R by an oblique crossvein (called MA in Chapter 1 of this Manual) (apomorphic)*

Hennig (1968: 5) stated that members of the Trichoceridae and Tipulidae (his Tipulomorpha or Polyneura) lacked a connection between the bases of M and R, although some members of the Tipulidae had a fold in that position. In contrast, this connection was present in, and might be a synapomorphy of, the rest of the Diptera (the Oligoneura). However, we believe that the distribution of this character needs further study. Members of the Trichoceridae lack this connection, as Hennig asserted, but some Tipulidae that we have examined (e.g. a species of *Hexatoma* Latreille) have a rather strong crossvein-like connection. In the Ptychopteridae, only a fold is present in species of *Ptychoptera* Meigen, whereas *Bittacomorpha clavipes* (Fabricius) seems to have a distinct crossvein. If this crossvein is really MA, its occurrence must be primitive, and its absence, as in the Trichoceridae, must be derived.

- *Venation with A_2 well developed and extending into wing blade, terminating in hind margin of wing (plesiomorphic) / A_2 rudimentary, not extending to hind margin of wing, and, in most Nematocera, not evident at all (apomorphic)*

The extent of A_2 in the Trichoceridae and Tipulidae is easily interpreted as being primitive, as Hennig did. Because the feature is a symplesiomorphy, however, its occurrence does not necessarily indicate a close relationship between these two families. In species of Trichoceridae that we have examined, the setae on its dorsal surface do not continue onto the abruptly bent apical portion, whereas in the Tipulidae that we have seen these setae continue to the apex of the vein. This observation suggests that, in the Trichoceridae, the apex of A_2 may be a crossvein and not homologous with what is assumed to be the same vein in the Tipulidae. Further study is needed. Additional evidence to corroborate that the Oligoneura form a monophyletic group seems to be lacking; in absence of such evidence we can postulate that loss of A_2 could have happened twice, but we hesitate to make any interpretation without further information.

- *Wing base broadening gradually, tapering evenly from base, thus without abrupt differentiation of wing into a stalk and blade (plesiomorphic) / wing base abruptly constricted near base, differentiating blade from stalk (apomorphic)*

Hennig (1968) made much of this character, believing that the plesiomorphic condition set the Polyneura apart from the rest of the Diptera (Oligoneura). However, the differences in width of the stalk between the Trichoceridae (his Fig. 2) and the Ptychopteridae (his Fig. 3) seem to us to be a result either of the presence in the Trichoceridae of a well-developed A_2, or of its absence in Ptychopteridae, as well as a result of the distance from the wing base of MA (discussed in the preceding item). Although Hennig did not couple these three character states, we postulate here that the extent of the stalk, the extent of A_2, and the position of MA may be functionally interrelated.

- *Transverse sutures of the scutum not meeting medially (plesiomorphic) / transverse sutures meeting middorsally, forming a V-shaped impression on the scutum (apomorphic)*

The distribution of the supposed apomorphic state of this character is puzzling; a truly distinct, V-shaped impression is present only in the Tipulidae, Blephariceridae, and Perissommatidae, and, much less distinctly, in the Ptychopteridae. In the Trichoceridae, which are supposed to share the apomorphic condition with the Tipulidae, the medial part of the impression is not well delineated. We believe that further study is needed to resolve the homology, distribution, and polarity of this character.

- *Mandibles present (plesiomorphic) / mandibles lacking (apomorphic)*

Mandibles have been lost in all the Bibionomorpha, Axymyiomorpha, Ptychopteridae, and Psychodomorpha, except some Psychodidae, and in at least some members of nearly every family of the Culicomorpha. Because the loss has occurred so many times we conclude that loss in the Tipulidae and Trichoceridae is doubtfully homologous.

- *Adult with a tergal depressor muscle of the trochanter (plesiomorphic) / adult lacking this muscle (apomorphic)*

As in the previous character, the tergal depressor muscle is apparently absent in several otherwise unrelated families, namely those with long legs and those that do not spring into the air when taking flight. Hennig pointed out that absence of this muscle could not be regarded as a synapomorphy. Although he held out hope that its absence in the Trichoceridae and Tipulidae could be a synapomorphy of the two families, evidence of this loss being a synapomorphy is as weak as it is for the previous character.

- *Aroleum present (plesiomorphic) / aroleum absent, replaced by a flap-like empodium (apomorphic)*

The sac-like aroleum at the apex of the tarsus was said to be "rudimentary or absent in most Diptera except Tipulidae" (Snodgrass 1935: 200). Hennig (1968), in attributing this conclusion to de Meijere (1901), suggested that the apomorphic condition might be a synapomorphy of the Oligoneura. We cannot see any fundamental difference between the pad-like structure at the apex of the tarsus in either the Tipulidae or the Trichoceridae, or in any dipteran in which an empodium is present, and we suggest that this character needs reexamination.

Proposed classification of the Tipulomorpha

In the arrangement we present here, we have relegated the Trichoceridae to the Psychodomorpha (see previous section and "Phylogeny of the Infraorder Psychodomorpha") and consider the Tipulidae *sensu lato* (Hennig's Tipuloidea) to be the only family belonging to the Tipulomorpha. In Volume 1 of this Manual the Tanyderidae were also included, but we have since found a more defensible position for them as the sister group of the Ptychopteridae (see previous section and "Phylogeny of the Infraorder Ptychopteromorpha"). Our proposed classification is as follows:

Infraorder Tipulomorpha
  Family Tipulidae
    Subfamily Tipulinae
    Subfamily Cylindrotominae
    Subfamily Limoniinae

Character states. Only one character state is used in interpreting the relationship of the Tipulomorpha, which here consists of the single family Tipulidae.

Larva

7. *Head capsule with well-defined posterior margin, not capable of complete withdrawal into thorax (plesiomorphic) / posterior margin of head capsule desclerotized and shallowly to deeply notched (Figs. 7.82, 7.87–88), with head capsule capable of complete withdrawal into thorax (apomorphic)*

The characteristic, fully retractible head, along with the desclerotized posterior margin of the head capsule, is one of the most distinctive synapomorphies of the family and is unique in the Nematocera. Larvae of the Brachycera have undergone a somewhat similar modification, although the desclerotization is in different areas. We assume the condition in the Brachycera is convergent because other concordant synapomorphies are not evident and because the similarity is probably only superficial.

Discussion. The presence in both the Tanyderidae and the Tipulidae of well-developed cervical sclerites and an elongate antepronotum gives adults of these two families a long-necked appearance. The feature was first thought, during preparation of Volume 1 of this Manual, to be a synapomorphy. However, no other character states have been found to substantiate such a relationship, and this arrangement was abandoned in favor of placing the Tanyderidae as the sister group of the Ptychopteridae on the basis of a shared synapomorphy in the legs of the male adult. Apomorphies shared by the Ptychopteromorpha (including the Tanyderidae) and Culicomorpha also support such a placement.

PHYLOGENY OF THE INFRAORDER BLEPHARICEROMORPHA

Hennig's classification

Hennig (1973) did not recognize a separate infraorder, the Blephariceromorpha, that we propose here. Instead, he included its members as two superfamilies, Blephariceroidea and Nymphomyioidea, in the Psychodomorpha as follows:

Superfamily Blephariceroidea
  Family Blephariceridae
    Subfamily Edwardsininae
    Subfamily Blepharicerinae
    Subfamily Paltostomatinae
    Subfamily Apistomyiinae
  Family Deuterophlebiidae
Superfamily Nymphomyioidea
  Family Nymphomyiidae

He included these three families in the Psychodomorpha, along with the Psychodidae, Ptychopteridae, and Tanyderidae, because they shared the coalescence of the mesomeron with the mesoepimeron. This character state had been recognized by Crampton (1925) as derived. Hennig (1968) expressed reservations about this character state as a synapomorphy and suggested that the affinities of the families possessing it were not understood. He regarded the placement of the Nymphomyiidae in this assemblage as particularly tenuous (Hennig 1973).

Hennig (1973) based the superfamily Blephariceroidea on the presence of a secondary network of folds in the wing, resulting from its development in an expanded condition within the pupal wing sheath before emergence of the adult. In all other insects, the wing is expanded only after emergence (see character 14).

Status of the Nymphomyiidae

Hackman and Väisänen (1982) recently resurrected and supported Rohdendorf's (1964) contention that the Nymphomyiidae (suborder Archidiptera) is the sister group of other extant Diptera (suborder Eudiptera). They reiterated evidence given by Rohdendorf (1964, 1977) and Cutten and Kevan (1970). We present here a more detailed analysis of these arguments.

Rohdendorf's original arguments for placing the Nymphomyiidae in the Archidiptera are unclear and we regard most of the character states he cited as autapomorphies. These are as follows: pupal head prognathous; adult with two large ocelli, each located behind the compound eye; adult with large elongate thorax; adult with two Malpighian tubules; and adult lacking a crop. The first four characters appear to be unique and are autapomorphic. The last one, however, requires further investigation. Adult Mecoptera and Siphonaptera also lack a crop, which, although not broadly investigated, is a character state apparently present in most Diptera, although it is lacking (presumably secondarily) in some Asilidae, in *Hippobosca* Linnaeus, and in some Oestridae (Hennig 1973). Hence, lack of a crop may represent the plesiomorphic condition; alternatively the crop may have been lost in extant Nymphomyiidae, which apparently do not feed.

Rohdendorf also included two fossil infraorders, the Dictyodipteromorpha and Diplopolyneuromorpha, known only from portions of wings, in the Archidiptera. The character state used to support this grouping was an elongate wing with parallel anterior and posterior margins. We fail to appreciate either the significance of this character state or the accuracy of the observations.

Cutten and Kevan (1970) provided a more detailed comparison of nymphomyiids with other Diptera. Of these comparisons, only a few, discussed here, seem significant to a phylogenetic placement of the Nymphomyiidae.

- *Larva with long, two-segmented prolegs on abdominal segments 1–8*

We consider the two-segmented condition to be only apparent, not representing true segmentation in the sense of thoracic legs of other insect larvae. Deuterophlebiidae larvae also exhibit a similar division of the leg; in both these groups the terminal portion, including the terminal hooks, can be withdrawn into the basal portion (Pulikovsky 1924), a condition not found in truly segmented legs. Cutten and Kevan (1970) also illustrated a muscle running the length of each leg, inserted at the apex of the extended leg; this muscle presumably pulls the end of the leg inside the basal portion, withdrawing the terminal hooks. Although Cutten and Kevan (1970) compared this abdominal proleg with those found in Zeugloptera (Lepidoptera) and Mecoptera, we regard this condition as not being homologous to that found in these outgroups. Instead, we consider the division between the two portions of the leg to be merely a flexion line to allow this movement.

The presence of abdominal prolegs, with apices that can be pulled inside a basal portion of the leg, seems, in the Nematocera, to be restricted to Nymphomyiidae and Deuterophlebiidae. However, similar-looking prolegs also occur in a few members of the Brachycera, e.g. *Atherix* Meigen (Fig. 32.7) and *Hemerodromia* Meigen (Fig. 47.56), a state that we regard as convergent. We do not know whether the apices of brachyceran prolegs can be pulled inside or how they function.

- *Pupa with articulated mandibles (decticous)*

This character state would possibly provide evidence that the Nymphomyiidae could be the sister group of all other Diptera, whose pupae are adecticous. All known pupae of Mecoptera are decticous, but those of Siphonaptera are not (see character 4 under "Monophyly and Phylogenetic Origins of the Diptera"). However, we have been unable to confirm Cutten and Kevan's statement, either from specimens of pupae of *Palaeodipteron walkeri* Ide, or from the literature, and we are unsure of the basis for their assertion.

- *Adult springing into the air on taking flight, by flexing the abdomen (observed by Tokunaga 1956, 1965)*

Adult Zeugloptera (Lepidoptera), Choristidae, and Nannochoristidae (Mecoptera) are also reported to take flight in this fashion. If this character state is not found in any other dipteran, it would be uniquely plesiotypic within the Diptera. Unfortunately, the method of taking flight is not well-enough studied in other taxa of Diptera to use in phylogenetic interpretation.

Other characters mentioned by Cutten and Kevan (1970) include: head capsule of larva with an epicranial suture; apex of larval antenna bifurcate; larval eyespot large; rostrum of larval head capsule projecting anteriorly; adult cerci possibly present; and a valvular ovipositor possibly present, reported in only one species of nymphomyiid, *Felicitomyia brundini* Kevan. None of these character states seems unique to the Nymphomyiidae, nor is any described in sufficient detail to be sure of homologies with other Diptera. For example, large larval eyespots of Nymphomyiidae were compared to those of Culicidae, but in the latter they are actually developing adult eyes and are not homologous. We feel that the larval eyespots of the Nymphomyiidae are not significantly different from those of, for example, simuliid larvae.

Another unusual feature of *Nymphomyia alba* Tokunaga, described in great detail by Tokunaga (1935, 1936) and interpreted by him as a primitive condition among Nematocera, is the presence in the adult male of eight separate, discrete, abdominal ganglia. This character state is apparently rare among adult Holometabola. Tokunaga did not comment on the female, and his drawing is of the male. Although neither Rohdendorf (1964) nor Cutten and Kevan (1970) mentioned this character, it may have influenced Rohdendorf when creating the Archidiptera.

In a series of six consecutive papers, Brandt (1880*a–f*) described briefly the nerve cords of a wide range of insects, especially of the orders Hymenoptera, Coleoptera, Lepidoptera, Diptera, and Siphonaptera (as part of Diptera). Although he dealt mostly with adults, he included some larvae. A few examples, showing differing degrees of coalescence of both thoracic and abdominal ganglia, were illustrated in considerable detail, showing segmental boundaries and lateral nerves and their destinations. Among them were six different Nematocera: the larva of *Culex* Linnaeus, and adults of *Bibio* Geoffroy, *Psychoda* Latreille, *Tipula* Linnaeus, *Chironomus* Meigen, and *Sciara* Meigen, as well as many Brachycera. The larva of *Culex* was shown with eight separate, abdominal ganglia, but in all the adults, of both sexes, the first abdominal ganglion was fused with the metathoracic ganglion, and the last two (7 and 8) were also fused. In *Sciara*, as in all Brachycera, there was further fusion of abdominal ganglia.

In contrast, in three species of fleas he studied, all eight ganglia were separate in the male, whereas in the female the first abdominal ganglion was fused with the metathoracic ganglion, and the remaining seven ganglia were separate.

In a study of several genera of Mecoptera, representing four families, Potter (1938) also demonstrated fusion of the first abdominal ganglion with the metathoracic ganglion as well as fusion of the seventh and eighth abdominal ganglia. In addition, she found sexual dimorphism similar to that reported by Brandt for fleas; females of some Mecoptera had one fewer abdominal ganglion than did males. In *Merope* Newman, she illustrated eight separate abdominal ganglia in the male and only seven in the female; in both, the first abdominal ganglion was located quite close to the metathoracic ganglion. We conclude that she identified the ganglia correctly from her statements (p. 482), that "The first abdominal ganglion lies in the metathorax, but innervates the first abdominal segment. The second ganglion lies in the first segment but supplies the second." However, for the female of *Panopa* Linnaeus, she illustrated six separate abdominal ganglia but stated (p. 471) that "The metathoracic ganglion innervates the spiracles of the first abdominal segment in addition to those of the metathorax," which suggests to us that the first abdominal ganglion has fused with the metathoracic ganglion. In *Bittacus* Latreille, she stated that the first abdominal ganglion is fused with the metathoracic ganglion yet innervates the first abdominal segment, and she illustrated this condition for both sexes.

Although it is difficult to develop a clear transformation series from the data available, evidently, at least in some Mecoptera and in female fleas, there is fusion between the first abdominal ganglion and the metathoracic ganglion in the adult. Brandt showed this fusion for the Nematocera he studied, as well as for all Hymenoptera, including a representative of Symphyta (Brandt 1880*c*), all Coleoptera (Brandt 1880*d*), and all Lepidoptera (Brandt 1880*e*). In contrast, the larvae of Hymenoptera and Lepidoptera that he illustrated had eight separate, abdominal ganglia. So widespread is this fusion throughout adult Holometabola that we are tempted to consider it as a part of the ground plan of members of this group.

Fusion of the seventh and eighth abdominal ganglia (and often more) is also widespread in all the holometabolous orders, even among their supposedly oldest groups, i.e. Symphyta (Brandt 1880*c*), Zeugloptera (Kristensen and Nielsen 1981). At the very least there is widespread homoplasy regarding fusion at the distal end of the nerve cord. As a result we cannot regard Nymphomyiidae as the sister group to the rest of the Diptera, as did Rohdendorf (1964, 1977), Cutten and Kevan (1970), and Hackman and Väisänen (1982), in spite of the presence of eight separate ganglia. Rather, we believe the condition in Nymphomyiidae is best interpreted as derived, possibly by neotenic retention of the larval condition.

Proposed classification of the Blephariceromorpha

Zwick (1977) provided an outstanding cladistic analysis of the Blephariceridae and we have followed his subfamilial classification. Our classification is as follows:

Superfamily Blephariceroidea
  Family Blephariceridae
    Subfamily Edwardsininae
    Subfamily Blepharicerinae (including Paltostomatini and Apistomyiini)
  Family Deuterophlebiidae
Superfamily Nymphomyioidea
  Family Nymphomyiidae

Character states. In the following, we analyze character states 8–19, used in interpreting relationships within the Blephariceromorpha.

Larva

8. *Abdominal segments without prolegs (plesiomorphic) / each abdominal segment with a pair of elongate prolegs, each with a protrusible apex surrounded by several concentric rings of hooklets or by a group of setae (apomorphic)*

Larvae of all instars of Nymphomyiidae and Deuterophlebiidae have paired, lateral, abdominal prolegs with the recurved hooklets at their apices. There are eight pairs in Nymphomyiidae but only seven pairs in Deuterophlebiidae, although a small pair of processes on the last abdominal segment in the latter may be homologous to an eighth pair. In the Blephariceridae, only the first instar larvae have prolegs and these, too, may terminate in recurved hooklets or protrusible setae (Zwick 1977: Figs. 23–28). These prolegs appear to be unique among the Nematocera. Although they bear a superficial resemblance to abdominal prolegs in the Mecoptera and Lepidoptera, we conclude that they are not homologous (see discussion under "Status of the Nymphomyiidae").

9. *Head and thorax each distinct and thorax separate from first abdominal segment (plesiomorphic) / head, thorax, and first abdominal segment fused into a compact mass, the "cephalic division" (Zwick 1977) (apomorphic)*

Fusion of head, thorax, and first abdominal segment, characteristic of all larvae of Blephariceridae, is one of the most complex and distinctive synapomorphies of the members of this family and is unique to them.

10. *Abdominal segments without suction cups (plesiomorphic) / first six abdominal segments each with midventral suction cup (Figs. 8.7–8) (apomorphic)*

Among other Nematocera, only larvae of the psychodid genus *Maruina* Müller possess ventral suction cups. These cups have a quite different structure and musculature from those of Blephariceridae, and we believe them to have been derived independently from an ancestor with a single, large, suction cup covering the ventral surface, as is known in extant larvae of *Horaiella* Tonnoir, by subdivision of the single, large cup.

11. *Antenna short and simple, not particularly elongate, nor bifurcate apically (plesiomorphic) / antenna exceptionally long and thick relative to body size, more than twice as long as head, bifurcating into two long, whip-like, transparent branches (Fig. 9.2) (apomorphic)*

Larvae of the Deuterophlebiidae are unique in having long, bifurcate antennae. This feature may be correlated with their rather rapid method of progressing over the substrate.

Pupa

12. *Pupa not directly cemented to its substrate by adhesive areas (plesiomorphic) / pupa flattened ventrally, closely appressed to its substrate, attached by three or four pairs of oval, adhesive areas (apomorphic)*

Lateral adhesive areas, for pupal attachment to the substrate, are apparently unique to pupae of the Blephariceridae and the Deuterophlebiidae. They are located laterally on abdominal segments 3–5 in the Deuterophlebiidae (Fig. 9.6), on segments 3–6 in the tribe Apistomyiini, and on segments 4–6 in other Blephariceridae (Figs. 8.14, 9.6). Although a comparative analysis has not been made, their appearance, function, and placement are sufficiently similar to suggest that they are homologous and are a synapomorphy of the two families. Pupae of the aquatic or subaquatic genus *Maruina* (Psychodidae) are also flattened ventrally and appressed to the substrate, but they lack the adhesive organs.

13. *Head hypognathous (plesiomorphic) / head prognathous (Figs. 10.8–10) (apomorphic)*

The prognathous pupal head of Nymphomyiidae is not known in any other group of Diptera, Siphonaptera, or Mecoptera and is, therefore, considered to be a synapomorphy of the family.

14. *Adult wing developing within pupal wing sheath as a shrivelled version of its final form, its margins matching those of wing sheath, with wing expanding only during and after emergence (plesiomorphic) / wing reaching its final size and shape before emergence, folded upon itself longitudinally two or three times to fit confines of wing sheath, and unfolding during emergence (apomorphic)*

This type of wing development, apparently unique to the Blephariceridae and Deuterophlebiidae, was mentioned by Hennig (1973) as a synapomorphy of the two families. We concur with his opinion. Edwards (1926: 120) included the Simuliidae with the Blephariceridae and Deuterophlebiidae because he incorrectly assumed that they, too, possessed fully expanded, folded wings before emergence. Wings of simuliids, however, are expanded during emergence.

Adult

15. *All femora and tibiae each comprising a single, undivided segment (plesiomorphic) / all femora and tibiae each divided into two parts by a membranous area (Fig. 10.1) (apomorphic)*

The apomorphic condition is unique to the Nymphomyiidae.

16. *Meron represented by a small, triangular sclerite (plesiomorphic) / meron large and rounded, as fully developed behind the mid coxa as is the katepisternum in front of the coxa (apomorphic)*

This character state may be a synapomorphy of the Deuterophlebiidae and Blephariceridae, but it is by no means unique to them; it is also characteristic of the Axymyiidae, Perissommatidae, and some Tipulidae. The thorax of the Nymphomyiidae is so modified that the meron apparently has not yet been identified.

17. *Male terminalia carried in various positions but usually not reflexed dorsally (plesiomorphic) / male terminalia reflexed dorsally, at least in preserved specimens, carried so that its morpholog-*

ically ventral surface is directed upwards (Fig. 9.1) (apomorphic)

Males of both blephariceridsand deuterophlebiids, at least as preserved specimens, carry their terminalia reflexed dorsally, and we suspect that this feature is another synapomorphy of the two families. However, males of at least one blepharicerid, *Bibiocephala grandis* Osten Sacken, carry their terminalia directed posteriorly when alive. Furthermore, this character state may not be unique among the Nematocera.

18. *Gonocoxites independent from each other and separate from sternite 9 (plesiomorphic) / gonocoxites and sternite 9 fused into a single sclerite (apomorphic)*

The Blephariceridae and Deuterophlebiidae share the apomorphic condition, in contrast with most other Nematocera in which the gonocoxites and sternite 9 are independent. However, this arrangement is not unique to these two families but is also found in other Nematocera; sometimes the two conditions even occur in the same family (e.g. Tipulidae, Simuliidae, and Bibionidae). Nor is it certain that we have interpreted polarity of this character state correctly, for the gonocoxites of the Mecoptera are also fused to one another, although they are separate from sternite 9. Many Brachycera have gonocoxites and hypandrium fused, but other groups have the hypandrium free. Stuckenberg (1973) has discussed this for some Brachycera.

19. *Aedeagus with a single, terminal opening corresponding to a single, external opening of the common spermathecal duct in the female, formed by the union of three spermathecal ducts from the three spermathecae (plesiomorphic) / aedeagus with three long, slender filaments, each with a terminal opening, correlated with three separate spermathecal openings in the female (apomorphic)*

Downes (1968) drew attention to the presence of a trifid aedeagus in the Tanyderidae, the Blephariceridae, and the tipulid subfamily Cylindrotominae (except species of *Liogma* Osten Sacken which have a bifid aedeagus) and suggested that this feature might represent the primitive condition in the Nematocera. Since then, however, a trifid aedeagus has been demonstrated in most Stratiomyidae, some Asilidae, and some Scenopinidae. The number of branches of the aedeagus is apparently associated with the number of external spermathecal openings in the previously mentioned taxa. A trifid aedeagus may have arisen independently in each of these. Hence, we treat it here as a synapomorphy of the Blephariceridae. Alternatively, the presence of a trifid aedeagus in the Tanyderidae, Blephariceridae, and Cylindrotominae could suggest that these three taxa shared a common ancestor, as implied by Downes (1968). We consider such a grouping unlikely in the light of other synapomorphies.

Discussion. We have been unable to elucidate the affinities of these three families with any great certainty. Although members of each family are quite distinct and differ from each other in all stages, we have offered here strong evidence of a sister-group relationship between the Blephariceridae and Deuterophlebiidae and have given reasons for a possible relationship between these two and the Nymphomyiidae. Nevertheless, we acknowledge that the position of the last family remains somewhat tenuous.

Some conflicting evidence may suggest a sister-group relationship between the Deuterophlebiidae and Nymphomyiidae.

Adults of the Nymphomyiidae have a small, median, membranous structure (possibly the labium), which is directed forward and bears a pair of minute papillae (possibly the labella as described in Chapter 10). Adults of the Deuterophlebiidae have a subcranial cavity but lack all traces of mouthparts. All other Nematocera have more or less well-developed mouthparts with at least maxillary palpi evident. We conclude that this condition can best be explained as a result of convergent loss.

Both sexes of members of the nymphomyiid *Palaeodipteron walkeri* emerge from the pupal case fully winged, but their wings are apparently shed soon afterward, for most specimens have been collected in a dealated condition, some *in copulo*, at considerable depths in the hyporheic zone (Cutten and Kevan 1970). Males of *Deuterophlebia* Edwards apparently retain their wings throughout their short lifespan, whereas females of *D. inyoensis* Kennedy, captured as they flew toward a swarm of males, shed them in captivity soon after their capture (Wood, personal observations). Pennak (1951) and Kennedy (1960) found only dealated females of *D. coloradensis* Pennak and *D. inyoensis*, respectively, which they assumed had lost their wings accidentally or from exertion. We suppose that deliberate shedding of wings is an adaptation for reentry into fast-flowing water, although there are apparently no observations on the oviposition habits of the Deuterophlebiidae.

These two character states suggest a possible sister-group relationship between the Nymphomyiidae and the Deuterophlebiidae, although we find it difficult to regard either one as being a synapomorphy, because of the conflicting evidence offered by the pupae. No blepharicerid is known that sheds its wings, although an ancestor of this family may have had the ability but had lost it. We have chosen to regard the shared character states of the Deuterophlebiidae and Blephariceridae presented here as synapomorphies and, hence, must consider the above two similarities between the Nymphomyiidae and the Deuterophlebiidae as convergences.

PHYLOGENY OF THE INFRAORDER AXYMYIOMORPHA

Hennig's classification

The systematic placement of the single, small family Axymyiidae has been debated ever since the type-genus *Axymyia* McAtee was originally described. First associated with the Bibionidae (Coquillett 1909), then with the Anisopodidae (Shannon 1921, Alexander 1942), the group has most recently been placed in the Pachyneuridae (Alexander 1965, Hennig 1973). Hennig regarded the Pachyneuridae (including the Axymyiidae) and Perissommatidae (which has extant representatives only in Australia and Chile) as comprising the Section Pachyneuriformia as a portion of the infraorder Bibionomorpha, as follows:

Infraorder Bibionomorpha
  Section Pachyneuriformia
    Family Perissommatidae
    Family Pachyneuridae (including Axymyiidae)

The assemblage was based on the simultaneous forked condition of R_{2+3} and the unbranched condition of R_{4+5}. In other members of the Bibionomorpha, R_{2+3} is unbranched, and the radial section, if it is forked at all, has only two branches, R_{2+3} and R_{4+5}.

Proposed classification

We propose classifying the Axymyiomorpha as containing the single family Axymyiidae, as follows:

Infraorder Axymyiomorpha
  Family Axymyiidae

Character states. In the following, we analyze character states 20–25, used in interpreting the monophyly of the Axymyiidae, the only family in our Axymyiomorpha.

Larva

20. *Labrum broad, flattened, usually moderately bilobate apically (plesiomorphic) / labrum in the form of a small, conical, setose projection (apomorphic)*

The apomorphic condition seems peculiar to the Axymyiidae and may be unique in the Nematocera. However, larvae of other families, e.g. the Blephariceridae and Thaumaleidae, have laterally compressed, finger-like labra, presumably as a result of the approximation of the mandibles medially.

21. *Mandible broad and stout, terminating in interlocking teeth (plesiomorphic) / mandible longer than broad, sharply pointed apically, with secondary tooth on inner margin (apomorphic)*

In the plesiomorphic condition, found in the Tanyderidae, Pachyneuridae, and Bibionidae, the mandibles appear to be adapted for crushing solid material, either for food or for burrowing. Although thorough studies have not been made, larvae of *Axymyia* spp. are apparently incapable of chewing the wood that they excavate to form their chamber. Their mandibles arise close together medially, and the extent of their movement is so limited that only the sharp apices meet medially. Their molar surfaces do not interact, and their mandibles are apparently only capable of enlarging the chamber. Differences in the structure, position, and action of the mandibles in pachyneurid and axymyiid larvae, as well as differences in food preferences and burrow shape (described in Chapters 11 and 12 of this Manual), have led us to conclude that the wood-boring habits of the two groups are not homologous.

22. *Maxilla well developed but having only a small palpus (plesiomorphic) / maxilla having a well-developed palpus, but with the remainder reduced to a finger-like setose lobe (apomorphic)*

The apomorphic condition appears to be unique to the Axymyiidae.

23. *Anal papillae, if present, no longer than one-third the total body length (plesiomorphic) / anal papillae about one-half the body length or longer (apomorphic)*

Axymyiid larvae are unique among Diptera in possessing such well-developed anal papillae. In most species the papillae are also branched, but in an undescribed species from western North America, the papillae are simple.

Pupa

24. *Abdominal tergite 8 lacking siphon (plesiomorphic) / abdominal tergite 8 with elongate siphon (Fig. 11.7) (apomorphic)*

Axymyiid pupae are the only Diptera known to have an abdominal siphon. We believe this organ is homologous with the larval siphon (Fig. 11.6) and may have a respiratory function. It bears apically a reduced set of spicules similar to that of the apex of the larval siphon.

Adult

25. *Surface of scutum variable, either glossy, pruinose, or a combination of these (plesiomorphic) / scutum with pair of dark, oval, shiny spots near middle (apomorphic)*

The apomorphic condition is unique to the Axymyiidae.

Discussion. We conclude that the forked condition of R_{2+3} is probably plesiomorphic and that other features of the Axymyiidae show no hint of a synapomorphy with the Pachyneuridae or Perissommatidae.

Hennig (1973) noted that an undescribed female from Alaska, known only from a single, damaged female in the

Canadian National Collection in Ottawa, shared a weakening of the base of the vein M with the Perissommatidae. We have reexamined this specimen in some detail and, though there is evidence of an affinity with the Axymyiidae, we fail to identify any synapomorphy shared with the Perissommatidae. Our investigations indicate a relationship between Perissommatidae and members of the Psychodomorpha where it is discussed in more detail.

We have proposed a new infraorder, the Axymyiomorpha, for the Axymyiidae because we cannot find any synapomorphies that show conclusively to what other taxa it may be related.

PHYLOGENY OF THE INFRAORDER BIBIONOMORPHA

Hennig's classification

Hennig (1973) included four sections, the Pachyneuriformia, Anisopodiformia, Bibioniformia, and Mycetophiliformia, in this infraorder. He presented only one character state, the reduction of the costa at the posterior margin of the wing, as a synapomorphy of the group. He also noted that no more than two spermathecae had been demonstrated in members of the Bibionomorpha. As the primitive number in the Diptera was presumed to be three, Hennig implied that this reduction might also be a synapomorphy of the group. Contrary to this hypothesis, however, bibionid females have three spermathecae whereas those of Anisopodidae have one to three. Changes in the number of spermathecae have occurred in so many lineages within so many families of Nematocera, that we do not consider this character state to be a trustworthy indicator of relationship. Finally, the belief that possession of three spermathecae is plesiomorphic in the Diptera may be questioned. Outgroup comparisons show that members of the Siphonaptera have one to two spermathecae and those of the Mecoptera have one. Within Nematocera, at least, some or all members of each of the following families have only one spermatheca: Tipulidae, Cecidomyiidae, Psychodidae, Anisopodidae, Scatopsidae, Dixidae, Corethrellidae, Culicidae, Simuliidae, and Ceratopogonidae.

Additional, uninterpreted features of the Bibionomorpha mentioned by Hennig include the absence of aquatic larvae and the loss of the blood-sucking habit, with concomitant loss of mandibles. Hennig's classification, based almost entirely on apparently derived conditions of the wing venation, was as follows:

Infraorder Bibionomorpha
- Section Pachyneuriformia
  - Family Perissommatidae
  - Family Pachyneuridae (including Axymyiidae)
- Section Anisopodiformia
  - Family Anisopodidae
  - Family Cramptonomyiidae
- Section Bibioniformia
  - Family Hesperinidae
  - Family Bibionidae
- Section Mycetophiliformia
  - Superfamily Scatopsoidea
    - Family Hyperoscelidae
    - Family Scatopsidae
  - Superfamily Cecidomyioidea
    - Family Lestremiidae
    - Family Cecidomyiidae
  - Superfamily Mycetophiloidea
    - Family Ditomyiidae
    - Family Diadocidiidae
    - Family Keroplatidae
    - Family Bolitophilidae
    - Family Mycetophilidae
    - Family Sciaridae
    - Family Baeonotidae (placed here provisionally)

Proposed classification

By our definition the infraorder Bibionomorpha contains most of the families included by Hennig. We have rearranged the elements, however, into three superfamilies, the Pachyneuroidea, Bibionoidea, and Sciaroidea, with the important exclusion of five families, the Anisopodidae, Synneuridae (which Hennig called Hyperoscelididae), Scatopsidae, Perissommatidae, and Axymyiidae. We have combined the Perissommatidae, Anisopodidae, Scatopsidae, and Synneuridae with the Psychodidae and Trichoceridae to constitute the Psychodomorpha on what we believe to be shared derived character states in the larval mouthparts. We have excluded the Axymyiidae simply because we lacked definite evidence of their affinities with the Bibionomorpha proper; we found no synapomorphies and deemed the similarities that seem to exist to be superficial.

We have treated the Cramptonomyiidae (*sensu* Krivosheina and Mamaev 1970 and Hennig 1973) as a subfamily of the Pachyneuridae, because complex details of the male terminalia, which we believe to be derived character states, are shared by all members of both taxa. We have combined the Hesperinidae with the Bibionidae, as a subfamily, because of the presence of synapomorphies in larvae of *Hesperinus* Walker, and other bibionids. For the same reasons, we have treated the Lestremiidae as a subfamily of the Cecidomyiidae. The Mycetophilidae *sensu lato*, however, present a more complex problem. Although treated here as a single family, in keeping with North American tradition, it is probably paraphyletic. Unfortunately the entire group is not yet well enough understood to interpret phylogenetic relationships in any detail. Some authors have regarded the Sciaridae as being merely a subfamily of the Mycetophilidae, whereas others have excluded it from the Mycetophilidae probably to avoid having to deal with it. Whether or not it is the sister group of the Cecidomyiidae will have repercussions on the ranking of the other subgroups of

Mycetophilidae. Some evidence is discussed later that suggests sciarids have arisen within the Mycetophilidae and that a sister-group relationship exists between sciarids and cecidomyiids. Because we have been unable to resolve the question of how the subfamilies of Mycetophilidae are related to one other, we have retained the Sciaridae here as a separate family.

We have placed, as an unjustified cladogram, the Bibionidae and Pachyneuridae as successive sister groups to the monophyletic Mycetophilidae, Sciaridae plus Cecidomyiidae lineage. This interpretation is based on what we perceive as a phenetic grade and, therefore, has no logical basis. Clearly, a major effort needs to be made to interpret the relationships between the families we include here in the Bibionomorpha.

In this Manual we have included the enigmatic genus *Baeonotus* Byers, given family rank by Byers (1969) and followed by Hennig (1973), in the Cecidomyiidae because of two synapomorphies, the rather regular arrangement of setae on the flagellomeres and the lack of tibial spurs. Our proposed classification, based on the phylogenetic analysis below, is as follows:

Infraorder Bibionomorpha
- Superfamily Pachyneuroidea
  - Family Pachyneuridae
    - Subfamily Pachyneurinae
    - Subfamily Cramptonomyiinae
- Superfamily Bibionoidea
  - Family Bibionidae
    - Subfamily Hesperininae
    - Subfamily Bibioninae
      - Tribe Pleciini (subfamilial rank in Ch. 13)
      - Tribe Bibionini
- Superfamily Sciaroidea (Mycetophiloidea a junior synonym)
  - Family Mycetophilidae
    - Subfamily Ditomyiinae
    - Subfamily Boletophilinae
    - Subfamily Diadocidiinae
    - Subfamily Keroplatinae
    - Subfamily Lygistorhininae
    - Subfamily Manotinae
    - Subfamily Sciophilinae
    - Subfamily Mycetophilinae
  - Family Sciaridae
  - Family Cecidomyiidae
    - Subfamily Lestremiinae
    - Subfamily Porricondylinae
    - Subfamily Cecidomyiinae

Character states. In the following, we analyze character states 26–36, used in interpreting relationships within the Bibionomorpha.

Larva

26. *Head capsule and mouthparts complete and of normal size (plesiomorphic) / head capsule minute, reduced to a tiny cone at the anterior end of the larva, with stylet-like mandible (Figs. 16.5, 16.7–9) (apomorphic)*

A reduced head capsule with minute styliform mandibles is one of the most distinctive synapomorphies of the Cecidomyiidae (see also discussion of postgenal lobes below). This character state is unique within the Nematocera. Although larvae of the Synneuridae also have a reduced head capsule, we believe that the feature is convergent because details differ and other character states suggest that this family belongs in the Psychodomorpha.

27. *Cardo free from anterior margin of head capsule (Fig. 3.2) (plesiomorphic) / cardo fused with, or closely appressed to, anterior margin of head capsule (apomorphic)*

Fusion, or at least close association, of the posterior edge of the cardo with the adjacent margin of the head capsule is found only in the Mycetophilidae (including Ditomyiinae) and the Sciaridae. Little information is available for the Cecidomyiidae. Petralia *et al.* (1979: Figs. 3, 4) show the base of the maxillae fused to the cranium in larvae of *Contarinia sorghicola* (Coquillett). We assume that this condition is widespread in the family. Otherwise the apomorphic condition appears to be unique in the Nematocera.

28. *Postgenal lobes more or less meeting midventrally to form a hypostomal bridge (Figs. 14.100, 14.102) (plesiomorphic) / lobes meeting in two places, enclosing a pyriform membranous area (Fig. 15.33) (apomorphic)*

This peculiar and apparently unique condition seems to be a synapomorphy of the Sciaridae.

29. *Thoracic prosternum unsclerotized (plesiomorphic) / a rod-like sclerite, the sternal spatula, present on ventral surface of prothorax (apomorphic)*

A sternal spatula is found in most Cecidomyiidae, although it is missing in a few derived members, and is unique to that family.

30. *Metathoracic spiracle present (plesiomorphic) / metathoracic spiracle absent (apomorphic)*

Among larvae of Diptera, only those of *Pachyneura* Zetterstedt and the Bibionidae (including *Hesperinus*) possess a metathoracic spiracle. If its loss is a reliable indicator of relationship, this feature could be a synapomorphy of the subfamily Cramptomyiinae and the Mycetophilid-sciarid-cecidomyiid lineage. Alternatively, as interpreted here, it could have been lost independently in the two groups as well as all other Diptera. A metathoracic spiracle is present in the Nannochoristidae.

31. *Intersegmental fissures each continuous around body (plesiomorphic) / intersegmental fissures not* *continuous laterally, rather, with dorsal and ventral fissures of each segment not aligned with one another (Figs. 13.12–14) (apomorphic)*

In nondipterous holometabolous insect larvae and in larvae of all other Diptera, each body segment is delineated by a continuous intersegmental fissure that encircles the body. The separation and displacement of the dorsal and ventral portions of each fissure is unique in the Bibionidae and, thus, is considered a synapomorphy of the family, one shared by *Hesperinus* and members of the Bibioninae.

32. *Somatic and germ cells with an equal number of chromosomes (plesiomorphic) / elimination of some chromosomes in the somatic cells, resulting in an unequal number in the germ and somatic cells (apomorphic)*

According to White (1973) some chromosomes are eliminated in the somatic cells of Sciaridae (about 20 species of *Sciara* Meigen and one species of *Plastosciara* Berg were studied). Matuszewski (1982), in his summary of the literature, listed 26 genera of cecidomyiids that exhibit chromosome elimination (all that were studied). Although such an elimination is not unique within the Nematocera in that it is also present in the distantly related chironomid subfamily Orthocladiinae, this feature probably is a synapomorphy of the Sciaridae and Cecidomyiidae.

Adult

33. *Compound eyes separated dorsally (plesiomorphic) / each compound eye with a medially directed extension meeting dorsally in the form of an eye bridge (Figs. 15.1, 16.1) (apomorphic)*

The shared presence of an eye bridge in the Sciaridae and the Cecidomyiidae (secondarily lost in a few) has led us to postulate a sister-group relationship between these two families. Among other Diptera, a similar eye bridge is only present in the Scatopsidae and Synneuridae. However, we believe that the condition has developed in those two families independently because of the presence of apomorphies suggesting a relationship between the Scatopsidae and members of the Psychodomorpha.

34. *Setae of flagellomeres arranged haphazardly (plesiomorphic) / setae of flagellomeres arranged in encircling rows or whorls (apomorphic)*

Although subject to much variation, the flagellomeres of almost all cecidomyiids have their setae arranged in rows; sciarids and mycetophilids usually do not. This character state is not unique to the cecidomyiids, however, for various families of Culicomorpha, particularly the males, also have setae arranged in whorls.

35. *Tibial spurs present (plesiomorphic) / tibial spurs absent (apomorphic)*

Tibial spurs are apparently present in all members of the Bibionomorpha but are lacking in the Cecidomyiidae. Therefore, their absence is considered here as being a synapomorphy of that family. The character is otherwise variable in other Nematocera and provides no further evidence for polarization.

36. *Gonostyli usually pointed and, when flexed, usually overlapping each other at midline; parameres and cerci not overhung by hood-like abdominal tergite 9; apices of gonostyli and apicoventral margin of tergite 9 with spicules and elongate setae (plesiomorphic) / gonostyli short and club-shaped, their apices meeting medially beneath rounded or truncate apex of abdominal tergite 9 that overhangs most of terminalia (Figs. 12.3–4); apices of gonostyli and apicoventral margin of tergite 9 studded with short setae (Figs. 12.3–4) (apomorphic)*

Other male Nematocera, including other members of the Bibionomorpha, have the plesiomorphic condition. A few members of the Sciaridae (Fig. 15.20) also have short, stout, clubbed gonostyli, but tergite 9 does not extend as far posteriorly and does not cover the cerci and parameres completely in dorsal view. Both the hood-like abdominal tergite 9 of Pachyneuridae, which overhangs all of the terminalia except for the gonostyli and the distal portions of the gonocoxites and which is studded ventroapically with dense short setae, and the short, clubbed gonostyli, which are also studded apically with short setae, are apparently peculiar to the Pachyneuridae and are regarded as a synapomorphy of that family. Although not studied carefully or compared widely to other members of the infraorder, the fused parameres of Pachyneuridae, which are lobed and folded complexly to form more than a single flange dorsal to the aedeagus, also appear to be unique to Pachyneuridae (Fig. 12.4).

Discussion. Although Hennig (1973, 1981) had no reservations about regarding this infraorder as being monophyletic, this hypothesis is not substantiated by convincing synapomorphies. The absence of mandibles in adult females is undoubtedly a derived character state but is subject to much homoplasy throughout the Nematocera and may not be meaningful in the case of this infraorder.

Neither have we been able to suggest a synapomorphy to support the concept of the Bibionomorpha. Although Pachyneuridae is suggested as being the sister group of the remaining Bibionomorpha and Bibionidae as the sister group of Sciaroidea, we have been unable to provide supporting synapomorphies. The arrangement reflects a phenetic similarity between the taxa that we intuitively believe to be correct.

The presence of eight pairs of abdominal spiracles in the larva has been assumed by all authors to be plesiomorphic. Although no one has demonstrated the likelihood of reappearance of abdominal spiracles after they become vestigial, this evolutionary step may not be impossible (Gould 1980). Most other aspects of the larva, such as the heavy, chewing mandibles, have also been considered plesiomorphic, but they may not be. A detailed comparison of larvae of the Bibionomorpha with those of Siphonaptera and Mecoptera might establish the former as the most plesiomorphic larval form in the Nematocera, but would, thereby, not support the hypothesis of monophyly of the Bibionomorpha.

We have attempted to interpret the phylogenetic relationships within Sciaroidea. Although we present the Mycetophilidae as the sister group of the Sciaridae plus Cecidomyiidae, some evidence suggests strongly that the Mycetophilidae are paraphyletic. The first character state supporting this conclusion is the unique, flattened, and heavily sclerotized maxilla of larvae of Sciaridae and all Mycetophilidae other than Ditomyiinae. The maxillae have a serrated anterior margin with the teeth directed opposite to those of the mandible. They are probably used to tear fungal hyphae. The Cecidomyiidae do not have serrate maxillae but their maxillae are very large (relative to other mouthparts) and, in this regard, are similar to Sciaridae and Mycetophilidae other than Ditomyiinae.

A second character suggesting paraphyly of the Mycetophilidae is the loss of the eighth abdominal spiracle of larvae of Sciaridae and all Mycetophilidae other than Ditomyiinae. All other Bibionomorpha including Cecidomyiidae have the eighth spiracle. We would assume that, if cecidomyiids are the sister group of sciarids, this character has reverted to the plesiomorphic condition.

These two character states indicate that the Ditomyiinae are the sister group of all other Mycetophilidae plus Sciaridae plus Cecidomyiidae. We have been unable to resolve further which specific group of the remaining mycetophilids may be the sister group of sciarids and cecidomyiids. Plachter (1979) provided a detailed morphological study of larvae of Mycetophilidae but unfortunately did not provide a rigorous phylogenetic analysis of his data. In one scheme he suggested that the Diadocidiinae and Mycomyiinae (treated as the tribe Mycomyini in Volume 1 of this Manual) were mostly closely related to Sciaridae.

In conclusion, we can only recommend a continued search for more character states that are synapomorphic, which can support or reject those already put forward.

PHYLOGENY OF THE INFRAORDER PSYCHODOMORPHA

Hennig's classification

Hennig's (1973) Psychodomorpha included four superfamilies, the Blephariceroidea, Nymphomyioidea, Ptychopteroidea, and Psychodoidea, an assemblage he admitted was of questionable monophyly. This assemblage was based on the shared coalescence, in the mesothorax, of the meron with the katepimeron, a character state that he believed was derived, although with reservations. He was particularly doubtful about the inclusion of the Blephariceroidea and Nymphomyioidea in this assemblage. Hennig's classification was as follows:

Infraorder Psychodomorpha
- Superfamily Blephariceroidea
  - Family Blephariceridae
    - Subfamily Edwardsininae
    - Subfamily Blepharicerinae
    - Subfamily Paltostomatinae
    - Subfamily Apistomyiinae
  - Family Deuterophlebiidae
- Superfamily Nymphomyioidea
  - Family Nymphomyiidae
- Superfamily Ptychopteroidea
  - Family Tanyderidae
  - Family Ptychopteridae
    - Subfamily Ptychopterinae
    - Subfamily Bittacomorphinae
- Superfamily Psychodoidea
  - Family Psychodidae
    - Subfamily Bruchomyiinae
    - Subfamily Phlebotominae
    - Subfamily Trichomyiinae
    - Subfamily Psychodinae

Proposed classification

Our composition of this infraorder, presented here for the first time, differs fundamentally from any other proposed arrangement. We have grouped the six families in our Psychodomorpha on the basis of what we interpret as apomorphic larval features. Such features have previously been regarded as plesiomorphic but little evidence exists for such an interpretation. Our classification of this infraorder is as follows:

- Superfamily Psychodoidea
  - Family Psychodidae
    - Subfamily Bruchomyiinae
    - Subfamily Phlebotominae
    - Subfamily Trichomyiinae
    - Subfamily Psychodinae
- Superfamily Trichoceroidea
  - Family Perissommatidae
  - Family Trichoceridae
  - Family Anisopodidae
    - Subfamily Mycetobiinae
    - Subfamily Anisopodinae
  - Family Scatopsidae
    - Subfamily Apistinae
    - Subfamily Scatopsinae
    - Subfamily Ectaetiinae
    - Subfamily Psectrosciarinae
  - Family Synneuridae

Character states. In the following, we analyzed character states 37–50, used in interpreting relationships within the Psychodomorpha.

Larva

37. *Head capsule and mouthparts complete, well developed, and strongly sclerotized (plesiomorphic) / head capsule greatly reduced and membranous, with mouthparts unsclerotized and indistinct (Fig. 21.6) (apomorphic)*

Hutson (1977: 94) listed the character states that are common to larvae of three of the four genera of Synneuridae (which Hutson treated as two families, Synneuridae and Canthyloscelidae), namely *Canthyloscelis* Edwards, *Hyperoscelis* Hardy and Nagatomi, and *Synneuron* Lundstrom. All three seem to be rather similar, and their most unusual feature is the reduction of the head capsule to a weakly sclerotized cephalic plate with rather prominent antennae and scarcely developed mouthparts. We strongly suspect that this complex, as well as considerable reduction, accompanied as it must be by specialized feeding habits, is a synapomorphy of the three genera and presumably of *Exiliscelis* Hutson as well, although the larva remains unknown. Hutson (1977) contended that the Scatopsidae are derived from *Synneuron* as a terminal branch of one of four lineages making up the Synneuridae plus Canthyloscelidae. If his contention is so, the head capsule of the larva must have been lost at least twice (thrice if the larva of *Exiliscelis* also lacks a fully formed head capsule), or else a fully formed head capsule must have been reacquired in the Scatopsidae. We consider both of these possibilities unlikely and have abandoned Hutson's cladogram (his Fig. 26) in favor of a sister-group relationship between the Scatopsidae and the Synneuridae.

Discussion of the other proposed synapomorphies that may be relevant to the sister-group argument between the Scatopsidae and the Synneuridae (Hutson 1977) follows.

38. *Labrum flattened dorsoventrally and usually bilobate (plesiomorphic) / labrum conical or wedge-shaped, rounded, or pointed anteroventrally, more or less covered apically and ventrally with backwardly pointing hairs (apomorphic)*

A conical setose labrum, found in all members of this infraorder except Synneuridae and the psychodid genera *Phlebotomus* Rondani and *Trichomyia* Curtis, may be a result of closer approximation of the mandibles; however, its setation and musculature suggest that it plays a role in feeding on particulate food. This feature is not unique to this infraorder; it is also characteristic of the Blephariceromorpha and Thaumaleidae, in which the mandibles are also closely approximated medially. We presume that the condition in the three taxa is convergent.

39. *Premandible in the form of a simple, small, unadorned sclerite, serving only as the point of insertion of the labral retractor muscles (plesiomorphic) / premandible in the form of a rounded or oval sclerite with a comb-like row of evenly spaced, anteriorly projecting teeth, presumably operated by the labral retractor muscles to comb the mandible or maxilla (Figs. 3.5, 114.3) (apomorphic)*

The presence of such a row of teeth on the anterior edge of a circular or oval premandible is apparently unique to the Psychodidae, Trichoceridae, Perissommatidae, Scatopsidae, and Anisopodidae and is one of the most convincing pieces of evidence of the monophyly of the infraorder here presented. Colless (1962) misidentified the premandibles of the Perissommatidae as parts of the hypopharynx. Careful examination has shown, however, that these structures are dorsal to the mandibles and appear to be on the ventral surface of the labrum.

Keilin and Tate (1940) found only a vestige of the premandible in the psychodid genus *Trichomyia*, a rather specialized wood-borer, in which the entire particulate food-gathering method of feeding presumably has been lost or modified.

40. *Torma fused with, and appearing as an extension from, the dorsal labral sclerite (plesiomorphic) / torma articulated with the dorsal labral sclerite (apomorphic)*

This articulation, which would allow greater mobility in the premandible, may be linked to the presence of its row of teeth. In the Culicomorpha, the torma is also articulated. In the Culicoidea, the torma is disassociated from the dorsal labral sclerite to permit mobility of the labral brushes; we presume this disassociation to be either convergent or a further step in a transformation series leading from a psychodomorph ancestor, through an ancestor of the Ptychopteridae, to the Dixidae. If the latter is true, this synapomorphy might group the Psychodomorpha, Ptychopteromorpha, and Culicomorpha.

We have been unable to identify a torma in the Perissommatidae, perhaps because of the weak sclerotization of the mouthparts in general.

41. *Mandibular movement in a more or less horizontal plane, with the mandibles striking each other apically when adducted (plesiomorphic) / movement in a nearly vertical plane as a result of a shift in position of the condyles (Fig. 3.11), with the mandibles striking the anterior edge of the hypostoma or the hypostomal teeth (apomorphic)*

Although there is a trend in the Blephariceromorpha, Ptychopteromorpha, and Culicomorpha for the mandibles to shift their plane of movement to the vertical, only in this infraorder is the shift so marked. The brachyceran mandible operates in the same way. Therefore, this character state may not be a synapomorphy of the Psychodomorpha alone but might also include the Brachycera.

The only exception to the apomorphic condition is in the Perissommatidae, in which the mandibles appear to move in an oblique plane. This character may suggest that the Perissommatidae may in fact be the sister group to the rest of the Psychodomorpha.

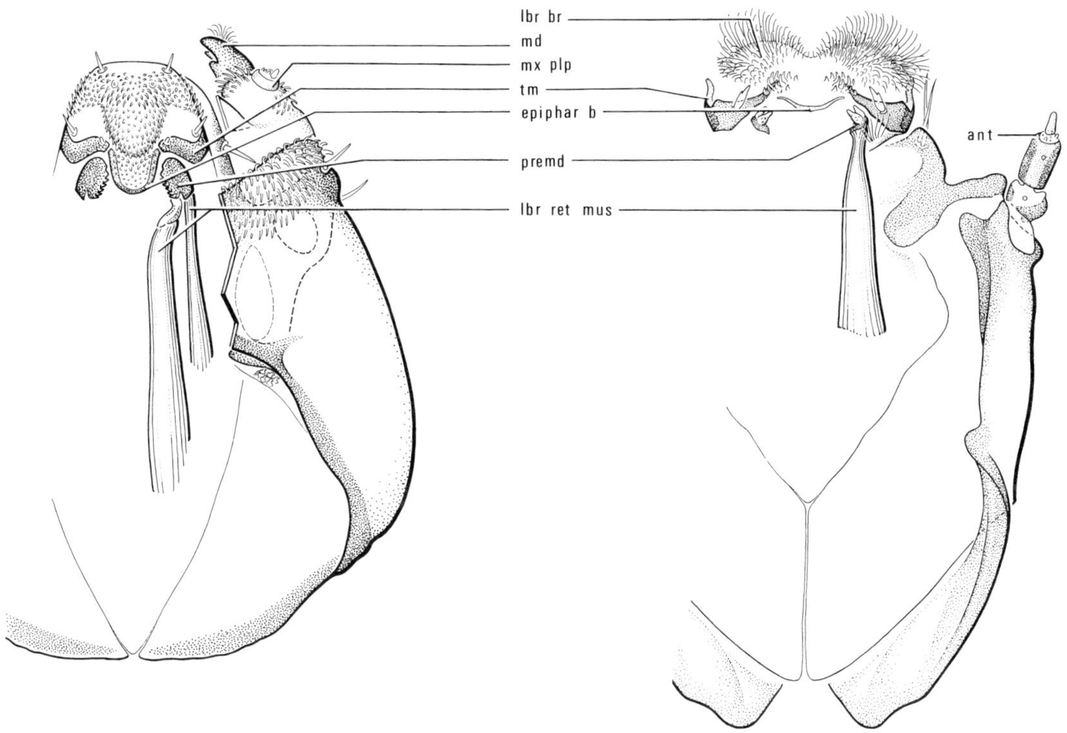

3. Trichoptera sp.

4. Protoplasa fitchii

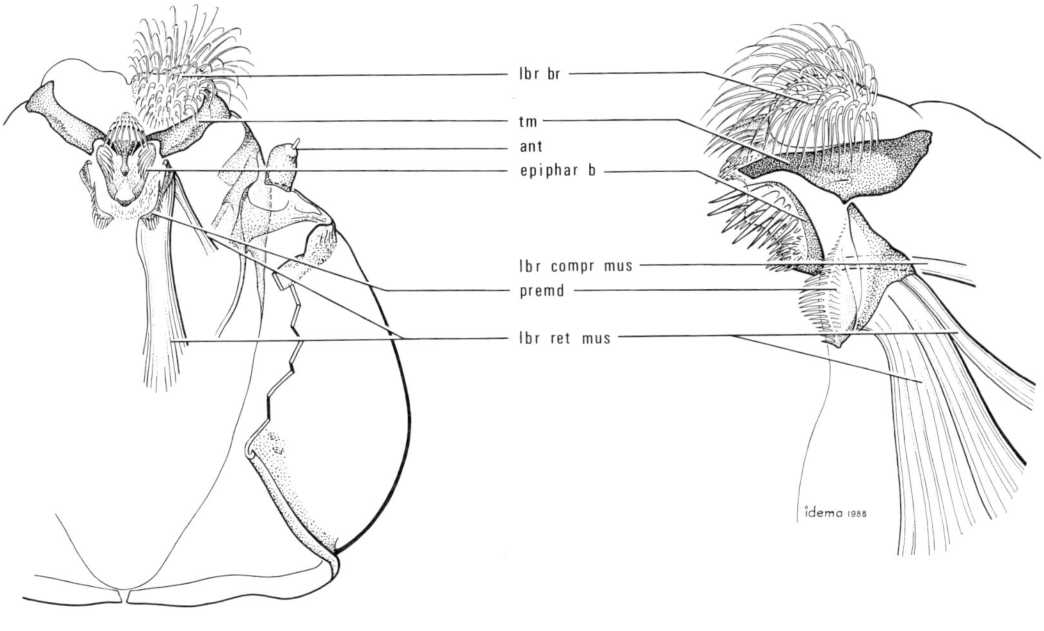

5. Bittacomorpha clavipes

6. Bittacomorpha clavipes

Figs. 114.3–6. Details of labra and associated muscles of the left side of the head capsule of various Nematocera (mid-ventral part of head capsule and all contents except labral muscles removed): ventral views of (3) *Trichocera* sp. (Trichoceridae); (4) *Protoplasa fitchii* Osten Sacken; and (5) *Bittacomorpha clavipes* (Fabricius) (Ptychopteridae); lateral view of (6) *B. clavipes*.

Abbreviations: ant, antennae; epiphar b, epipharyngeal bar; lbr br, labral brush; lbr compr mus, labral compressor muscle; lbr ret mus, labral retractor muscle; md, mandible; mx plp, maxillary palpus; premd, premandible; tm, torma.

42. *Mandible with teeth confined to the apex (plesiomorphic) / mandible in the shape of a chela (Fig. 3.4) with a small, subbasal, thumb-like projection curving toward the apex of the mandible (apomorphic)*

This peculiarly shaped mandible, here considered a synapomorphy of the infraorder, seems to be unique to this group. The Perissommatidae also have a subbasal projection on the mandible but this is directed medially.

43. *Toothed apical portion of mandible solidly fused with base (plesiomorphic) / apical portion of mandible separated from the base by a less strongly sclerotized line of weakness (apomorphic)*

Both Anthon (1943) and Hennig (1973) regarded the line of weakness to be evidence of segmentation and thus plesiomorphic. However, we strongly contest such an interpretation. No other mandibulate pterygote has such a condition and certainly the pterygote mandible is primitively unsegmented. The transverse division found in the Trichoceridae, Scatopsidae, and Anisopodidae must be secondary.

A similar condition is also found in the Ptychopteridae. If homologous, it could be a synapomorphy of these three families plus the Ptychopteridae (as discussed under the section "Phylogeny of the Infraorder Ptychopteromorpha"); therefore, these three families of Psychodomorpha would form a paraphyletic group. Although we have presented an alternative hypothesis, more study is needed of the possible relationship of the Ptychopteridae to these three families.

Because of inadequate material, we were unable to assess this character for the Perissommatidae. However, there may be a line of weakness near the base of the toothed, medial, mandibular projection.

Otherwise the apomorphic character state is unknown elsewhere in the Nematocera.

44. *Cardo in the form of a clearly delineated sclerite (plesiomorphic) / cardo reduced, mainly membranous (apomorphic)*

In the Trichoceridae, Perissommatidae, Scatopsidae, and Anisopodidae the maxilla is mainly membranous and the area is covered with backwardly pointing setae forming a characteristic pattern, unique in the Diptera. This feature is, therefore, considered to be a synapomorphy of the four families.

45. *Maxillary palpus well developed and palpiform (plesiomorphic) / maxillary palpus reduced to a disc-like sclerotized ring (apomorphic)*

In the Trichoceridae, Perissommatidae, Scatopsidae, and Anisopodidae the maxillary palpus is flush with the surface of the maxilla, recognizable only by a ring of chitin surrounding the sensilla. This character, which appears to be unique in the Nematocera, may be intimately correlated with the previous one, as a result of the way in which the larva presses the ventral surface of the head to the substrate while feeding.

46. *Hypostomal area fully sclerotized, forming a hypostomal bridge (plesiomorphic)/hypostomal area membranous (apomorphic)*

Larvae of nearly all Nematocera have some degree of closure of the head capsule ventrally between the foramen magnum and the labiohypopharynx, and psychodid larvae have this connection well developed. Larvae of other families, however, notably the Tipulidae and the families of the Culicomorpha, also have a well-developed hypostomal bridge crowned anteriorly with a row of hypostomal teeth, against which the mandibles strike when adducted. The interpretation of this region is controversial. Some authors consider the bridge to incorporate portions of the labium (e.g. the mentum), whereas others (for example Anthon 1943) regard it as a fusion of the subgenal lobes, displacing the labium anteriorly (see discussion in Ch. 3, p. 68). It is even more difficult to establish whether such a well-developed hypostomal bridge crowned with a row of teeth as it appears in the Tipulidae, Psychodidae, and Culicomorpha represents the primitive condition in the Diptera, whether this feature is an intermediate development that might be a synapomorphy representing a particular lineage, or whether it encompasses two or three independent developments.

In forms with a well-developed hypostomal bridge, the tentorium is usually reduced, even vestigial, and the transverse tentorium is usually absent. The reduction of the tentorium of most Nematocera may have occurred because the bridge serves to strengthen the head capsule and perhaps replaces the transverse tentorium, which renders it obsolete. In support of this idea, larvae of the Anisopodidae, which have no hypostomal bridge at all, have a remarkably well-developed transverse tentorium. However, the presence of both a well-developed hypostomal bridge and transverse tentorium in the Perissommatidae may refute this concept.

Anthon (1943) concluded that the presence of a transverse tentorium represented the primitive condition because it was the usual condition in primitive insects. Larvae of Nannochoristidae and Siphonaptera also have transverse tentoria. This hypothesis, that the anisopodid larva represents the most primitive type of dipterous head capsule because of the presence of a fully sclerotized transverse tentorium, is difficult to refute. Nevertheless, we hesitate to agree with this conclusion. Instead, we prefer to suppose that the anisopodid and perissommatid transverse tentoria are a *de novo* development, a reformation and resclerotization of a transverse connection, a condition that is apparently not present in the other families of Diptera.

Among the families of the Psychodomorpha, the hypostomal bridge is most extensive in the Psychodidae and the tentorium is correspondingly vestigial. In the Trichoceridae the bridge is reduced to a transverse band joining the posterior corners of the head capsule (superficially resembling a transverse tentorium), and although the anterior and posterior tentorial arms are complete, a transverse tentorium is lacking. In the Scatopsidae, which appear to lack a tentorium, and in the Anisopodidae, which have the most extensive and heavily sclerotized tentorium of all, the hypostomal bridge is nonexistent. We have interpreted reduction of the bridge to be a derived feature.

Adult

47. *Antennal flagellomeres with only hair-like pigmented sensilla and lacking membranous sensilla (plesiomorphic) / antennal flagellomeres with membranous sensilla (Figs. 17.2–9) (apomorphic)*

Almost all members of Psychodidae have membranous sensilla on at least some of their flagellomeres and those lacking these sensilla must have lost them secondarily. The character state is nearly unique within Diptera but at least many Cecidomyiidae also exhibit similarly modified sensilla (e.g. Figs. 16.46–49).

48. *Two or more palpal segments present (plesiomorphic) / only one palpal segment present (apomorphic)*

Although Scatopsidae are the only members of our Psychodomorpha to have only one palpal segment, some other Nematocera also exhibit the condition. In some families, such as the Cecidomyiidae, Tipulidae, and Mycetophilidae, only a few members have one palpal segment, which must have resulted from secondary loss. In other families, for example the Nymphomyiidae and Deuterophlebiidae, in which the palpus is lacking, absence of the palpus may be an autapomorphy of the family. We believe these examples to be convergent with those of the Scatopsidae, although the possibilities of parallelism and the potential weakness of the character state as an indication of monophyly are evident.

49. *Ommatidia forming two groups on either side of head (plesiomorphic) / ommatidia divided laterally into two dorsal and two lateral groups by a band of integument (apomorphic)*

The only other Nematocera to exhibit the apomorphic condition are some members of the Cecidomyiidae (e.g. *Campylomyza* Meigen, *Trisopsis* Kieffer, and *Odontodiplosis* Felt). Considering the synapomorphies of and within this family, the occurrence of divided eyes in Cecidomyiidae is certainly convergent with that of the Perissommatidae.

50. *Costa continuous along anterior margin of wing (plesiomorphic) / costa with break near base (Figs. 17.10–13) (apomorphic)*

Within the Nematocera the apomorphic condition is unique to the Psychodidae. Some psychodids have more than one break in the costa.

Discussion. In one of the earliest comparative studies presenting details of the head capsule of Nematocera, Keilin (1912) drew attention to the similarity between the head capsules of the Trichoceridae and Anisopodidae, and their dissimilarity to those of the Tipulidae. Goetghebuer (1925) noted further the similarity in the structure of the labrum and especially in the shape of the premandibles of the Anisopodidae, Trichoceridae, Psychodidae, and Scatopsidae. These similarities in structure were confirmed by Anthon (1943) who expanded and elaborated on the work of Goetghebuer by showing that many other details of the head capsules of two anisopodids, *Sylvicola* Harris and *Mycetobia* Meigen, the trichocerid *Trichocera* Meigen, and the psychodids, *Philosepedon* Eaton and other genera, were constructed on the same basic plan. Edwards (1926) believed that the type of head capsule found in the Anisopodidae, Trichoceridae, Psychodidae, and Scatopsidae represented the primitive condition. Anthon, too, concluded that several features of this plan, such as the vertical orientation of the mandibles and their subdivision into two articulating "segments," were primitive. We have reached the opposite conclusion.

We think that the primitive nematoceran mandible most likely operated horizontally, or nearly so, as in the Tipulomorpha, Bibionomorpha, Nannochoristidae, and Siphonaptera, and that it was most likely undivided. A truly divided or segmented mandible is otherwise unknown in Pterygota. Consequently, an obliquely or vertically operating mandible, apically crowned with a brush of setae for harvesting food and with an unsclerotized transverse band that gives it a "segmented" appearance, along with most other details of its anatomy, such as the peculiar premandibles, must be derived and are here considered synapomorphies of the infraorder Psychodomorpha. Although Anthon did not study the larva of *Scatopse* Geoffroy, he noted that its head capsule also conformed to the same basic plan. The Synneuridae (Hyperoscelidae of Hennig, and including the Canthyloscelidae of Hutson 1977) are, on adult characters, apparently most closely related to the Scatopsidae. Synneurid larvae have a greatly reduced head capsule, so no direct comparison with larvae of the other families can be made easily.

A recent challenge to the placement of the Trichoceridae close to the Anisopodidae, supporting instead its traditional position adjacent to the Tipulidae, has come from developmental studies by Dahl (1980). She suggested that portions of the male adult terminalia of an anisopodid, *Sylvicola cincta* (Fabricius), were not homologous to supposedly homologous parts of the terminalia of tipulids (based on a study of *Limonia nubeculosa* (Meigen)) and trichocerids (based on three species of *Trichocera*). Male terminalia of *Sylvicola* developed entirely from imaginal discs, whereas those of *Limonia* Meigen and *Trichocera* developed from both imaginal discs and ectodermal tissue of the pupa. An alternate interpretation to one of nonhomology, which we prefer, suggests a unique developmental change in the Tipulidae and Trichoceridae. The terminalia of all other known holometabolous insects develop from imaginal discs, so that partial development of terminalia from pupal ectoderm would be interpreted as a synapomorphy of the Tipulidae and Trichoceridea. However, we prefer to await confirmation from additional research before abandoning our hypothesis that the Trichoceridae are more closely related to the Anisopodidae and Psychodidae (as the infraorder Psychodomorpha presented here) than they are to the Tipulidae.

The apparent synapomorphies in the larval mouthparts and head capsule are the basis for inclusion of these five families in a presumably monophyletic group, the Psychodomorpha of this Manual. Adult structures are not supportive of this arrangement; however, neither do they detract from it. In a cladistic analysis, lack of similarity is insufficient grounds for separating taxa that can otherwise be grouped on the basis of shared derived character states. In the past, similarities between adults of the Trichoceridae and Tipulidae, and between the Anisopodidae and some members of the Bibionomorpha, have led to quite different classifications. However, we have concluded that similarity between adult anisopodids and some members of the Bibionomorpha, or between adults of the Trichoceridae and Tipulidae, is a result either of symplesiomorphy or of homoplasy.

We have not been able to identify a synapomorphy of the Scatopsidae and Synneuridae. Nevertheless, on the basis of marked similarity between adults of the Scatopsidae and some Synneuridae (e.g. *Synneuron* Lundstrom), we have intuitively placed the two families as sister groups.

The peculiar southern hemisphere Perissommatidae (also known as a Jurassic fossil from Siberia (Kalugina and Kovalev 1985)) are placed in this infraorder on the basis of our interpretation of the larval mouthparts. We believe that the larva of *Perissomma fusca* Colless shares derived characters of some other Psychodomorpha.

We have taken a cautious approach to an interpretation of the relationships between the Perissommatidae, Trichoceridae, Anisopodidae, Scatopsidae, and Synneuridae. The following character states can be interpreted as synapomorphies but give conflicting evidence.

A prostheca on the medial surface of the mandible is characteristic of most Nematocera larvae and is conspicuous on the mandible of the larvae of Nannochoristidae as well. Therefore, its presence is probably plesiomorphic within the Diptera. Its absence in the Anisopodidae and Perissommatidae may be considered a synapomorphy of those two families; however, this absence is not unique for it also occurs in some other families as well, e.g. the Chaoboridae, Mycetophilidae, and Cecidomyiidae.

The Anisopodidae and Perissommatidae are the only Nematocera to possess well-developed tentoria, which could be interpreted as a synapomorphy of the two families (however, see discussion under character 46 above).

The larvae of the Trichoceridae, Perissommatidae, and most Anisopodidae (lacking in *Olbiogaster* Osten Sacken) have elongate spicules on the anteroventral margin of the head capsule. We know of no other Nematocera, Siphonaptera, or Mecoptera with this condition.

PHYLOGENY OF THE INFRAORDER PTYCHOPTEROMORPHA

Hennig's classification

Hennig (1973, 1981) did not recognize a separate infraorder by this name, but instead included the two constituent families as the superfamily Ptychopteroidea, in the infraorder Psychodomorpha, as follows:

Superfamily Ptychopteroidea
  Family Tanyderidae
  Family Ptychopteridae

Proposed classification

We propose a classification that is essentially the same as that of Hennig, except that we have elevated the entire taxon to the rank of infraorder. We chose to do so because we believe that the Ptychopteromorpha, or possibly only a part of it alone, the Ptychopteridae, is the sister group of the Culicomorpha and is not closely related to the Psychodidae or to any other member of Hennig's Psychodomorpha.

Character states. In the following we analyze character states 51–53, used in interpreting relationships within the Ptychopteromorpha.

Pupa

51. *Respiratory organs, if present, equal in size and shape (plesiomorphic) / respiratory organ of one side much longer than that of other side (Fig. 22.7) (apomorphic)*

The apomorphic condition is unique to the Ptychopteridae, although a few Tipulidae have respiratory organs that differ slightly in length (Rogers 1949).

Adult

52. *Last tarsomere of male not capable of being folded proximally against the penultimate tarsomere and without interlocking tufts of setae on tarsomere 4 (plesiomorphic) / last tarsomere of male capable of being folded forward against tarsomere 4; tarsomere 4 with basal swelling opposed by tarsal claws and with interlocking tufts of setae (apomorphic)*

This character state was the only one suggested by Hennig (1973) as a synapomorphy linking the Tanyderidae and the Ptychopteridae. It is not present in *Bittacomorpha* Westwood or *Bittacomorphella* Alexander, however, but members of these genera have quite modified legs, and it may be secondarily lost. Although its distribution within the Nematocera has not been fully explored, it is evidently not present in any other Nematocera.

Hennig (1968) suggested that this modification assists the male in holding the female during copulation.

53. *Halter lacking a prehalter (plesiomorphic) / halter with a basal appendage called the prehalter (Fig. 2.71) (apomorphic)*

The presence of a prehalter is unique within the Diptera.

Discussion. The presence, in the Ptychopteridae, of a more complex torma than that found in *Protoplasa* Osten Sacken and *Protanyderus* Handlirsch (Tanyderidae), along with invagination of the premandible for greater surface area in support of associated musculature, suggests that the Ptychopteridae alone, rather than Ptychopteromorpha, may be the sister group of the Culicomorpha (see under "Phylogeny of the Infraorder Culicomorpha").

PHYLOGENY OF THE INFRAORDER CULICOMORPHA

Hennig's classification

The monophyly of the Culicomorpha, along with the relationships between its component families, is one of the least contentious issues in the phylogenetic interpretation of the Nematocera. Hennig (1973) listed at least nine character states that he considered to be synapomorphies of the infraorder: adult ocelli vestigial or absent; pedicel enlarged, with specially developed Johnston's organ (secondarily reduced in Simuliidae); wing without a discal cell; M_3 absent; radial sector with only three branches; sperm pump absent, with sperm being transferred within a spermatophore; first abdominal spiracle of adult absent (vestigial in Simuliidae); larva eucephalous, with hypostomal bridge dentate at anterior margin; and leg sheaths of third pair of pupal legs curved below wing sheaths to allow mobility of abdomen. The condition of the pedicel, sperm pump, larval head, and pupal leg sheaths are discussed below. The remaining character states are not unique to the Culicomorpha but are subject to considerable homoplasy in various groups of Nematocera.

Hennig (1973) divided the infraorder into two sister groups, the superfamilies Culicoidea and Chironomoidea. As synapomorphies for the Culicoidea, he offered the following four characters: antepronotum and postpronotum of adult separate, the latter appearing as part of the pleuron; third palpal segment of adult without sensory vesicle (actually present in some Chaoboridae); larva metapneustic, with anterior spiracles reduced and posterior spiracles surrounded by a characteristic, five-lobed structure; and antenna of larva single-segmented. Hennig further subdivided the Culicoidea into two family groups: the Dixidea, with only one family, the Dixidae; and the Culicidea, with two families, the Chaoboridae and Culicidae. The superfamily Chironomoidea shared the following five synapomorphies: costa of wing not continuing beyond apex of wing, ending at apex of R_{4+5}, leaving hind margin of wing membranous (except in Thaumaleidae, in which there is a suggestion of thickening); anterior branch of radial sector (which Hennig considered to be R_2) ending in R_1; A_1 not reaching hind margin, usually lying close to CuP; larva basically amphipneustic; prothorax and terminal segment of larva each with paired or unpaired parapodium (proleg). Further, the Chironomoidea were subdivided into two family groups, the Thaumaleidea, containing only the Thaumaleidae, in which the costa encircled the hind margin of the wing to some degree, and the Chironomidea, containing the Simuliidae, Ceratopogonidae, and Chironomidae, in which the hind margin of the wing was entirely membranous. Hennig's classification was as follows:

Infraorder Culicomorpha
- Superfamily Culicoidea
  - Family group Dixidea
    - Family Dixidae
  - Family group Culicidea
    - Family Chaoboridae
    - Family Culicidae
- Superfamily Chironomoidea
  - Family group Thaumaleidea
    - Family Thaumaleidae
  - Family group Chironomidea
    - Family Simuliidae
    - Family Ceratopogonidae
    - Family Chironomidae

Proposed classification

Our classification is similar to that of Hennig, with the single exception of the addition of the family Corethrellidae, which was formerly considered to be a subfamily or tribe of the Chaoboridae. Justification for the recognition of this taxon at the family level is provided below. Our proposed classification is as follows:

Infraorder Culicomorpha
- Superfamily Culicoidea
  - Family Dixidae
  - Family Corethrellidae
  - Family Chaoboridae
  - Family Culicidae
- Superfamily Chironomoidea
  - Family Thaumaleidae
  - Family Simuliidae
  - Family Ceratopogonidae
  - Family Chironomidae

Character states. In the following we analyze character states 54–84, used in interpreting relationships within the Culicomorpha.

Larva

54. *Anterolateral margin of head capsule complete (plesiomorphic) / anterolateral margin of head capsule with groove in which antennae lies when adducted (apomorphic)*

The apomorphic feature is unique to the Corethrellidae.

55. *Labral brush relatively simple, composed of rather few rows of short, well-spaced, curved setae and capable only of slow closure, and with a rather small, labral retractor muscle originating in the middle of the cephalic apotome (plesiomorphic) / labral brush complex, with numerous, tightly packed, long setae or a single row of long, sickle-shaped setae and capable of rapid closure, and with an elongate, labral retractor muscle clearly divided into two portions, each originating farther back on the enlarged cephalic apotome (apomorphic)*

Members of the Ptychopteridae and Tanyderidae have simple labral brushes capable only of slow movement. These brushes bear only a few rows of curved setae. Much more complex labral brushes are present in all members of the Dixidae and in almost all the Culicidae and Simuliidae. We regard the simple labral brushes found in the Ptychopteridae (and possibly also in the Tanyderidae) as being homologous with those found in the Dixidae, Culicidae, and Simuliidae and that the simple brushes represent the plesiomorphic condition from which the more complex structures developed. Therefore, we regard the possession of labral brushes as probably being present in the ancestor of the Culicomorpha and, hence, as a synapomorphy of the infraorder, even though members of the Corethrellidae, Chaoboridae, Thaumaleidae, Ceratopogonidae, and Chironomidae lack these structures.

56. *Labral brush consisting of numerous rows of closely packed, long setae arising from head surface, with both brushes closing simultaneously to entrap food particles from standing water (plesiomorphic) / labral brush elevated from head surface on apex of stalk to form a labral or cephalic fan consisting of a single row of extremely long setae (Fig. 27.76); each fan with an arc-like base when structure is extended fan-like in current of water to extract food particles; brushes closed alternately (apomorphic)*

The plesiomorphic condition is characteristic of the Dixidae and Culicidae (secondarily reduced in the Toxorhynchitinae). The apomorphic condition, although secondarily lost in a few members, is one of the most distinctive synapomorphies of Simuliidae. Although the occurrence of vestigial fans, or their absence, in *Crozetia crozetensis* (Womersley), *Simulium oviceps* Edwards, and in all members of *Twinnia* Stone & Jamnback and *Gymnopais* Stone has been interpreted by Rubtzov (1956), Davies (1965), and others as plesiomorphic, we believe that the ancestor of all simuliids possessed labral brushes or cephalic fans of some sort and that absence of fans is a derived condition (Wood 1978).

57. *Torma in the form of a band-like sclerite, with point of attachment of premandible at the surface of cuticle occurring at posteromedial corner of the torma (plesiomorphic) / torma folded on itself, invaginated, with the premandible intimately associated with the torma and internal and dorsal to body of torma (Figs. 114.7–8, 11) (apomorphic)*

The plesiomorphic condition is found in the Tanyderidae (Fig. 114.4) and all Psychodomorpha (Fig. 114.3), both outgroups of the Culicomorpha. An infolded torma, on which the point of attachment of the premandible is internal, is found in all members of the Dixidae, Culicidae, and Simuliidae. An internal premandible, which serves as the insertion point of the labral retractor muscle, presumably allows torsion of the torma, apparently increasing the sweep of the labral brush. Because it occurs in some members of both the Culicoidea and the Chironomoidea, we regard this infolding as being a synapomorphy of the Culicomorpha. Alternatively, infolding may have happened twice, in the Culicoidea and in the Simuliidae.

58. *Premandible mainly an external sclerite, with a well-developed external comb of setae and a small, invaginated apodeme for insertion of the labral retractor muscles (plesiomorphic) / premandible mainly internal as a result of invagination, serving almost entirely as a point of insertion of the labral retractor muscle, with only a small remnant of the external sclerite (apomorphic ') / premandible entirely internal, apparently without any remnant of an external sclerite (apomorphic ")*

In the plesiomorphic condition, found in the Ptychopteromorpha (Figs. 114.4–6), the premandible can move only in a posteromedial direction, pulling the torma in the same direction, thus imparting scarcely any rotational or other twisting motion to the torma. The premandibular comb apparently assists in combing food particles from the mandible and is itself cleaned by the prostheca. A transformation series, in which the external, comb-bearing part of the premandible either is greatly reduced relative to the internal, invaginated part, as in the Dixidae, or is not evident at all, as in the Culicidae and the Simuliidae, will be described in detail by Wood (in preparation). We consider the large external premandible of some Chironomidae as a superficial reversal to the plesiomorphic condition.

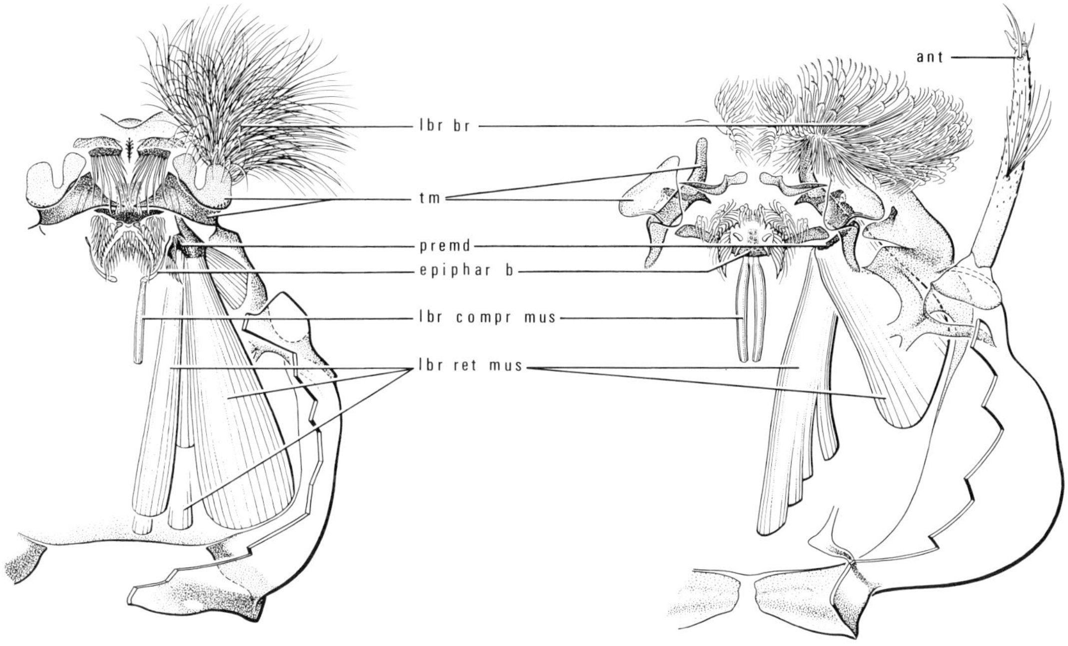

7. Dixa sp.

8. Aedes communis

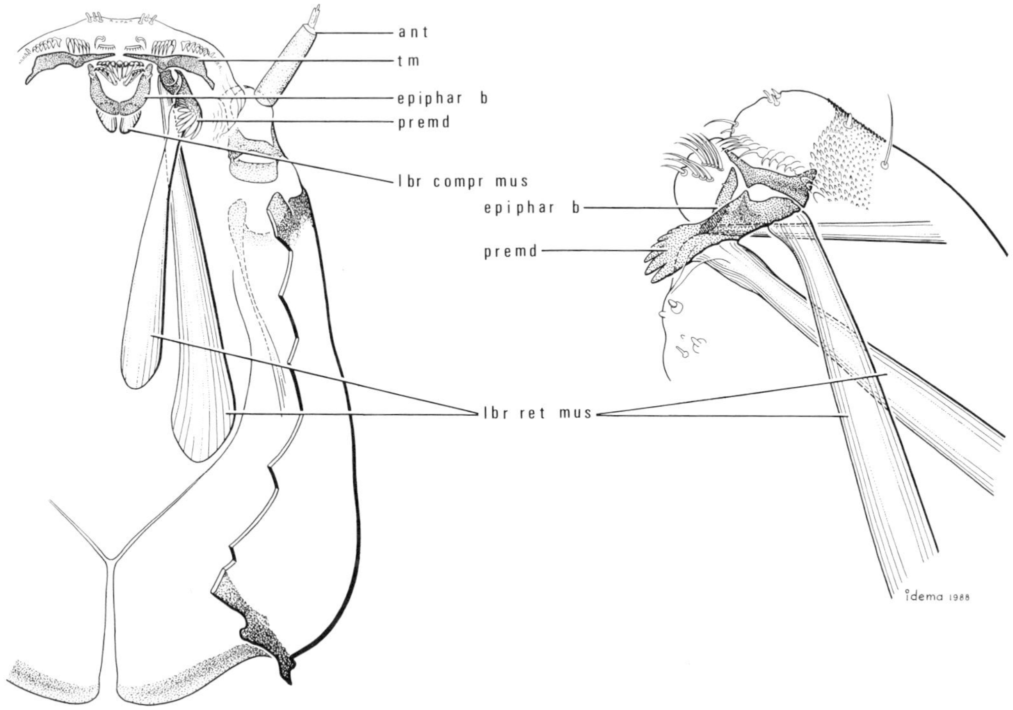

9. Diamesa sp.

10. Diamesa sp.

Figs. 114.7–10. Details of labra and associated muscles of the left side of the head capsule of various Nematocera (mid-ventral part of the head capsule and all contents excepts labral muscles removed): ventral views of (7) *Dixa* sp. (Dixidae); (8) *Aedes communis* (De Geer) (Culicidae); and (9) *Diamesa* sp. (Chironomidae); lateral view of (10) *Diamesa* sp.

Abbreviations: ant, antenna; epiphar b, epipharyngeal bar; lbr br, labral brush; lbr compr mus, labral compressor muscle; lbr ret mus, labral retractor muscle; md, mandible; mx plp, maxillary palpus; premd, premandible; tm, torma.

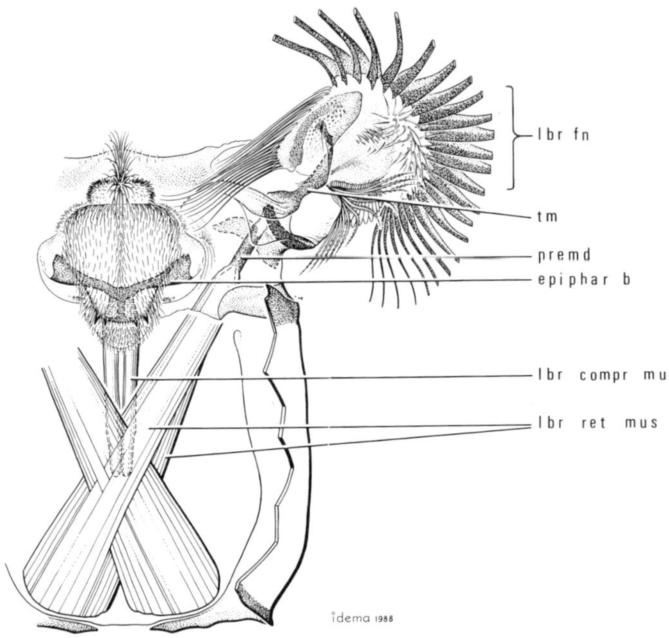

11. Prosimulium sp.

Fig. 114.11. Details of labra and associated muscles of the left side of the head capsule of *Prosimulium* sp. (Simuliidae) (mid-ventral part of head capsule and all contents except labral muscles removed).
Abbreviations: epiphar b, epipharyngeal bar; lbr compr mus, labral compressor muscle; lbr fn, labral fan; lbr ret mus, labral retractor muscle; premd, premandible; tm, torma.

59. *Premandible mainly an external sclerite, with well-developed external comb of setae and with a small, invaginated apodeme for insertion of labral retractor muscles (plesiomorphic) / premandible with tongs-like or spike-like external portion and with invaginated internal portion bearing two separate insertions of the two parts of the labral retractor muscles; premandible lever-like, moveable in an anteroposterior direction as a result of the separate insertions of the labral retractor muscles and capable of forceps-like action resulting from contraction of the labral compressors pulling the intertorma in a dorsal direction (Figs. 114.9–10) (apomorphic)*

In the Chironomidae (secondarily lost in some), the external portion of the premandible is not comb-like, but is used in an entirely novel way, as a pair of tongs for grasping food particles (Wood, personal observations). Contraction of the labral compressor muscles pulls the epipharyngeal bar dorsally, bringing the distal ends of the premandibles together. Simultaneous contraction of the labral retractor muscles conveys food particles to the pharynx. The premandibles of the Forcipomyiinae (Ceratopogonidae) also act as levers. They, too, have two opposing insertions of the labral retractor muscles as in the Chironomidae, but they are exceedingly minute, and their function while feeding is not understood. The apomorphic character state is unique within the Diptera and is not known in other Holometabola.

60. *Maxillary palpus reduced or barrel-shaped (plesiomorphic) / maxillary palpus markedly elongate (Fig. 23.10) (apomorphic)*

The apomorphic feature is unique to Dixidae.

61. *Labiohypopharynx not connected to paraclypeal phragma (plesiomorphic) / dorsolateral corner of labiohypopharynx connected via cibarial bar (Harbach and Knight 1980) to paraclypeal phragma, thus increasing its stability for use as an anvil for pounding food particles (apomorphic)*

The presence of a connection between the labiohypopharynx and paraclypeal phragma is apparently unique to the Culicoidea and is here considered a synapomorphy of that group. This feature is present in the Dixidae, Corethrellidae, and Culicidae, but not in the Chaoboridae. We assume that its absence in the Chaoboridae represents a secondary loss, related to the swallowing of large prey.

62. *Pharyngeal filter present (plesiomorphic) / pharyngeal filter absent, replaced by a muscular pharynx (apomorphic)*

A pharyngeal filter, which consists of rows of backwardly pointing hairs for straining food particles from expelled water (Snodgrass 1959), is widespread in the Nematocera. This feature is present in at least some members of all the infraorders except the Bibionomorpha, and we presume that it was present in the ancestor of the

Culicomorpha as well. It is well developed in the Dixidae, Culicidae, and Thaumaleidae, but not in the Chaoboridae or Corethrellidae in which it was presumably lost secondarily as a result of swallowing whole prey. It is also absent in the filter-feeding Simuliidae, which apparently engulf indiscriminately all material captured by the labral fans and transfer this material to the pharynx (Craig and Chance 1982), and in the Ceratopogonidae and Chironomidae, which we believe to be selective rather than indiscriminate feeders. We regard the presence of a muscular pharynx in the Simuliidae, Ceratopogonidae, and Chironomidae, instead of a pharyngeal filter (which is present in Thaumaleidae) as being a synapomorphy of these three families.

63. *Pharyngeal apparatus absent to somewhat developed (plesiomorphic) / pharyngeal apparatus markedly developed with two strongly diverging arms and rows of combs (apomorphic)*

All known larvae of Ceratopogonidae have a characteristic, well-developed pharyngeal complex (Borkent *et al.* 1987). Although this character state appears to be unique, we are unsure of homologies in other Nematocera. The larvae of *Bittacomorpha* (Ptychopteridae) (but not *Ptychoptera*) also exhibit a markedly developed pharyngeal structure that works with a hammer and anvil motion but looks quite different and does not appear to be homologous.

64. *Antenna slender, tapering apically, and usually rather short (plesiomorphic) / antenna large and stout (apomorphic)*

The antenna of larvae of the Nematocera is, in general, rather small and slender. In the Culicoidea, however, it is long, stout, and massive relative to the body size of the larvae and is usually armed apically with setae. The apomorphic condition is unique within the Diptera, Siphonaptera, and Mecoptera.

65. *Antenna capable of relatively little movement, terminating in several straight setae (plesiomorphic) / antenna prehensile, terminating in claw-like setae, capable of holding prey captured by the mandibles as well as of grasping the substrate (apomorphic)*

Prehensile antennae have been considered a synapomorphy of the Chaoboridae plus Corethrellidae, which have traditionally been considered as members of a single family, the Chaoboridae; this tradition is maintained in Volume 1 of this Manual. In the Chaoboridae *sensu stricto* (excluding Corethrellidae), the antennae are used for holding prey while it is being devoured. In *Eucorethra* Underwood and *Mochlonyx* Loew, the antennal bases are widely separated (Fig. 24.10), whereas in *Chaoborus* Lichtenstein they are adjacent medially. In addition, in *Chaoborus* and perhaps in the other genera as well, the larva can anchor itself by grasping objects on the bottom with its antennae. However, in the Corethrellidae, the antennae arise close to each other near the midline, fold back into a groove along the side of the head when at rest, and project directly forward when extended. Although their function is apparently unknown, our observations of feeding *Corethrella* Coquillett (two species from Ecuador) and manipulation of preserved specimens indicate that *Corethrella* larvae do not use their antennae for feeding as do the Chaoboridae. The antennae appear incapable of being bent down toward the mouth and, therefore, seem to be of no use in feeding. Perhaps they are used for gripping some substrate in their environment, a feature that could prevent them from being washed out of precarious habitats (such as bromeliads) during storms. These observations suggest that the prehensile antennae of the Corethrellidae and Chaoboridae are not homologous.

The apomorphic condition is otherwise unique within Diptera.

66. *Prothoracic proleg absent (plesiomorphic) / prothoracic proleg present, crowned apically with rows of hooklets (apomorphic)*

A prothoracic proleg (bifurcate and appearing as a paired structure in Chironomidae) is present in most members of the four families of the Chironomoidea, namely all Thaumaleidae, all Simuliidae, members of the subfamily Forcipomyiinae and first instar *Culicoides* Latreille (Ceratopogonidae), and virtually all Chironomidae; but it is unknown outside these families. It was regarded by Hennig (1973) and by Wood (1978) as a synapomorphy of the Chironomoidea and is accepted as such here. The proleg is used for locomotion and, in the Simuliidae and Chironomidae, for grasping and manipulating silk (Barr 1984).

67. *Anterior thoracic (mesothoracic) spiracle present (plesiomorphic) / anterior thoracic spiracle absent or vestigial (apomorphic)*

We regard the amphipneustic condition (Ch. 3, p. 83), found in the Tanyderidae, Axymyiidae, most Psychodomorpha, and most Brachycera, as being relatively plesiomorphic compared with the metapneustic condition, in which the anterior spiracles are absent. Among the Culicomorpha, only the Thaumaleidae have anterior spiracles, i.e. are amphipneustic. Absence of the anterior spiracles in the remaining Chironomoidea (members of the transantarctic genus *Archaeochlus* Brundin have posterior spiracles only, Brundin 1966: 295) may be a synapomorphy of these three families, but this feature is probably of dubious value as an indicator of relationship.

68. *Thoracic segments distinct, not appreciably wider than first abdominal segment (plesiomorphic) / all three thoracic segments enlarged and fused to each other (Figs. 24.10, 25.33–34) (apomorphic)*

Distinct thoracic segments are found in all insect larvae except in all members of three families of the Culicoidea, namely the Corethrellidae, Chaoboridae, and Culicidae. In these, the three segments are fused and swollen into a distinct thoracic region. We regard this condition as being a synapomorphy of the three families.

69. *Abdominal spiracles flush with surface (plesiomorphic) / abdominal spiracles elevated on a conical siphon (apomorphic)*

In the Dixidae, as in most larvae of Nematocera, the posterior abdominal spiracles are flush with the surface of the body, whereas in the remaining families of the Culicoidea, namely the Corethrellidae, Chaoboridae, and Culicidae, these spiracles are more or less elevated on a siphon. The siphon is shortest and least developed in *Eucorethra* Underwood (Chaoboridae) and in *Anopheles* Meigen (Culicidae), both of which are surface feeders like the Dixidae. As both these genera are probably the most primitive members of their respective families, an exceptionally short siphon may represent the primitive condition. The development of a more elongate siphon may have occurred independently in the Corethrellidae, Chaoboridae, and Culicidae. Regardless of whether it is short or long, we consider the presence of a siphon as being a synapomorphy of the Corethrellidae plus Chaoboridae plus Culicidae. In some members of the Psychodinae (Psychodidae) and in Perissommatidae, the spiracles are also elevated on a siphon, superficially similar to that found in the Culicoidea.

70. *Posterior abdominal spiracles surrounded by two pairs of flaps, namely one small pair flanking the spiracles and arising on abdominal segment 8 and a second, much larger pair posterior to them, arising on a separate and distinct segment 9 (Fig. 23.10) (plesiomorphic)/posterior pair of flaps reduced, elevated to apex of siphon (Figs. 25.28–32) (apomorphic)*

Homologies of the structures surrounding the posterior abdominal spiracles of most amphipneustic and metapneustic dipterous larvae are not well understood. Usually there are two pairs of flaps, one lateral to the spiracles and the other posterior to them. Often one middorsal lobe, or even a pair of lobes, occurs anterior to, and between, the spiracles. There is so much variation among the families of Diptera that it is difficult to deduce the plesiomorphic condition. However, in larvae of the Dixidae, the lateral flaps (postspiracular processes, Fig. 23.10) that are intimately associated with the spiracles seem to arise, with the spiracles, on segment 8, whereas the posterior flaps (posterolateral processes), which are much larger, appear to arise on segment 9.

We have taken this condition to represent the plesiomorphic condition in the Culicoidea. In the Corethrellidae, Chaoboridae, and Culicidae, the spiracles and both pairs of flaps are borne at the apex of the siphon; there appears to be no segment 9, and we have concluded, along with previous students of the Culicidae, that it has been incorporated into the siphon, which appears to arise from segment 8. This condition is presented as a synapomorphy of the three families.

71. *Posterior abdominal spiracles surrounded by two pairs of flaps, namely one small pair flanking the spiracles and arising on abdominal segment 8 and a second, much larger pair posterior to them, arising on a separate and distinct segment 9 (Fig. 23.10) (plesiomorphic) / abdominal spiracles having the posterior pair of flaps modified into cylindrical procerci (Figs. 26.4, 29.136–137) (apomorphic)*

Among the Chironomoidea, only members of the Thaumaleidae and the chironomid genus *Archaeochlus* (Brundin 1966: Fig. 404) possess abdominal thoracic spiracles. In the Thaumaleidae, a pair of fringed flaps arises just behind these spiracles; both flaps and spiracles appear to be on abdominal segment 8, but are followed by only one more segment, which we assume to be segment 10. In *Archaeochlus*, in contrast, the flaps (called procerci by chironomid specialists), which we suggest are homologous to the posterior flaps (posterolateral processes) of the Culicoidea as well as to those of the Thaumaleidae, arise on a clearly defined segment 9 and are of a characteristic cylindrical shape. In other chironomid larvae, the procerci also arise in the same position, even though the spiracles are absent. In our scheme we assume that the posterior flaps have been lost in the Simuliidae and Ceratopogonidae. We know of no other Diptera with such a modification.

72. *Lobes surrounding abdominal spiracles immovable (plesiomorphic) / lobes movable, folding together when the larva submerges (apomorphic)*

In Dixidae, as in other larvae of Nematocera possessing spiracular lobes, the lobes are immovable, and we consider this feature to be the plesiomorphic condition. The ability of the lobes to fold together is considered a synapomorphy of the Corethrellidae, Chaoboridae, and Culicidae.

73. *Anal papillae retractible (plesiomorphic) / anal papillae permanently exerted, nonretractible (apomorphic)*

In *Microchorista* Byers (Nannochoristidae, Pilgrim 1972), the Tanyderidae (Exner and Craig 1976), Ptychopteridae, Dixidae, Ceratopogonidae (Wirth and Grogan 1979), and Simuliidae, the anal papillae can be extended from, or retracted into, the anus. In the Corethrellidae, Chaoboridae, and Culicidae, however, they remain permanently exerted. In chironomid larvae they are also nonretractable. This condition may be a synapomorphy of the Corethrellidae, Chaoboridae, and Culicidae; however, further observations are needed, not only on the Thaumaleidae and Chironomidae, but also on nonculicomorphan Nematocera.

74. *Last abdominal segment (segment 10) without fan-like row of long setae (plesiomorphic) / last* *abdominal segment with fan-like row of long setae on midventral surface, each seta having a transverse, T-shaped base (apomorphic)*

Larvae of the Corethrellidae, Chaoboridae, and Culicidae, all free-swimming, are the only larvae of Nematocera to possess this anal fan, which is regarded as being a synapomorphy of the three families. The entire anal fan acts as a fin for swimming when the abdomen is lashed from side to side, and the T-shaped base on each seta presumably minimizes sideways movement (Figs. 24.14, 25.28–32).

75. *Pigment of adult eye appearing in the pupa, after the larval stage (plesiomorphic) / pigment of adult eye developing precociously, becoming conspicuous as early as the second instar, and always well developed by the last instar, located anterior to the pigment-containing cells characterizing the larval eye or stemma (apomorphic)*

Precocious development of the adult eye, early in the life of the larva, presumably enhances the larva's ability to detect changes in light intensity and, thereby, to avoid predators. This condition is known only in the Chaoboridae and Culicidae, although this pigmentation occurs only along the hind margin of the eye in *Mansonia perturbans* (Walker). This feature is considered to be a synapomorphy of these two families.

Pupa

76. *Metathoracic leg sheath extending beyond wing sheath, parallel to sheaths of other two legs (plesiomorphic) / metathoracic leg sheath bent in an S-shape, concealed beneath wing sheath, ending beside apex of mesothoracic leg (apomorphic)*

This character state, which presumably allows greater mobility of the pupal abdomen, was one of the synapomorphies of the Culicomorpha proposed by Hennig (1973). Although not applying to all members (some chironomid pupae have straight metathoracic leg sheaths), the character state is unique to the Culicomorpha.

77. *Apex of abdomen terminating in a pair of immovable lobes (plesiomorphic) / apex of abdomen terminating in a pair of articulated, membranous paddles, each with supporting midrib (apomorphic)*

The apomorphic condition is unique to the Culicidae and Chaoboridae. By rapid flexion of the abdomen, as in a lobster or crayfish, these paddles enable the aquatic pupae to swim rapidly to the bottom of their water body to escape predators. Members of the chironomid subfamily Tanypodinae also swim in a similar fashion, although they lack articulated paddles.

Adult

78. *Proboscis no longer than two or three times head capsule length (plesiomorphic) / proboscis, composed of labrum, mandibles, laciniae, and hypopharynx, all ensheathed within the labium, markedly elongate (apomorphic)*

The elongate mouthparts of adult Culicidae, and the manner in which the stylets are used to probe for and enter capillaries without damaging superficial tissues of their host, are one of the most characteristic features of the family. The condition is unique within the Diptera and is only superficially approximated by some other nonbloodfeeding taxa (e.g. *Elephantomyia* Osten Sacken and *Toxorhina* Loew of Tipulidae).

79. *Pedicel not especially enlarged, about the same diameter as scape, and male flagellum not markedly plumose nor noticeably different from that of female (plesiomorphic) / pedicel enlarged, especially in males, usually globular and much wider than scape as a result of a greatly enlarged Johnston's organ, and male flagellum plumose, with much longer, denser setae than those of female (apomorphic)*

Specialization of the male antenna for sound reception, a feature that enables the male to locate a female by detecting the sound of her wing beats, is unique to the Culicomorpha and is characteristic of most members of the Corethrellidae, Chaoboridae, Culicidae, Ceratopogonidae, and the Chironomidae. A few species of the Culicidae, Ceratopogonidae, and Chironomidae, which mate on the ground without a prior mating flight, do not have the long setae of the flagellum (presumably lost) but still possess an enlarged pedicel. The pedicel of adult thaumaleids is large relative to the flagellomeres but otherwise their antennae resemble those of simuliids in structure. However, no simuliid has a globular pedicel or a long, setose flagellum. Males of the Simuliidae (except for a few small-eyed species that mate on the ground) have enlarged eyes and capture females in flight. We concur with Hennig (1973) in concluding that the specialized antenna has been lost in this family. We suppose that the same is true with regard to the Thaumaleidae.

Although an enlarged pedicel is mainly limited to the Culicomorpha, some cecidomyiids (*Micromya* Rondani, *Anarete* Haliday Figs. 16.36–37) exhibit a similar condition. We regard this as an example of convergence.

80. *Wing with vein M_3 present and with a discal cell (dm) (plesiomorphic) / wing with neither M_3 nor a discal cell (apomorphic)*

Vein M_3 and the discal cell are usually present together in most of the infraorders of Nematocera, namely the Tipulidae (Tipulomorpha), the Anisopodidae and Trichoceridae (Psychodomorpha), the Pachyneuridae (Bibionomorpha), the Tanyderidae (Ptychopteromorpha), and the lower Brachycera. The wing of members of the Psychodidae has M_3 but lacks the discal cell. Loss of either or both of these character states is undoubtedly derived; however, both are lacking and presumed lost in the members of many other families of Nematocera, and we consider it unreliable as a measure of relationship. Nevertheless, no member of the Culicomorpha possesses either M_3 or a discal cell; this loss may be a synapomorphy of the infraorder.

81. *Radial sector of wing with more than three branches (plesiomorphic) / radial sector of wing with three or fewer branches (apomorphic)*

Few families of Nematocera have a four-branched radial sector; those that do, the Tipulidae, Tanyderidae, Psychodidae, and Ptychopteridae, are usually considered to be relatively primitive. Only in the Tanyderidae and Psychodidae do all four branches reach the wing margin. No member of the Culicomorpha has more than three branches of the sector. In the Culicoidea R_2 and R_3 are separate distally but have a common stem, whereas R_{4+5} is unbranched, which results in a total of three branches of the sector. This situation is uncommon in the Chironomoidea and is found only in a few (presumably more primitive) Chironomidae. In the remaining Chironomoidea, the sector is either unbranched, or two-branched, consisting of R_{2+3} and R_{4+5}. Reduction to three branches or fewer, although probably a synapomorphy of the Culicomorpha, is also found in the Axymyiidae, some Anisopodidae (*Sylvicola* Harris), and Pachyneuridae (*Pachyneura* Zetterstedt). Other Bibionomorpha have only two branches or fewer. The lower Brachycera also have three branches, but R_{2+3} is unbranched whereas R_{4+5} is branched, the reverse of the Culicoidea. This reduction probably cannot be considered alone as being a reliable indicator of relationship.

82. *R_{2+3} straight or slightedly curved (plesiomorphic) / R_{2+3} strongly arched (Fig. 23.4) (apomorphic)*

The apomorphic feature is unique to Dixidae.

83. *Sperm transferred as a liquid or amorphous mass, often by sperm pump (plesiomorphic) / sperm transferred by a complex, two-chambered, symmetrical spermatophore that is formed within the male before or during ejaculation (apomorphic)*

Although Hennig presented this character state as being a synapomorphy of the Culicomorpha, Wood (1978) showed that it is not present in the Culicoidea but suggested it may be a synapomorphy of the Chironomoidea, provided that it is present in the Thaumaleidae. Spermatophore formation and deposition was described in greatest detail for two species of *Culicoides* Latreille by Pomerantzev (1932) and Linley (1981). Linley (1981) showed that a two-chambered envelope secreted by the accessory gland was filled with two packets of sperm just before ejaculation. Identical structures are present in the Simuliidae and Chironomidae (discussed by Wood 1978), but the exact process of spermatophore formation is not yet understood. Presence of a spermatophore has yet to be established in the Thaumaleidae.

Leppla *et al.* (1975) reported the presence of a spermatophore in a bibionid, *Plecia nearctica* Hardy. Pollock (1970) described a spermatophore in *Glossina austeni* Newstead. Spermatophores have not otherwise been reported in the Diptera. Mickoleit (1974) described the spermatophore of *Boreus westwoodi* Hagen but regarded it as being a convergent development within the Mecoptera.

Discussion. Although we have presented the Thaumaleidae as the sister group of the rest of the Chironomoidea, some evidence indicates that the Simuliidae is the sister group of the rest. A dorsal mandibular brush is present in larvae of the Ptychopteridae, Culicoidea, and Simuliidae but not in those of the Tanyderidae. It is absent in Thaumaleidae, Ceratopogonidae, and Chironomidae, which may be a synapomorphy of these three families. The dorsal mandibular brush is used for combing food particles from the labral brush while the mandible is adducted, bringing the mandibular brush hairs through those of the labral brush just before it is extended.

Present classifications place *Corethrella* in a separate tribe or subfamily within the Chaoboridae. As already discussed, the only apparent synapomorphy that might suggest this placement is the presence of prehensile antennae in the larvae, which we believe are not homologous in *Corethrella* and other Chaoboridae. We regard the precocious development of the adult eye within the larva and the presence of a pair of movable lobes, or paddles, at the apex of the pupal abdomen, as synapomorphies of the Culicidae plus Chaoboridae, not including *Corethrella*. Thus, the Chaoboridae, as previously recognized, would be a paraphyletic group. To avoid this situation, *Corethrella* is, therefore, considered as being a separate and newly recognized family, the Corethrellidae.

In this presentation we did not list all apparent synapomorphies for the morphologically distinct Simuliidae. Further adult synapomorphies have been discussed by Wood and Borkent (1982).

CONCLUSIONS AND SUGGESTIONS FOR FUTURE RESEARCH

Our analysis of the families of the Nematocera provides a fresh outlook on their phylogenetic relationships. In some instances general conclusions by previous workers have been supported (e.g. Culicomorpha), whereas for other taxa we have proposed quite different arrangements (e.g. Psychodomorpha). Previous schemes have relied heavily on adult characters. Our reinterpretation of these, plus consideration of added characters from immature stages, have resulted in different phylogenetic conclusions.

As noted in our analyses, many of our hypotheses of polarity of character states are tenuous. The Nematocera are a remarkably diverse group of organisms that exhibit marked structural variation in each life stage. This diversity means not only that characters are often of unsure homology but also that, once homologous characters have been established, homologous character states may be difficult to recognize.

A common problem of phylogenetic analysis of Nematocera is in the types of comparisons made by some workers. Too often the only comparisons made have been between a few distantly related groups, which has resulted in questionable statements of homologies and a failure to express the distribution of character states. It seems to us that a stable phylogenetic scheme for the Nematocera will depend on the detection of many character states that form series of nested sets (rather than discovering many character states at a particular node of the cladogram). Such a scheme will depend on future workers studying and describing character states in a large variety of taxa, and on more careful and detailed comparison being made between closely related groups.

In particular, we consider the following areas of future study to be particularly fruitful for revealing synapomorphies: first-instar head capsule and mouthparts, larval pharyngeal filter, pupal respiratory organ and the manner of its attachment to the spiracle of the pharate adult (see, for example, Coffman 1979), and details of female terminalia. We have attempted to note in the text some characters that are in particular need of reexamination. In addition, a wide array of structures that have not been previously examined remain to be studied.

References

Alexander, C. P. 1927. The interpretation of the radial field of the wing in the nematocerous Diptera, with special reference to the Tipulidae. Proc. Linn. Soc. N.S.W. 52: 42–72.

Alexander, C. P. 1942. Guide to the insects of Connecticut. Part VI. The Diptera or true flies of Connecticut. First Fascicle. Family Anisopodidae. Bull. Conn. St. geol. nat. Hist. Surv. 64: 192–198 [reprinted 1966].

Alexander, C. P. 1965. Family Pachyneuridae. Page 196 *in* A. Stone, C. W. Sabrosky, W. W. Wirth, R. H. Foote, and J. R. Coulson, eds. A catalog of the Diptera of America north of Mexico. Agric. Handb. 276.

Anthon, H. 1943. Der Kopfbau der Larven einiger nematoceran Dipterenfamilien: Rhyphidae, Trichoceridae, Psychodidae und Ptychopteridae. Spolia zool. Mus. haun. 3: 1–61; 11 plates.

Barr, W. B. 1984. Prolegs and attachment of *Simulium vittatum* (sibling IS-7) (Diptera: Simuliidae) larvae. Can. J. Zool. 62: 1355–1362.

Borkent, A. 1984. The systematics and phylogeny of the *Stenochironomus* complex (*Xestochironomus*, *Harrisius*, and *Stenochironomus*) (Diptera: Chironomidae). Mem. ent. Soc. Can. 128: 1–269.

Borkent, A., W. W. Wirth, and A. L. Dyce. 1987. The newly discovered male of *Austroconops* (Ceratopogonidae: Diptera) with a discussion of the phylogeny of the basal lineages of the Ceratopogonidae. Proc. ent. Soc. Wash. 89: 587–606.

Boudreaux, H. B. 1979. Arthropod phylogeny with special reference to insects. John Wiley & Sons Canada Limited, Rexdale, Ont.

Brandt, E. 1880*a*. Vergleichend-anatomische Skizze des Nervensystems der Insekten. Trudy̆ russk. ént. Obshch. 15: 3–19; pls. 3, 4.

Brandt, E. 1880*b*. Ueber die Metamorphosen des Nervensystems der Insekten. Trudy̆ russk. ént. Obshch. 15: 20–30; pls. 5, 6.

Brandt, E. 1880*c*. Vergleichend-anatomische Untersuchungen über das Nervensystem der Hymenopteren. Trudy̆ russk. ént. Obshch. 15: 31–50; pls. 7–10.

Brandt, E. 1880*d*. Vergleichend-anatomische Untersuchungen das Nervensystems der Kafer (Coleoptera). Trudy̆ russk. ént. Obshch. 15: 51–67; pls. 11–13.

Brandt, E. 1880*e*. Vergleichend-anatomische Untersuchungen über das Nervensystems der Lepidopteren. Trudy̆ russk. ént. Obshch. 15: 68–83; pl. 14.

Brandt, E. 1880*f*. Vergleichend-anatomische Untersuchungen über das Nervensystems der Zweiflügler (Diptera). Trudy̆ russk. ént. Obshch. 15: 84–101; pls. 15–18.

Brauer, F. 1863. Monographie der Oestriden. Wien.

Brauer. F. 1869. Kurze Charakteristik der Dipteren-Larven. Verh. zool.-bot. Ges. Wien 19: 843–852.

Brundin, L. 1966. Transantarctic relationships and their significance, as evidenced by chironomid midges with a monograph of the subfamilies Podonominae and Aphroteniinae and the Austral Heptagyiae. K. svenska Vetensk-Akad. Handl. 11: 7–472.

Byers, G. W. 1969. A new family of nematocerous Diptera. J. Kans. ent. Soc. 42: 366–371.

Chiswell, J. R. 1955. On the last instar larva of *Tipula livida* Van der Wulp (Diptera, Tipulidae) with notes on the fronto-clypeal region of larval Tipulinae and caterpillars. Proc. R. ent. Soc. Lond., Ser. A, 30: 127–136.

Coffman, W. P. 1979. Neglected characters in pupal morphology as tools in taxonomy and phylogeny of Chironomidae (Diptera). Ent. scand. Suppl. 10: 37–46.

Colless, D. H. 1962. A new Australian genus and family of Diptera (Nematocera: Perissommatidae). Aust. J. Zool. 10: 519–535.

Cooper, K. W. 1974. Sexual biology, chromosomes, development, life histories and parasites of *Boreus*, especially of *B. notoperates*. A southern California *Boreus*. II. (Mecoptera: Boreidae). Psyche (Camb.) 81: 84–120.

Coquillett, D. W. 1901. The Nematocera not a natural group of Diptera. Ann. ent. Soc. Am. 8: 93–98.

Coquillett, D. W. 1909. Rediscovery of the bibionid genus *Eupeitenus*. Ent. News 20: 106.

Craig, D. A., and M. M. Chance. 1982. Filter feeding in larvae of Simuliidae (Diptera: Culicomorpha): aspects of functional morphology and hydrodynamics. Can. J. Zool. 60: 712–724.

Crampton, G. C. 1924. Remarks on the phylogeny and interrelationships of nematocerous Diptera. Psyche (Camb.) 31: 238–242.

Crampton, G. C. 1925. A phylogenetic study of the thoracic sclerites of the non-tipuloid nematocerous Diptera. Ann. ent. Soc. Am. 18: 49–74.

Crampton, G. C. 1926. A phylogenetic study of the thoracic sclerites of the psychodoid Diptera, with remarks on the interrelationships of the Nematocera. Ent. News 37: 33–38, 65–70; 2 plates.

Cranston, P. S., D. H. D. Edward, and D. H. Colless. 1987. *Archaeochlus* Brundin: A midge out of time (Diptera: Chironomidae). Syst. Ent. 12: 313–334.

Crowson, R. A. 1981. The biology of the Coleoptera. Academic Press, Toronto, Ont.

Cutten, F. E. A., and D. K. McE. Kevan. 1970. The Nymphomyiidae (Diptera), with special reference of *Palaeodipteron walkeri* Ide and its larva in Quebec, and a description of a new genus and species from India. Can. J. Zool. 48: 1–24.

Dahl, C. 1980. Comparison of postembryonic organization of the genital segments in Trichocridae, Tipulidae, and Anisopodidae (Diptera, Nematocera). Zool. Scr. 9: 165–185.

Davies, L. 1965. The structure of certain atypical Simuliidae (Diptera) in relation to evolution within the family, and the erection of a new genus for the Crozet Island black-fly. Proc. Linn. Soc. Lond. 176: 159–180.

de Meijere, J. C. H. 1901. Uber das letzte Glied der Beine bei den Arthropoden. Zool. Jb. (Abt. Anat. Ontog. Tiere) 14: 417–476.

de Meijere, J. C. H. 1916. Beitrage zur Kenntnis der Dipteren-larven und puppen. Zool. Jb. (Syst. Okol. Geogr. Tiere) 40: 177–322.

Downes, J. A. 1968. Notes on the organs and processes of sperm-transfer in the lower Diptera. Can. Ent. 100: 608–617.

Edwards, F. W. 1926. The phylogeny of nematocerous Diptera: a critical review of some recent suggestions. Verh. III int. Kongr. Ent. (Zurich) 1: 111–130.

Eldredge, N., and J. Cracraft. 1980. Phylogenetic patterns and the evolutionary process. Method and theory in comparative biology. Columbia University Press, New York, N.Y.

Exner, K., and D. A. Craig. 1976. Larvae of Alberta Tanyderidae. Quaest. Ent. 12: 219–237.

Gassner, G. 1963. Notes on the biology and immature stages of *Panorpa nuptialis* Gerstaecker (Mecoptera: Panorpidae). Tex. J. Sci. 15: 142–154.

Goetghebuer, M. 1925. Contribution à l'étude des "prémandibules" chez les larves des Diptères nématocères. Pages 142–152 *in* Encycl. ent. B. 2. Diptères. 1 (3).

Gould, S. J. 1980. Hen's teeth and horse's toes. Nat. Hist., N.Y. 89 (7): 24–28.

Griffiths, G. C. D. 1975. Book Review. Rohdendorf, B. B. 1975. The historical development of Diptera. Quaest. Ent. 11: 3–5.

Hackman, W., and R. Väisänen. 1982. Different classification systems in the Diptera. Annls. zool. Fenn. 19: 209–219.

Harbach, R. E., and K. L. Knight. 1980. Taxonomists' glossary of mosquito anatomy. Plexus Publishing, Inc., Marlton, N. J.

Hennig, W. 1948. Die Larvenformen der Dipteren. Eine Ubersicht über die bisher bekannten Jugendstadien der zweiflügeligen Insekten. 1. Teil. Akademie-Verlag, Berlin.

Hennig, W. 1950. Die Larvenformen der Dipteren. Eine Ubersicht über die bisher bekannten Jugendstadien der zweiflügeligen Insekten. 2. Teil. Akademie-Verlag, Berlin.

Hennig, W. 1952. Die Larvenformen der Dipteren. Eine Ubersicht über die bisher bekannten Jugendstadien der zweiflügeligen Insekten. 3. Teil. Akademie-Verlag, Berlin.

Hennig, W. 1953. Kritische Bemerkungen zum phylogenetischen System der Insekten. Beitr. Ent. 3: 1–85.

Hennig, W. 1954. Flügelgeäder und System der Dipteren unter Berücksichtigung der aus dem Mesozoikum beschriebenen Fossilien. Beitr. Ent. 4: 245–388.

Hennig, W. 1965. Phylogenetic systematics. A. Rev. Ent. 10: 97–116.

Hennig, W. 1966. Phylogenetic systematics. *Translated by* D. D. Davis and R. Zangerl. University of Illinois Press, Urbana, Ill.

Hennig, W. 1968. Kritische Bemerkungen über den Bau der Flügelwurzel bei den Dipteren und die Frage nach der Monophylie der Nematocera. Stuttg. Beitr. Naturk. 193: 1–23.

Hennig, W. 1969. Die Stammesgeschichte der Insekten. Senckenbergischen Naturforschenden Gesellschaft. W. Kramer, Frankfurt am Main. Senkenberg-Buch 49.

Hennig, W. 1973. Ordnung Diptera (Zweiflügler). Handb. Zool. 4 (2) 2 / 31 (Lfg. 20): 1–337.

Hennig, W. 1981. Insect phylogeny. *Translated and edited by* A. C. Pont (revisionary notes by D. Schlee). John Wiley & Sons, New York.

Hinton, H. E. 1971. Some neglected phases in metamorphosis. Proc. R. ent. Soc., Ser. C, 35: 55–64.

Hinton, H. E. 1981. Biology of insect eggs in three volumes. I. Pergamon Press, Oxford, England.

Hutson, A. M. 1977. A revision of the families Synneuridae and Canthyloscelidae (Diptera). Bull. Br. Mus. nat. Hist. (Ent.) 35 (3): 65–100.

Imms, A. D. 1944. On the constitution of the maxillae and labium in Mecoptera and Diptera. J. microsc. Sci. n. ser. 85: 73–96.

Kalugina, N. S., and V. G. Kovalev. 1985. Jurassic Diptera of Siberia [in Russian]. U.S.S.R. Academy of Sciences, Moscow.

Keilin, D. 1912. Structure du pharynx en fonction du régime chez les larves de diptères cyclorhaphes. C. r. hebd. Séanc. Acad. Sci. Paris 155: 1548–1550.

Keilin, D. 1944. Respiratory systems and respiratory adaptations in larvae and pupae of Diptera. Parasitology 36: 1–66.

Keilin, D., and P. Tate. 1940. The early stages of the families Trichoceridae and Anisopodidae (= Rhyphidae) (Diptera: Nematocera). Trans. R. ent. Soc. Lond. 90: 39–62.

Kennedy, H. D. 1960. *Deuterophlebia inyoensis*, a new species of mountain midge from the alpine zone of the Sierra Nevada range, California (Diptera: Deuterophlebiidae). Trans. Am. microsc. Soc. 79: 191–210.

Klemm, N. 1966. Die Morphologie des Kopfes von *Rhyacophila* Pict. (Trichoptera). Zool. Jb. (Abt. Anat. Ontog. Tiere) 83: 1–50.

Knab, F. 1915. The Nemocera not a natural group of Diptera. Ann. ent. Soc. Am. 8: 93–98.

Kristensen, N. P. 1968. The anatomy of the head and alimentary canal of adult Eriocraniidae (Lep., Dacnonypha). Ent. Meddr. 36: 239–315.

Kristensen, N. P. 1975. The phylogeny of hexapod "orders." A critical review of recent accounts. Z. zool. Syst. Evolut. Forsch. 13: 1–44.

Kristensen, N. P. 1981. Phylogeny of insect orders. A. Rev. Ent. 26: 135–157.

Kristensen, N. P., and E. S. Nielsen. 1981. Abdominal nerve cord configuration in adult non-Ditrysian Lepidoptera. Int. J. Insect Morph. Embryol. 10: 89–91.

Krivosheina, N. P., and B. M. Mamaev. 1970. Morphology, ecology and phylogenetic relationships of the Cramptonomyiidae (Diptera, Nematocera), a family new to the fauna of the U.S.S.R. [in Russian]. Ént. Obozr. 49: 886–898 (English translation in Ent. Rev., Wash. 49: 541–548).

Lafon-Cazal, M. 1983. Ch. 13. Cephalic neurohemal organs in Hymenoptera. Pages 422–432 *in* A. P. Gupta, ed. Neurohemal organs of arthropods: Their development, evolution, structures, and functions. C. C. Thomas, Springfield, Ill.

Lameere, A. 1906. Notes pour la classification des Diptères. Mém. Soc. r. ent. Belg. 12: 105–140.

Latreille, P. A. 1802. Histoire naturelle, générale et particulière, des Crustaces et des Insectes. Tome troisième. Pages 1–468 *in* C. S. Sonnini, ed. Histoire naturelle par Buffon. Paris.

Leppla, N. C., T. C. Carlysle, and R. H. Guy. 1975. Reproductive systems and the mechanics of copulation in *Plecia nearctica* Hardy (Diptera: Bibionidae). Int. J. Insect Morph. Embryol. 4: 299–306.

Lindner, E. (ed.) 1949. Die Fliegen der palaearktischen Region. Schweizerbart, Stuttgart.

Linley, J. R. 1981. Ejaculation and spermatophore formation in *Culicoides melleus* (Coq.) (Diptera: Ceratopogonidae). Can. J. Zool. 59: 332–346.

Macquart, J. 1834. Histoire naturelle des insectes. Vol. 1. Pages 1–578 + 12 plates *in* N. E. Roret, ed. Collection des suites à Buffon. Paris.

Malloch, J. R. 1917. A preliminary classification of Diptera, exclusive of Pupipara, based upon larval and pupal characters, with keys to imagines in certain families. Part 1. Bull. Ill. St. Lab. nat. Hist. (1918) 12: 161–409; plates 28–57.

Matuszewski, B. 1982. Diptera I. Cecidomyiidae. Animal Cytogenetics. Vol. 3: Insecta 3.

Mickoleit, G. 1971. Das Exoskelet von *Notiothauma reedi* MacLachlan, ein Beitrag zur Morphologie und Phylogenie der Mecoptera (Insecta). Z. Morph. Tiere 69: 318–362.

Mickoleit, G. 1974. Ueber die Spermatophore von *Boreus westwoodi* Hagen (Insecta, Mecoptera). Z. Morph. Tiere 77: 271–284.

Mickoleit, G. 1981. Pages 332–333 *in* W. Hennig. Insect phylogeny. *Translated and edited by* A. C. Pont (revisionary notes by D. Schlee). John Wiley & Sons, New York.

Nelson, G. J. 1974. Classification as an expression of phylogenetic relationships. Syst. Zool. 23: 344–359.

Osten Sacken, C. R. 1891. On the characters of the three divisions of Diptera: Nemocera vera, Nemocera anomala, and Eremochaeta. Berl. ent. Z. 37: 417–465.

Patterson, C., and D. E. Rosen. 1977. Review of ichthyodectiform and other Mesozoic teleost fishes and the theory and practice of classifying fossils. Bull. Am. Mus. nat. Hist. 158: 81–172.

Pennak, R. W. 1951. Description of the imago of the mountain midge *Deuterophlebia coloradensis* Pennak. Am. Mus. Novit. 1534: 1–11.

Petralia, R. S., A. L. Wuensche, G. L. Teetes, and A. A. Sorensen. 1979. External morphology of the mouthparts of larvae of sorghum midge, *Contarinia sorghicola*. Ann. ent. Soc. Am. 72: 850–855.

Pilgrim, R. L. C. 1972. The aquatic larva and the pupa of *Choristella philpotti* Tillyard, 1917 (Mecoptera—Nannochoristidae). Pacif. Insects 14: 151–168.

Plachter, H. 1979. Zur Kenntnis der Präimaginalstadien der Pilzmucken (Diptera, Mycetophiloidea). Teil II: Eidonomie der Larven. Zool. Jb. (Abt. Anat. Ontog. Tiere) 101: 271–392.

Pollock, J. N. 1970. Sperm transfer by spermatophores in *Glossina austeni* Newstead. Nature, Lond. 225: 1063–1064.

Pomerantzev, B. I. 1932. On the morphology and anatomy of the genitalia of *Culicoides* (Diptera, Nematocera) [in Russian]. Parazit. Sb. 3: 183-214.

Potter, E. 1938. The internal anatomy of the order Mecoptera. Trans. R. Ent. Soc. Lond. 87: 467–501.

Pulikovsky, N. 1924. IV. Metamorphosis of *Deuterophlebia* sq. [sic] (Diptera, Deuterophlebiidae Edw.). Trans. R. ent. Soc. Lond. 1924: 45–62.

Rogers, J. S. 1949. The life history of *Megistocera longipennis* (Macquart) (Tipulidae, Diptera) a member of the neuston fauna. Occ. Pap. Mus. Zool. Univ. Mich. 521: 1–14; 2 plates.

Rohdendorf, B. B. 1964. The historical development of two-winged insects [in Russian]. Trudy̆ paleont. Inst. 100: 1–311.

Rohdendorf, B. B. 1974. The historical development of Diptera [translated from Russian]. B. Hocking, H. Oldroyd, and G. E. Ball, eds. University of Alberta Press, Edmonton, Alta.

Rohdendorf. B. B. 1977. The classification and phylogeny of Diptera. Pages 81–88 *in* O. A. Scarlato and K. B. Gorodkov, eds. Systematics and evolution of the Diptera (Insecta) [in Russian]. A collection of scientific works. Zool. Inst. Acad. Sci., Leningrad, U.S.S.R.

Rothfels, K. H., and G. F. Mason. 1975. Achiasmate meiosis and centromere shift in *Eusimulium aureum*. Chromosoma 5: 111–124.

Rothschild, M. 1975. Recent advances in our knowledge of the order Siphonaptera. A. Rev. Ent. 20: 241–259.

Rubtzov, I. A. 1956. Fauna S.S.S.R. Insects Diptera. Black flies (Fam. Simuliidae) [in Russian]. Zool. Inst. Acad. Sci. U.S.S.R. (N.S. 64) 6 (6): 1–860.

Schiner, J. R. 1864. Ein neues System der Dipteren. Verh. zool.-bot. Ges. Wien 14: 201–212.

Shannon, R. C. 1921. Another anomalous dipteron added to the Rhyphidae. Proc. ent. Soc. Wash. 24: 50–51.

Snodgrass, R. E. 1935. Principles of insect morphology. McGraw Hill, New York, N.Y.

Snodgrass, R. E. 1946. The skeletal anatomy of fleas (Siphonaptera). Smithson. misc. Collns 104 (18).

Snodgrass, R. E. 1959. The anatomical life of the mosquito. Smithson. misc. Collns. 139 (8): 1–87.

Stuckenberg, B. R. 1973. The Athericidae, a new family in the lower Brachycera (Diptera). Ann. Natal. Mus. 21: 649–673.

Tillyard, R. J. 1935. The evolution of the scorpion flies and their derivatives (Order Mecoptera). Ann. ent. Soc. Am. 28: 1–45.

Tokunaga, M. 1935. A morphological study of the nymphomyiid fly. Philipp. J. Sci. 56: 127–214.

Tokunaga, M. 1936. The central nervous, tracheal, and digestive systems of a nymphomyiid fly. Philipp. J. Sci. 59: 189–216.

Tokunaga, M. 1956. 4499. *Nymphomyia alba* Tokunaga. Page 1567 *in* T. Shiraki *et al.*, Diptera, *in* T. Esaki *et al.*, eds. Incongraphia Insectorum Japonicorum. Edito Secunda Reformata.

Tokunaga, M. 1965. Nymphomyiidae. Page 184 *in* S. Asahina, T. Ishihara, and K. Yasumatsu, eds. Iconographia Insectorum Japonicorum. Colore naturali edita. Hokuyurkan, Tokyo 3.

White, M. J. D. 1949. Cytological evidence on the phylogeny and classification of the Diptera. Evolution 3: 252–261.

White, M. J. D. 1973. Animal cytology and evolution. 3rd ed. Cambridge University Press, England.

Wiggins, G. B. 1977. Larvae of the North American caddisfly genera (Trichoptera). University of Toronto Press, Toronto, Ont.

Wiley, E. O. 1981. Phylogenetics. The theory and practice of phylogenetic systematics. John Wiley & Sons, New York.

Willmann, R. 1981. Das Exoskelett der männlichen Genitalien der Mecoptera (Insecta) II. Die phylogenetischen Beziehungen der Schnabelfliegen-Familien. Z. zool. Syst. Evolut.-Forsch. 19: 153–174.

Willmann, R. 1987. The phylogenetic system of the Mecoptera. Syst. Ent. 12: 519–524.

Wirth, W. W., and W. L. Grogan. 1979. Natural history of Plummers Island, Maryland. XXIV. Biting midges (Diptera: Ceratopogonidae). 2. The species of the tribes Heteromyiini and Sphaeromiini. Proc. biol. Soc. Wash. 91: 847–903.

Wood, D. M. 1978. Taxonomy of the Nearctic species of *Twinnia* and *Gymnopais* (Diptera: Simuliidae) and a discussion of the ancestry of the Simuliidae. Can. Ent. 110: 1297–1337.

Wood, D. M., and A. Borkent. 1982. Description of the female of *Parasimulium crosskey* Peterson (Diptera: Simuliidae), and a discussion of the phylogenetic position of the genus. Mem. ent. Soc. Wash. 10: 193–210.

Zwick, P. 1977. Australian Blephariceridae. Aust. J. Zool., Supp. 46: 1–121.

PHYLOGENY AND CLASSIFICATION OF THE "ORTHORRHAPHOUS" BRACHYCERA 115

N. E. Woodley

INTRODUCTION

Although classification of the Diptera has been an integral part of the taxonomy of the order almost since formal taxonomic publication was initiated by Linnaeus, interest in the evolution and phylogeny of flies was post-Darwinian and was really only pursued during this century. A rigorous methodology for hypothesizing phylogenies of organisms is quite recent. The subject received its first full treatment by Hennig in 1950, a work that has since become widely recognized.

Williston (1908) and Lindner (1949) summarized the early works that dealt significantly with the classification of Diptera. In this chapter, I discuss primarily our knowledge of the evolutionary branching sequences of taxa at the family level and above in the suborder Brachycera, i.e. the cladistic relationships of the major groups. Rendering this information into a formal classification is only a secondary concern. Because the hypotheses concerning the cladistics of the Brachycera are mostly based on a rather small amount of evidence, I prefer to make as few changes as possible in formal classification. Otherwise, the development of new knowledge will only result in instability of nomenclature and frequent changes in a classification used widely beyond the specialized field of systematics.

Works containing evolutionary information on the classification of Diptera include those of Williston (1908), Malloch (1917), Hardy (1927), Mackerras and Fuller (1942), Steyskal (1953, 1974), Colless and McAlpine (1970), Nagatomi (1977), and Hackman and Väisänen (1982). Works by Oldroyd (1977) and Rohdendorf (1974) have not utilized methodology acceptable to most modern dipterists, and their conclusions have not been followed by subsequent workers. Relatively few students of the Brachycera have utilized cladistic methodology. Most notable of those who have is, of course, Hennig (1973, 1976, 1981, and most of his specialized papers). Others include Stuckenberg (1973), Griffiths (1972), Nagatomi (1981), and Chvála (1983), as well as some recent authors dealing with intrafamilial relationships.

Early clarification is also required for the term "orthorrhaphous" Brachycera. This group includes all Brachycera exclusive of the Muscoidea as recognized in this chapter (= Muscomorpha of Ch. 116). However, the group cannot be recognized formally because it is paraphyletic, as the Muscoidea cannot be considered the sister group to the "Orthorrhapha." Thus it is a grouping of convenience only.

As with Chapter 114 on Nematocera, my treatment of the Brachycera is based on Hennig's numerous works (summarized in 1973, 1981), to which I have added further information from the literature, as well as a few of my own observations.

ORIGINS OF THE BRACHYCERA

Identification of the nematocerous sister group of the Brachycera is a matter that remains unsettled. Hennig (1973) considered the most likely candidate within the Nematocera to be the Bibionomorpha. He placed the Pachyneuriformia, Anisopodiformia, Bibioniformia, and Mycetophiliformia in his Bibionomorpha, which was equivalent to the Pachyneuroidea + Axymyiomorpha, Anisopodoidea, Bibionoidea, and Sciaroidea + Scatopsoidea of Chapter 114. He based his theory on two characters, which he considered possibly synapomorphic: the enlargement of the katatergite ("die starke Vergrösserung des 2. Laterotergites") and the undivided postphragma of the thorax. Two additional characters, which have sporadic distribution within the Nematocera, were mentioned as also being possibly synapomorphic: the enlarged lower calypter ("squamula thoracicalis") and the two right-angled bends in the pleural suture between the episternum and epimeron.

Hennig's hypothesis has not been improved upon or modified by any subsequent work. A thorough survey of these characters is needed in the Nematocera, as the distribution of character states within the subordinate taxa of Hennig's Bibionomorpha is poorly known. Only when we have such information at hand can Hennig's hypothesis be refined and the possible sister group of the Brachycera be narrowed to a superfamily or perhaps a family of Nematocera. As the Brachycera is not the sister group of the Nematocera as a whole, the latter is a paraphyletic group (Hennig 1968).

In looking for a putative sister group to the Brachycera within the Nematocera, certain features found in Brachycera, which are plesiomorphic relative to many extant Nematocera, should be considered. These are the presence of mandibles in the adult, wing venation with R_4 and M_3 present as well as a closed discal cell, and the presence of three spermathecae. Only six families of Nematocera have M_3 present: the Tanyderidae, Tipulidae, Pachyneuridae, Trichoceridae, Psychodidae, and Anisopodidae. Interestingly, the Anisopodidae have some members with three spermathecae and have a wing venation that is very similar to the ground-plan venation of

Brachycera proposed by Hennig (1973, Fig. 112). Both have a small discal cell near the middle of the wing, with veins M_1, M_2, and M_3 emitted from the cell and with CuA_1 closely approximate to the posterobasal corner of the cell. This venation is possibly synapomorphic, being derived from the ground-plan venation of the Nematocera in an ancestor of the Anisopodidae + Brachycera, with subsequent loss of vein R_4 in the Anisopodidae. Also, larval mandibles are closest to moving in a vertical plane in the Anisopodiformia within the Nematocera; therefore, this feature is the nearest to the apomorphic condition found in the Brachycera (see Ch. 114, under Psychodomorpha). The long history of the two suborders since the probable time of divergence makes the problem of finding a sister group to the Brachycera within the Nematocera difficult. Ample time has passed for the nematocerous sister group to have changed considerably in morphology and, thereby, to have lost the more plesiomorphic features cited above.

The Brachycera probably arose during the Triassic, because flies with well-developed brachycerous characters are known from the lower Jurassic: the first rich assemblage known is from the Malm period. The best summaries of data on these flies are two papers by Kovalev (1981, 1982). He describes new fossils in detail, and discusses in depth his reexaminations of material already described. He criticizes the numerous genera and families proposed in various papers by Rohdendorf (see summary in Rohdendorf 1974), as often being based on fragments and artifactual impressions, thereby leading to misinterpretation of wing venation. More important, however, he describes new forms that clearly differ little from some extant Rhagionidae. Some of these from the upper Dogger to lower Malm deposits of Transbaikalia exhibit character states that are apomorphic relative to some extant rhagionids, such as the reduction of antennal flagellomeres to fewer than eight and the loss of vein M_3 (Kovalev 1982). No fossils that can be conclusively assigned to any other extant families are known from these early periods. Therefore, the Rhagionidae are possibly closest to the most primitive Brachycera among the families of present-day fauna.

CHARACTERIZATION OF THE BRACHYCERA

Several features are present in Brachycera that are synapomorphic for the suborder. These have been discussed briefly by Hennig (1973). In this and following discussions of characters, the italicized words following a character number describe the apomorphic state of the character being considered. The character states are numbered consecutively 1–38 throughout the chapter and the numbers correspond to those used in the three cladograms.

1. *Posterior portions of larval head capsule elongated posteriorly into thorax*

In the Brachycera, the posterior elements of the larval head capsule are elongate and form an internalized structure, coupled with a tendency toward retraction of the entire head capsule into the thorax. These skeletal elements are in the form of a single dorsal plate composed of the cranium and the elongate tentorial rods. The cranium may be further modified to form articulated metacephalic rods in some Asiloidea and Empidoidea. The internal portion of the head capsule becomes the only conspicuous element in the Muscoidea and is known as the cephalopharyngeal skeleton. In Nematocera, the tentorium is contained within the completely exposed head capsule (Fig. 3.5), and the cranium is not retracted into the thorax.

2. *Larval mandibles moving in a vertical plane*

Mandibles of larvae of all known Brachycera are sickle shaped and move in a parallel fashion, vertically. Larval mandibles in Nematocera move in a horizontal plane, in an opposable fashion. The members of Anisopodidae among the Nematocera have larval mandibles that most nearly approximate the brachycerous condition, although these still move in a more or less horizontal plane.

3. *Reduction of antennal flagellomeres to eight*

The basic number of antennal flagellomeres in the most plesiomorphic Brachycera is eight. This number is found in at least some members of all families of the Xylophagomorpha and Stratiomyomorpha, as well as in all Tabanomorpha except the Athericidae. In Muscomorpha this number is reduced even further, as these flies never have more than four flagellomeres. Most Nematocera have more than eight flagellomeres, with eight or fewer being found only in the Deuterophlebiidae (4), Nymphomyiidae (3–4), Bibionidae (7–10), Scatopsidae (5–10), Cecidomyiidae (7, 8), and Simuliidae (7–9). In the genus *Rachicerus* Walker (Xylophagidae), the number of flagellomeres is much greater than eight, ranging from 13 to 39 (Nagatomi 1970). This form of flagellum is unique in the Brachycera. It is thought to be of secondary origin (Hennig 1967, 1973), rather than plesiomorphic.

4. *Maxillary palpus two-segmented*

All brachycerous flies have two palpal segments or less. There are no known taxa with more than this basic number, the various reports to the contrary in the literature being erroneous. This condition is clearly derived from the nematocerous ground-plan number of five. No known major taxon of Nematocera has two or fewer palpal segments, although in isolated taxa of the Tipulidae, Mycetophilidae, Sciaridae, Cecidomyiidae, and Scatopsidae two or fewer palpal segments are known. The palpi are small and vestigial in the Deuterophlebiidae and Nymphomyiidae. The palpi of Brachycera are usually stiff in appearance, projecting forward, whereas they are normally pendant in Nematocera.

5. *Veins CuA_2 and A_1 with apices approximate, forming a nearly closed cell cup (anal cell)*

This character was mentioned by Hennig in his work of 1954 but not subsequently in his 1973 monograph. The most primitive condition in the Brachycera is found in flies such as *Pseudoerinna* Shiraki (Fig. 30.2, as *Bequaertomyia* Brennan) and some Bombyliidae (Figs. 45.44–48), in which the cell cu*p* is slightly open at the wing margin. The vast majority of Brachycera have this cell closed at the wing margin, and in most Muscoidea the cell is quite reduced, with its apex far removed from the wing margin. In other taxa, such as leptogastrine Asilidae and a few Bombyliidae (such as *Mancia* Coquillett, Fig. 45.43), the rather widely open cell cu*p* may be a result of the reduction in the anal area of the wing. In the Nematocera, the anal cell is most similar to the brachycerous condition in *Protoplasa* Osten Sacken (Fig. 6.2), but in virtually all other taxa cell cu*p* is widely open. The nematocerous ancestors of the Brachycera may possibly have approached the Brachycera in configuration of cell cu*p*. However, the anal region of the wing in many nematocerous taxa is modified with great reduction in the venation, so that this possibility is difficult to resolve.

Hennig (1954) mentioned a few other venational features that he thought to be apomorphic for the Brachycera. Most notable of these is the fusion of R_2 and R_3, which he also mentioned as being a synapomorphy of the Brachycera in his 1973 work. This fusion is, however, so widespread in the Nematocera that it is difficult to determine if it is synapomorphic for a clade larger than just the Brachycera. The same is true for the reduction of vein A_2 and the fusion of Sc_2 with R_1, both mentioned by Hennig (1954) as being apomorphic for the Brachycera. These characters cannot be considered cladistically useful until much more is understood about character-state homologies in the Nematocera.

PHYLOGENY OF THE INFRAORDERS OF THE BRACHYCERA

As emphasized by Hennig (1973, 1981), the phylogeny of the taxa within the Brachycera is not well resolved. Traditionally, the Brachycera have been separated into two or three major groups, with various names at various ranks. Some classifications recognize two major groups, the Orthorrhapha and the Cyclorrhapha (as in Colless and McAlpine 1970); other schemes, such as those in Volume 1 of this Manual and in Hennig (1973), recognize three infraorders, the Tabanomorpha, Asilomorpha, and Muscomorpha. Although ranked differently in these two systems, the Cyclorrhapha and Muscomorpha (= Muscoidea in the sense of this chapter) are synonymous and represent an undoubted monophyletic group (see Ch. 116). The Orthorrhapha is equivalent to the Tabanomorpha + Asilomorpha of Volume 1 (= Xylophagomorpha + Stratiomyomorpha + Tabanomorpha + Muscomorpha, exclusive of Muscoidea in this chapter). It is essentially a "negative" grouping, having the plesiomorphic character state opposed to the apomorphic mode by which the Muscoidea emerge from their puparium. There is no known synapomorphy for the Orthorrhapha to indicate that it is a monophyletic assemblage. Thus cladistically, the Orthorrhapha is a paraphyletic group in much the same way as is the Nematocera. The subordinate groups of Brachycera as recognized by Hennig (1973) and as in Chapter 1 of this Manual are also questionably monophyletic, at least according to our present, very limited, knowledge of character systems within the Brachycera. Careful survey work is needed throughout the Brachycera, such that character distributions can be delimited and used in a cladistic fashion.

The Tabanomorpha of Hennig (1973) was characterized by three character states: macrochaetae absent, empodium pulvilliform, and tibial spurs pilose. He stated that these were not clearly polarized, given the knowledge of homologies at the time. The presence of both a pulvilliform empodium and pilose tibial spurs in scattered Nematocera, including Bibionidae and Anisopodidae, indicate that these two character states are plesiomorphic in the Brachycera. The absence of macrochaetae is also inconclusive. I have seen apparent macrochaetae on the mesonotum in some species of *Chrysopilus* Macquart (Rhagionidae), and they are distinctly present on the tibiae of *Atherimorpha* White (Rhagionidae). Therefore the absence of macrochaetae is probably not uniform within Hennig's Tabanomorpha. Accordingly, present knowledge does not provide support for the Tabanomorpha *sensu* Hennig as being a monophyletic taxon.

Fig. 115.1 presents a cladogram that outlines our present knowledge of the relationships of families of Brachycera, exclusive of the Muscoidea, which are treated in detail in Chapter 116. The characters are numbered sequentially, and the point at which the apomorphic character state evolved is indicated on the figure by the appropriate number. From this summary of relationships, which should be regarded as tentative, future studies can expand to support or refute these ideas.

The following list summarizes the classification of the Brachycera in a manner that reflects the cladogram in Fig. 115.1. In this system I have tried to utilize categories that correspond with those in Volume 1 and that are more or less consistent with our knowledge of cladistic relationships of these flies.

Infraorder Xylophagomorpha
  Family Xylophagidae
  Family Pantophthalmidae, *incertae sedis*
  Genus *Exeretoneura* Macquart, *incertae sedis*
  Genus *Heterostomus* Bigot, *incertae sedis*
Infraorder Stratiomyomorpha
  Family Xylomyidae
  Family Stratiomyidae
Infraorder Tabanomorpha
  Family Rhagionidae
  Family Pelecorhynchidae
  Family Athericidae
  Family Tabanidae
  Family Vermileonidae, *incertae sedis*
Infraorder Muscomorpha
 Superfamily Nemestrinoidea
  Family Nemestrinidae
  Family Acroceridae

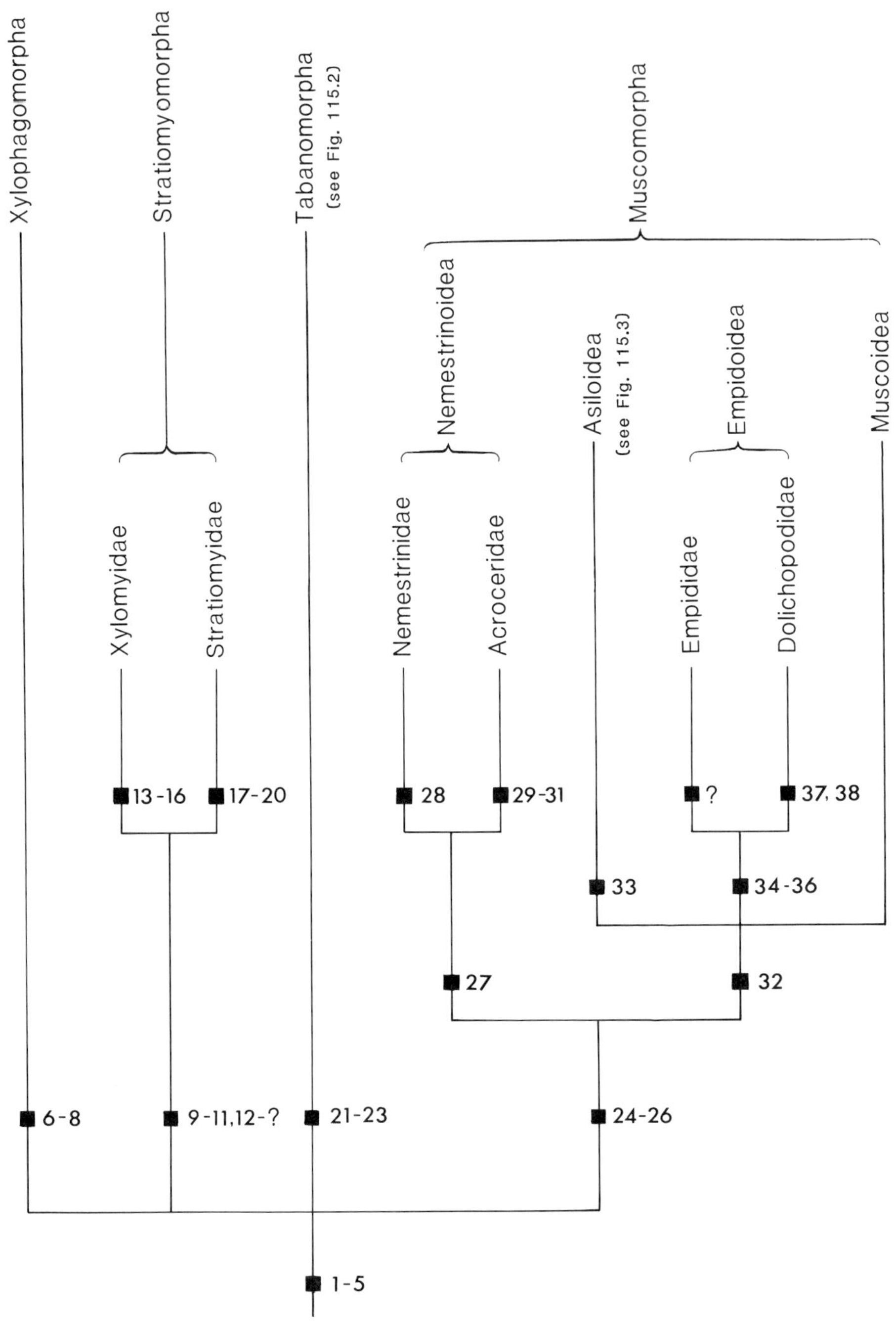

Fig. 115.1. Cladogram showing relationships of major taxa of the suborder Brachycera. Numbers indicate points of origin of apomorphic character states and correspond to discussions in the text.

Superfamily Asiloidea
Family Bombyliidae
Family Therevidae
Family Scenopinidae
Family Mydidae
Family Apioceridae
Family Asilidae
Superfamily Empidoidea
Family Empididae, *sensu lato*
Family Dolichopodidae
Superfamily Muscoidea

(See Ch. 116 for details of classification.)

The organization of the remaining discussion follows this classification. As stated previously, characters are numbered sequentially, and the apomorphic character state immediately follows the number.

Four infraorders of Brachycera are recognized. Unfortunately, knowledge of the relationships between these infraorders remains obscure, and they are shown as an unresolved polychotomy in Fig. 115.1. Each lineage has apparent autapomorphic character states defining it, but no characters that provide a basis for groupings of the infraorders are known at present.

PHYLOGENY OF THE INFRAORDER XYLOPHAGOMORPHA

As recognized here, this infraorder is more restrictive than the analagous taxon used in Volume 1 of this Manual. Volume 1 included in the Stratiomyoidea the Xylophagidae as recognized here, as well as the Xylomyidae and Stratiomyidae. The latter two families are excluded here from the Xylophagomorpha because there is at present no conclusive evidence that the three families form a monophyletic group.

Monophyly of the Xylophagomorpha. Characters 6–8 support the monophyly of the Xylophagomorpha.

6. *Larval head capsule externally elongate, cone-shaped, and strongly sclerotized (Figs. 34.11–15)*

Although the head capsule is quite variable within the Brachycera, that found in the Xylophagomorpha is unique and therefore autapomorphic for the infraorder.

7. *Larval head capsule with internal portion of cranium divided, forming a pair of metacephalic rods (Hennig 1973)*

The divided internal portion of the cranium is found in all known xylophagid larvae (Webb 1979, 1983*a*, 1984; Webb and Lisowski 1983; Krivosheina 1967). Nearly all other Brachycera have the posterior portion of the cranium composed of a single, undivided plate. Only in a few Asilidae (Fig. 42.79) and in the Empididae (Fig. 47.58) and Dolichopodidae (Figs. 48.41, 48.43) are paired metacephalic rods also found. In these three families, the rods are fundamentally different; they are articulated to the anterior, exposed portion of the cranium, rather than being continuous with it. Furthermore, many asilids have the plesiomorphic, undivided cranium. Subdivision of the cranium has likely evolved more than once, although Teskey (Ch. 3) has suggested that a possible precursor to the hinged condition found in Asiloidea and Empidoidea is found in *Xylophagus*.

8. *Anal segment of larva with sclerotized dorsal plate that surrounds the spiracles, ending in a pair of hook-like processes (Figs. 34.11–12; Hennig 1973)*

This character state is unique to the Xylophagomorpha and is found in all known larvae of the group (see references for character 7). It is undoubtedly autapomorphic for the lineage.

Discussion. Only the family Xylophagidae, as recognized in Volume 1 of this Manual, can be placed in this infraorder with certainty. There are 111 species of xylophagids included in the following genera: *Rachicerus* Walker (66 species), *Gymnorhachicerus* Frey (1 species; Oriental), *Paleorachicerus* Nagatomi (1 species; Oriental), *Xylophagus* Meigen (19 species), *Anacanthaspis* Röder (1 species; Palearctic), *Arthropeas* Loew (4 species), *Coenomyia* Latreille (3 species), *Dialysis* Walker (12 species), *Napemyia* Webb (1983*b*, possibly = *Dialysis*, 1 species; Nearctic), and *Odontosabula* Matsumura (5 species; Palaearctic). I disagree with authors who subdivide the Xylophagidae into three smaller families (Hennig 1973; Nagatomi 1977; Webb 1983*a*,*b*), recognizing the Coenomyiidae and Rachiceridae as well as the Xylophagidae. These authors leave the Xylophagidae with only a single genus (*Xylophagus*) and the Rachiceridae with three (*Rachicerus* and its two small segregates, *Gymnorhachicerus* and *Paleorachicerus*). Given the autapomorphic features that characterize the Xylophagidae as recognized here, these small segregates seem unwarranted. All but one of the features utilized by Nagatomi (1977) to characterize the Xylophagidae and Rachiceridae are probably plesiomorphic. The exception is the unusual pectinate, multisegmented antennal flagellum of *Rachicerus* and its allies, which is probably autapomorphic (Hennig 1967, 1973). Some evidence, such as the reduced alula of the wing (Hennig 1973) and the fused first-cercal segments in adult females (Nagatomi and Iwata 1976, Figs. 4–6), which are probably apomorphic, indicates a possible sister-group relationship between *Xylophagus* and *Rachicerus* and allies. Larvae are unknown for *Gymnorhachicerus*, *Paleorachicerus*, *Anacanthaspis*, *Odontosabula*, and *Napemyia*, but, owing to their general similarity to other genera within the Xylophagidae, their placement seems relatively certain.

Three taxa should be included here as *incertae sedis*. They are the family Pantophthalmidae and the isolated genera *Exeretoneura* Macquart and *Heterostomus* Bigot. They are all exotic to the Nearctic region but have been associated with the Xylophagomorpha by various authors; therefore brief comments seem appropriate here.

Family Pantophthalmidae. This small family of strictly Neotropical flies consists of two genera, *Opetiops* Enderlein (1 species) and *Pantophthalmus* Thunberg (19 species) (Val 1976). Hennig (1973) suggested that the family is the sister group of the Coenomyiidae, Xylophagidae, and Rachiceridae combined (i.e. Xylophagidae as recognized here). However, he did not present evidence to support this opinion. The larvae of Pantophthalmidae have none of the derived features of the Xylophagomorpha outlined above, but these features are possibly apomorphic for the Xylophagidae only, and undiscovered characters may eventually support the relationship of pantophthalmids to xylophagids. The adults have lost spurs on both the front and hind tibiae and have a closed wing cell m_3. The larvae have a large sclerotized plate dorsally on the anal segment, but the plate does not have a forked posterior process, and the spiracles are enclosed in a deep ventral cleft. The larvae also have lateral gill-like structures at the foremargin of the anal segment. All these features are probably autapomorphic for the family. Larval structures are summarized by Greene and Urich (1931). These flies are very large in size, among the largest dipterans, and have a unique general appearance. Further evidence is necessary before their relationships to other primitive Brachycera can be elucidated. However, the idea of Mackerras and Fuller (1942: 31), that the Pantophthalmidae arose from within the Stratiomyidae, is clearly untenable.

Genus Heterostomus. This genus is monotypic, with the single species, *H. curvipalpis* Bigot, known only from Chile. This fly has a general appearance not unlike some of the larger xylophagid genera. Unfortunately, the larva is not known. Malloch (1932), quoted later by Hennig (1972), stated that *Heterostomus* had a general habitus very similar to that of *Austroleptis* Hardy (Rhagionidae). In my opinion, the two genera are very different in general appearance. *Heterostomus* is much larger than *Austroleptis* and has many structural differences, including the wing venation and terminalia of both sexes. Although Malloch was right in excluding *Heterostomus* from the pangoniine Tabanidae, where it had been placed by Kröber (1930), it cannot be considered a rhagionid. Hennig (1972) suggested that the genus was possibly a pelecorhynchid, based on similarities of wing venation. This theory can also be refuted, based on present knowledge. *Heterostomus* lacks the bulbous clypeus found in the Tabanomorpha, which includes both the Rhagionidae and Pelecorhynchidae. I prefer to place *Heterostomus* in the Xylophagomorpha as *incertae sedis*, admittedly based only on its similarity in appearance to some members of the Xylophagidae. Nagatomi (1977) considered the genus to belong possibly to the Coenomyiidae, and he later (1981, 1982*c*) erected a separate family for it. Positive placement of the genus will probably be difficult until the larvae are discovered.

Genus Exeretoneura. This small genus contains four species from eastern Australia and Tasmania. Macquart (1846) originally placed the genus in the Leptidae (= Rhagionidae), and most authors until recently have considered it to belong to the Nemestrinidae because of its peculiar wing venation (see Bernardi 1972 for review). Paramonov (1953), in the last taxonomic treatment of the genus, included it in the Nemestrinidae, but he noted the doubtfulness of this placement and mentioned similarities between *Exeretoneura* and *Coenomyia* and related genera. After studying the world genera of Nemestrinidae, Bernardi (1972) excluded *Exeretoneura* from that family, presenting evidence that similarities in wing venation were probably convergent rather than synapomorphic. Nagatomi (1977) again capsulized the similarities between *Exeretoneura* and his concept of Coenomyiidae and proposed the family Exeretoneuridae for the genus, but he stated "it may possibly be better, however, to relegate *Exeretoneura* into the Coenomyiidae." As in *Heterostomus*, the larvae are not known, and their discovery will be instrumental in placing the genus for certain. The unique features of the wing venation, and the peculiar overall appearance of the genus, may indicate an isolated position for *Exeretoneura* within the more primitive Brachycera; perhaps it may be reasonable to maintain *Exeretoneura* in its own family. At present I prefer to treat it as *incertae sedis* in the Xylophagomorpha, as I feel that proposal of new families for small segregate genera obscures their systematic position and is warranted only when good cladistic information supports such action.

PHYLOGENY OF THE INFRAORDER STRATIOMYOMORPHA

Hennig (1973), Steyskal (1974), and Nagatomi (1977) have all placed the Stratiomyomorpha nearest the Xylophagomorpha in putative phylogenies of Brachycera. However, none of these authors presented character evidence that supports the sister-group relationship of the two groups in a cladistic fashion. The infraorder as recognized here contains two families, the Xylomyidae and Stratiomyidae. In another work (Woodley 1983) I have presented much of the information included here, as well as some comments on the relationships of subordinate taxa within the Stratiomyidae.

Monophyly of the Stratiomyomorpha. Characters 9–12 support the monophyly of the Stratiomyomorpha.

9. *Pupation within the last larval integument*

Pupation within a puparium formed from the last larval integument is unique to the Stratiomyomorpha within the more primitive Brachycera and is clearly apomorphic. Such a puparium is absent in all other Diptera except the Muscoidea (= Cyclorrhapha, Muscomorpha), in which it has evolved independently. Austen (1899) presented the best early review of this character, and since that time the Xylomyidae have been considered close relatives of the Stratiomyidae, although this similarity had been noted at least as early as Westwood's (1840) general work. In my opinion, this character is the most obvious and conclusive synapomorphy linking two families of brachycerous flies.

10. *Larval cuticle encrusted with "warts" of calcium carbonate*

Again, this character state is unique to the Stratiomyomorpha. It is not found in any other Diptera and is, therefore, very strong evidence for the monophyly of the infraorder. McFadden (1967) briefly reviewed earlier papers discussing the morphology of these cuticular deposits.

As the larvae of the Stratiomyomorpha are so unusual and unique within the Diptera, a detailed study would almost certainly reveal many additional autapomorphies for the infraorder. Excellent summaries of larval morphology, which include many illustrations, have been presented by Rozkošný (1973, 1982, 1983).

11. *Loss of tibial spurs on front legs*

No Stratiomyomorpha are known to have tibial spurs on the front legs, and I consider this autapomorphic. The xylomyid genus *Coenomyiodes* Brunetti was originally described (Brunetti 1920) as having one spur on the front tibia, but Nagatomi (1982*c*) reported that there was none. My subsequent examination of the holotype has confirmed that no spur is present. Within the primitive Brachycera, only the Pantophthalmidae have lost front tibial spurs, apparently independently, although it cannot be completely precluded that this feature is synapomorphic between Stratiomyomorpha and this family.

12. *Abbreviation of costal vein*

Hennig (1967, 1973) has considered the termination of the costal vein at M_2 or before as autapomorphic for the Stratiomyomorpha. In most Xylophagomorpha, the costa is clearly circumambient, continuing around the wing. In *Xylophagus*, it continues to vein CuA_2 and is faint beyond that point, but it is still visible in larger specimens. In *Rachicerus*, Nagatomi (1977) stated that the costa ends "at or beyond wing tip." In specimens of some larger species that I have examined in the genus (e.g. *R. obscuripennis* Loew), the costal vein gradually becomes weak beyond R_5 but is visible at least to $CuA_1 + M_3$. In other, small species (e.g. *R. fulvicollis* Walker), the costa becomes more quickly reduced beyond R_5 and is apparently absent beyond M_1 in some species. In Xylomyidae, the costa ends abruptly at M_2 in *Coenomyiodes*, *Xylomya* Rondani (Fig. 35.3) and in some species of *Solva* Walker, and at M_1 in some *Solva* (Fig. 35.2) and in *Arthropeina* Lindner. In nearly all stratiomyids, the costa ends between R_5 and M_1, but at least in a few Chiromyzinae that I have seen it may continue weakly beyond that point. If the end of the costa diminishes gradually, it is sometimes difficult to ascertain where it actually stops, especially as the bases of the fine hairs that are found along the posterior margin of the wing may cause the edge to appear thickened.

Because of the ambiguity in interpreting where the costa actually ends in some stratiomyids and in *Rachicerus*, at present it is impossible to determine whether reduction of the costa is synapomorphic for the Stratiomyomorpha + *Rachicerus* or whether it represents homoplasy. Thus, it is rather weak evidence for the monophyly of the Stratiomyomorpha, as proposed by Hennig.

Hennig (1967) considered the sclerotized bridge between the prosternum and proepisternum found in many stratiomyids and xylomyids to be synapomorphic. However, the lack of such fusion of these sclerites in the primitive stratiomyids *Allognosta* Osten Sacken (Nagatomi 1977) and *Parhadrestia* James (Woodley 1986) casts doubt on this character.

Monophyly of the Xylomyidae. The Xylomyidae is a small family of flies (107 described species) present in all zoogeographic regions. I recognize four genera in the family: *Arthropeina* Lindner (1 species; Neotropical), *Coenomyiodes* Brunetti (1 species; Oriental), *Solva* Walker (73 species), and *Xylomya* Rondani (32 species). Characters 13–16 support the monophyly of the Xylomyidae.

13. *Presence of only two spermathecae in female*

Females of all xylomyids I have examined have only two spermathecae, whereas all Xylophagomorpha and nearly all Stratiomyidae have three, the number believed to be plesiomorphic for Diptera (Downes 1968, Hennig 1973). Webb (1984) reported only two spermathecae for *Rachicerus*, but all specimens that I have examined have three. Loss of one spermatheca in xylomyids is, therefore, strong evidence for monophyly of the family. I have observed a few stratiomyids in which only two spermathecae are present (e.g. members of the genus *Euparyphus* Gerstäcker), but in these cases loss of one spermatheca can certainly be attributed to homoplasy, as closely related genera in all known cases have three.

14. *Gonocoxal apodemes of male terminalia very reduced or absent*

All xylomyids that I have seen have their gonocoxal apodemes greatly reduced or absent. The "dorsal lobe of basistyle" (Nagatomi and Tanaka 1971) may be homologous with gonocoxal apodemes, as implied by Rozkošný (1973), who called them "dorsal processes of the synsternite." Because these lobes are quite variable in size and shape (see figures in Nagatomi and Tanaka 1971), it is difficult to determine whether they are indeed homologous to the gonocoxal apodemes. Regardless of whether the gonocoxal apodemes are absent or merely very reduced, this region of the male terminalia is highly modified in a manner unique to xylomyids and can be considered autapomorphic for the family.

15. *Spermathecal ducts very long*

All xylomyids that I have examined have very long spermathecal ducts, longer to much longer than the abdomen. In *Xylomya*, the ducts are about twice as long as the abdomen and are folded once near the middle; hence, the spherical spermathecae themselves are near the posterior end of the abdomen. In *Solva*, the ducts are even longer and are coiled several times within the abdomen. In most Xylophagomorpha and in Stratiomyidae, the ducts are shorter than the abdomen and are more or less straight. In *Xylophagus*, the ducts are extremely long and are coiled within the abdomen, but unlike xylomyids, they end in generally membranous spermathecae. Webb's (1979) statement that this genus lacks terminal spermathecae is incorrect. In *Rachicerus*, the ducts are not as long as the abdomen, are not coiled, and end in well-sclerotized spermathecae. At present I hypothesize that very long spermathecal ducts have evolved independently in xylomyids and *Xylophagus*, and therefore the character state is autapomorphic for Xylomyidae. However, very long spermathecal ducts could also be considered synapomorphic for *Xylophagus* and xylomyids, in which case they would have had to become shorter again in stratiomyids. Careful study of the spermathecal apparatus in more primitive Brachycera could well provide a suite of characters useful in cladistic analysis of these flies.

16. *Wing cell m_3 closed before margin (Figs. 35.2–3)*

This character state has been used by most authors to differentiate between xylomyids and stratiomyids. It is considered a derived state, in comparison with a widely open cell m_3, as found in *Xylophagus* (Fig. 34.4). Although a closed cell m_3 is possibly synapomorphic between xylomyids and some subgroup of Xylophagomorpha, its distribution within the Xylophagomorpha is not uniform. Without corroborating information, the usefulness of this feature remains ambiguous. If xylomyids and the sister group of the Stratiomyomorpha share a closed cell m_3, then stratiomyids likely exhibit wing venation that is even more derived, rather than being plesiomorphic as thought by Hennig (1973). Therefore, the open cell m_3 found in stratiomyids may be secondarily evolved. This theory may not be as unlikely as it seems at first. The ends of veins M_1, M_2, M_3, and CuA_1 are often weak in stratiomyids, especially M_3. In fact, in the majority of stratiomyids, M_3 is abbreviated. In the most primitive forms (*Parhadrestia*, Chiromyzinae, and Beridinae) it is abbreviated in length in all genera, except in the beridine genus *Exaeretina* Enderlein in which the end of the vein is weak. An open cell m_3 could have been formed simply by the loss of the distal portion of the cell combined with the migration of the radial and medial veins toward the costal margin of the wing.

Monophyly of the Stratiomyidae. Hennig (1973) has remarked that autapomorphic character states for the stratiomyids have not been clearly worked out. Because these flies are so distinctive, however, there has been relatively little controversy over the included subgroups. The family is a large one, with about 400 currently recognized genera and well over 2000 described species. Characters 17–20 are considered to be autapomorphic for the family.

17. *Loss of spurs on hind tibiae*

No known stratiomyids have spurs on the hind tibiae. This loss is considered autapomorphic, as nearly all xylomyids and Xylophagomorpha have two apical spurs on the hind tibiae, with the following exceptions. The Pantophthalmidae all lack hind tibial spurs, which probably represents an independent loss. In the Xylomyidae, all species of *Arthropeina* and *Xylomya* known to me have two hind tibial spurs. In most species of *Solva*, the hind tibiae have two spurs, but some possess either one or none (Daniels 1976, Nagatomi 1981). I have examined some species that have two hind tibial spurs, both of which are greatly reduced in size. The loss of one or both hind tibial spurs in *Solva* has clearly occurred independently of that in stratiomyids.

18. *Female cerci separated by abdominal tergite 10*

In all known stratiomyids, the female cerci are widely separated at their bases by abdominal tergite 10. This feature has been illustrated for a number of taxa by Nagatomi and Iwata (1978), who used the character state as a diagnostic feature for the family in their key. In xylomyids and all Xylophagomorpha except Pantophthalmidae, the cerci are virtually contiguous at their bases. In the pantophthalmids, the female cerci are somewhat separated by tergite 10, but they are much closer together than are those of stratiomyids and tergite 10 is notably smaller. This similarity thus seems superficial and likely is the result of convergence.

19. *Radial veins crowded toward costal margin of wing (Figs. 36.27–37)*

This character state has been used frequently to characterize the stratiomyids in keys, and I believe it is autapomorphic for the family. The best indicator of this feature is that vein R_5 ends before the wing apex. In the more primitive members of the family (*Parhadrestia*, Chiromyzinae, and Beridinae), the radial veins are more widely spaced than in more derived taxa, but R_5 still ends distinctly before the wing apex except in a few Beridinae. In these unusual taxa, an example of which is *Macromeracis philippii* (Rondani), R_5 ends only slightly before the wing apex. In these cases, the anteroapical region of the wing has probably become secondarily enlarged, as cells r_{2+3} and r_4 are abnormally large, even though vein R_{2+3} is crowded toward the costa. In xylomyids and nearly all Xylophagomorpha, vein R_5 ends at the wing tip or well beyond it (Figs. 34.2–3; 35.2–3). In a few *Xylophagus* that I have seen, R_5 ends slightly before the wing tip. Otherwise, only in *Exeretoneura* does R_5 end before the wing apex, and in this unusual genus the venation is quite peculiar and the vein homologies are not clearly worked out (Bernardi 1972). The character state found in this genus is most probably not synapomorphic with stratiomyids, especially because the apex of R_5 curves distinctly toward the anterior margin of the wing. Otherwise, Nagatomi (1981) has hypothesized that both the states of R_5 ending well before and well beyond the wing tip are derived, and that the plesiomorphic state is one in which R_5 ends near the wing apex. This putatively plesiomorphic state is found in xylomyids, *Xylophagus*, *Rachicerus*, and *Heterostomus*.

20. *Discal cell reduced in size*

I consider the small size and more-or-less rounded shape of the discal cell to be autapomorphic for stratiomyids (Figs. 36.27–37). The exact nature of this character state is difficult to describe, as it is somewhat variable within the family. Nevertheless, this feature of the wing venation is distinctive and easily recognizable by those familiar with soldier flies. The discal cell of stratiomyids is smaller in relation to the total area of the wing than it is in most Xylophagomorpha. In xylomyids, *Xylophagus*, and *Rachicerus*, the discal cell is commonly a little smaller than it is in other Xylophagomorpha, but its shape is quite different than that found in stratiomyids. It is elongate along the longitudinal axis of the wing, more than two times longer than high, with the anterior and posterior edges nearly parallel sided. It is located near the middle of the wing, in the anterior–posterior axis. In the Stratiomyidae in which the discal cell is largest (e.g. in some Chiromyzinae and Antissinae), the cell is shorter and broader, the anterior and posterior edges are not nearly parallel, and the length is at most about twice the height (but usually the length and height are nearly equal). The discal cell is also closer to the anterior edge of the wing. I regard this unique form of the discal cell to be autapomorphic, as none of the potential outgroups of the Stratiomyomorpha, nor the Xylomyidae, exhibit such conformation.

PHYLOGENY OF THE INFRAORDER TABANOMORPHA

Four families can definitely be assigned to this infraorder, and the placement of these taxa together has been accepted by most recent authors (Hennig 1973, Stuckenberg 1973, Nagatomi 1977). The families Tabanidae and Athericidae (since the latter was recognized by Stuckenberg 1973) have not been controversial, whereas the limits of the Rhagionidae and Pelecorhynchidae have changed repeatedly (see Chs. 30, 33; Nagatomi 1977, 1982*a*; Krivosheina 1971). I prefer to place the Vermileonidae in the Tabanomorpha, its traditional placement, in contrast to its assignment in the Asiloidea (Ch. 39) for reasons discussed below. However, it is placed here as *incertae sedis*.

Monophyly of the Tabanomorpha. Characters 21–23 apparently support the monophyly of the Tabanomorpha as recognized here.

21. *Larval mandible with mandibular brush near its base*

Hennig (1973) mentioned this feature as the sole character he was aware of that supported the recognition of the Tabanoidea (= Tabanomorpha as used here) as a monophyletic group. The character is unique to the group (see Ch. 3, Figs. 3.6–7). Hennig (1967, 1973) pointed out that the larvae of many of the more primitive genera of Rhagionidae are unknown, so there is some uncertainty that the character is found throughout the Tabanomorpha.

22. *Larval head retractile*

In all known larvae of Tabanomorpha the head is retractile into the thorax. The head is permanently exerted in Xylophagomorpha and Stratiomyomorpha, as well as in some Asiloidea (e.g. Scenopinidae and Therevidae). Therefore a retractile head is apparently autapomorphic for the Tabanomorpha.

23. *Adult with convex, bulbous clypeus*

Nagatomi (1981) thought that a swollen, bulbous face was derived from a flat, plate-like facial area, as found in Xylophagomorpha and Stratiomyomorpha. Such a facial morphology is found throughout the Tabanomorpha (Figs. 30.4, 31.2–4, 32.2–3, 33.3) and seems to be autapomorphic for the infraorder. The only apparent exception is the genus *Austroleptis* Hardy (found in Australia, Tasmania, Chile, and Argentina), which has a flat clypeus. Although the genus is traditionally placed in the Rhagionidae (Hennig 1972), its assignment to the family is questionable (Nagatomi 1982*a*) and will probably remain in doubt until its larvae are known.

Relationships between families of Tabanomorpha. The four families definitely ascribed to the Tabanomorpha have been studied by various authors, and characters that are of value in defining their relationships are available. Fig. 115.2 presents a cladogram with character-state distributions that I believe show the most support for these relationships.

As pointed out several times by Hennig (1967, 1972, 1973), good, derived features that may be used to define the Rhagionidae unambiguously are not known. Hennig (1972) suggested one, the supposed sharp separation of the first antennal flagellomere from those distal to it. The value of this character, however, seems dubious, as certain genera assigned to the rhagionids, such as *Arthroceras* Williston and *Atherimorpha* White, have the basal flagellomere little differentiated from the others (see figures in Nagatomi 1982*a*). Those features used to characterize the Rhagionidae by Nagatomi (1977) seem to me to be plesiomorphic or ambiguous at present. The family is of moderate size, with about 520 species placed in the following genera: *Austroleptis* Hardy, *Alloleptis* Nagatomi & Saigusa, *Archicera* Szilády, *Arthroceras* Williston, *Arthroteles* Bezzi, *Atherimorpha* White, *Bolbomyia* Loew, *Chrysopilus* Macquart, *Desmomyia* Brunetti, *Litoleptis* Chilcott, *Neorhagio* Lindner, *Ptiolina* Zetterstedt, *Rhagina* Malloch, *Rhagio* Fabricius, *Schizella* Bezzi, *Solomomyia* Nagatomi, *Spania* Meigen, *Spaniopsis* White, *Spatulina* Szilády, *Stylospania* Frey, and *Symphoromyia* Frauenfeld. Nagatomi (1982*a*) presented the most recent review of the world genera, but he includes *Glutops* Burgess and *Pseudoerinna*, which are treated here as Pelecorhynchidae. As mentioned previously, Kovalev (1981, 1982) treated some fossil species

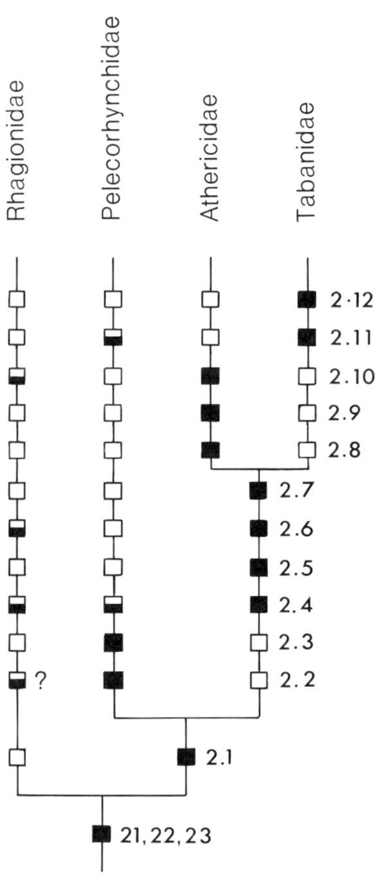

Fig. 115.2. Cladogram showing distributions of character states, showing relationships of families in the infraorder Tabanomorpha. Open squares indicate plesiomorphic character states; solid squares indicate apomorphic states; and half-filled squares indicate characters in which both states are found in a single lineage. Numbers correspond to discussions in the text.

of rhagionids from the Jurassic, which have wing structure little different from present-day members of the family. They represent the oldest fossil Brachycera that are definitely assignable to an extant family.

Character 2.1 supports the monophyly of the group Pelecorhynchidae + Tabanidae + Athericidae.

2.1. *Larval mandible with poison canal*

Tabanidae (Ch. 31), Athericidae (Webb 1977), and Pelecorhynchidae (Teskey 1970, for *Glutops*), based on what we know of their larvae, all possess a canal in the center of each mandible, used for injecting poison into their prey. Known larvae of Rhagionidae (Ch. 33) have no such poison canal. This apomorphic feature indicates a monophyletic origin of the three families that possess the poison canal. Stuckenberg (1973: 669) gives a brief summary of this character.

The Pelecorhynchidae is a small family, which is composed of three genera: *Glutops* Burgess (11 species), *Pelecorhynchus* Macquart (36 species, from Australia, Tasmania, and Chile), and *Pseudoerinna* Shiraki (2 species; *Bequaertomyia* Brennan, treated in Chapter 30, is a synonym according to Nagatomi 1982*a*,*b*). Nagatomi (1982*a*) treated *Glutops* and *Pseudoerinna* as rhagionids, and Krivosheina (1971) erected the Glutopidae for *Glutops*, a move rejected by Kovalev (1981). *Glutops* and *Pseudoerinna* are more similar in general appearance to Rhagionidae than is *Pelecorhynchus*, but, based on the extreme similarity of the general body form of larvae of *Pelecorhynchus* and *Glutops* (Teskey 1970) and on the difference between these and known rhagionid larvae, the two are likely closely related. The larvae are not known for *Pseudoerinna*.

Characters 2.2 and 2.3 support the monophyly of the Pelecorhynchidae.

2.2. *First segment of female cercus with strong lateral process*

All three genera included in the Pelecorhynchidae here have a strong lateral process on the first segment of the female cercus (illustrated by Nagatomi and Iwata 1976: Figs. 11, 17, 18). A similar morphology is found only in a few rhagionids (Nagatomi and Iwata 1976: Figs. 14, 20,

21), but the lateral lobe is not as strongly attenuate apically in these taxa. This character state is therefore apparently autapomorphic for the Pelecorhynchidae. If the similar character state in the rhagionids was found to be synapomorphic with that of pelecorhynchids, the latter could be treated as rhagionids. But, because of strong evidence (character 2.1) supporting the monophyly of the Pelecorhynchidae + Tabanidae + Athericidae, the similarity of the female cerci in pelecorhynchids and rhagionids seems more likely the result of convergence.

2.3. *Larva with stout spines laterally on labrum and apically on maxilla*

This character state is apparently unique in the Pelecorhynchidae (Mackerras and Fuller 1942, Teskey 1970) and is therefore autapomorphic for the family.

Characters 2.4–2.7 provide evidence for the sister-group relationship between Pelecorhynchidae and Athericidae + Tabanidae by supporting the monophyly of the latter group.

2.4. *Male terminalia with hypandrium fused to gonocoxites*

Stuckenberg (1973) noted that Athericidae have the hypandrium fused to the gonocoxites, which is also the case in Tabanidae (Ch. 31). Many rhagionids, in contrast, have the hypandrium free of the gonocoxites (as in Fig. 33.14), as do *Glutops* and *Pseudoerinna* in the Pelecorhynchidae. Some rhagionids, such as some *Chrysopilus* and *Spaniopsis* (Stuckenberg 1973) and *Pelecorhynchus* (Mackerras and Fuller 1942), have the hypandrium partly to completely fused to the gonocoxites. Thus it appears that, although this character state is apomorphic for Athericidae + Tabanidae, it has evolved independently in other lineages.

2.5. *Male terminalia with long aedeagal tines present*

Stuckenberg (1973) first mentioned this character in terms of the relationship of the Athericidae and Tabanidae. The presence of large aedeagal tines (sometimes called endophallic tines) is limited to these two families (see Fig. 32.6; Stuckenberg 1973: Figs. 5, 11–13), although Stuckenberg mentioned possible "precursor" structures in some *Rhagio* and *Atherimorpha* in the Rhagionidae (see his Fig. 8). These tines are not nearly as strongly developed as are those in the Athericidae and Tabanidae.

2.6. *Female cercus one-segmented*

In all known Athericidae and Tabanidae the cerci of adult females are one-segmented. In some more primitive tabanids, there is evidence of the two original segments (as in Fig. 31.42), but fusion is present. In rhagionids and pelecorhynchids, the cerci are two-segmented without evidence of fusion (see Nagatomi and Iwata for various figures). The only exception is the enigmatic *Austroleptis*, which has one-segmented cerci (Nagatomi and Iwata 1976: Fig. 34). As already mentioned, its placement in the Rhagionidae is uncertain, but it certainly cannot be placed in either the Athericidae or Tabanidae.

2.7. *Metathorax with postspiracular scale*

Stuckenberg (1973) discussed this character and its significance and provided figures of the scale in both the Athericidae and Tabanidae (see his Figs. 36–39). Presence of the scale is unique to the two families.

The Athericidae is a small family of about 90 species placed in seven genera: *Atherix* Meigen, *Atrichops* Verrall (Palaearctic), *Dasyomma* Macquart (Neotropical, Australian), *Pachybates* Bezzi (Afrotropical), *Suragina* Walker, *Trichacantha* Stuckenberg (Afrotropical), and *Xeritha* Stuckenberg (Neotropical). Previously considered rhagionids, they were given family status by Stuckenberg (1973) when their closer relationship to Tabanidae rather than Rhagionidae was demonstrated. The characterization of the family was discussed in some detail in Stuckenberg's paper, based on adult morphology.

Characters 2.8–2.10 support the monophyly of the family.

2.8. *Larva with abdominal segments one to seven with paired ventral prolegs armed with apical crochets*

The known larvae of athericids are very distinctive in general habitus. One of the most striking features is the series of abdominal prolegs along the venter, each of which has a circle of apical claw-like crochets (Fig. 32.7). Larvae are known for *Atherix*, *Atrichops*, and *Suragina*, and figures are available for Japanese members of these genera (Nagatomi 1961). These structures are unlike any others in the Diptera and are certainly autapomorphic for the Athericidae.

2.9. *Wing with vein R_{2+3} shortened, ending near R_1 (Fig. 32.4)*

Stuckenberg (1973) noted the significance of this character, which is found throughout athericids but not in other Tabanomorpha, with the exception of a few species of the rhagionid genus *Chrysopilus* (see Stuckenberg 1973: Fig. 47). In *Chrysopilus* the condition has probably arisen by convergence, as most of the species of the genus have typical rhagionid venation in which R_{2+3} is not shortened.

2.10. *Antennal flagellum stylate (Figs. 32.2–3)*

As the previous characters seem to provide good evidence that the phylogeny of the Tabanomorpha is as presented in Fig. 115.2, the stylate antennal flagellum is probably autapomorphic for Athericidae. The Pelecorhynchidae and Tabanidae have a more plesiomorphic, less modified flagellum than the Athericidae. One must then assume that the stylate condition as found in some rhagionids (Figs. 33.5–7) has resulted from convergence.

There is a tendency toward the evolution of a thick, compact base to the antennal flagellum with an apical style or arista. This trend occurs in numerous, independent lineages of more primitive Brachycera, such as *Dialysis* (Xylophagidae), many Stratiomyidae, Rhagionidae, and Athericidae.

The Tabanidae is one of the larger families of Diptera, worldwide in distribution, with 126 recognized genera containing 2965 species (Moucha 1976). The monophyly of the family has not been doubted, although occasionally genera of Pelecorhynchidae and *Heterostomus* have been included as aberrant genera in the family, a conclusion that has been discounted by recent authors. A thorough search should reveal many autapomorphic features; only two of the most conspicuous are treated here.

2.11. *Fork of vein R_{4+5} encompassing wing tip*

This feature has long been used to characterize the Tabanidae in identification keys. The only taxon within the Tabanomorpha but outside the Tabanidae with the character state is the genus *Pelecorhynchus*, which has led some earlier authors to consider that genus to be a member of the Tabanidae. A vaguer similarity exists between *Pseudoerinna* and tabanids, but in this genus the fork is not nearly as divergent as in tabanids (compare Fig. 30.2 with Figs. 31.28–38). Nagatomi (1981) gives a general survey of the character states of vein R_{4+5} and their polarity and distribution within the primitive Brachycera.

2.12. *Lower calypter much enlarged*

Another conspicuous feature of the Tabanidae, unique to the family within the Tabanomorpha, is the enlarged lower calypter. It is without doubt autapomorphic for the family.

Family Vermileonidae, *incertae sedis*. The Vermileonidae is a small family composed of four genera: *Lampromyia* Macquart (19 species; Palaearctic, Afrotropical), *Vermileo* Macquart (8 species; widespread), *Vermiophis* Yang (1 species; China), and *Vermitigris* Wheeler (3 species; Oriental). No one has questioned the monophyly of this group of genera. Adults have autapomorphic features, which include the reduction of the wing base and alula and the slender, clavate abdomen (Hennig 1967). Larvae exhibit many derived features, such as the reduced mouthparts, the ventral proleg on the first abdominal segment, and behavior that is amazingly convergent to that of antlions (Neuroptera: Myrmeleontidae). Larvae of vermileonids and antlions both construct sand pits as pitfall traps for their ant prey.

The placement of the family within the Brachycera has been unsettled recently. Hennig (1973) recognized the vermileonids as a subfamily of Rhagionidae, their traditional placement. Nagatomi (1977) first gave the group family rank, but noted "the true position of this family is still obscure." Teskey (Ch. 39) regarded the group as being related to the Asiliformia (*sensu* Vol. 1), based on the resemblance of their male terminalia to those of Therevidae and on the similarity of the mouthparts of the larvae to those of Empidoidea.

I believe that the family cannot be related to Asiloidea, because at least some vermileonids have eight antennal flagellomeres and all possess true tibial spurs. Also, the females notably do not possess acanthophorites, despite the fact that they oviposit in sand. Finally, Vermileonidae have a pulvilliform empodium, which is plesiomorphic relative to Asiloidea.

The family is placed in the Tabanomorpha here, but as *incertae sedis*; although the larvae have a retractile head, they lack the mandibular brush (which could have been lost, however, with reduction of the mouthparts). The adults generally do not have a bulbously developed clypeus, although in *Lampromyia* the face is protuberant (Stuckenberg 1960: Figs. 1, 25, 46) but in a manner different from typical Tabanomorpha. This development is probably related to the musculature used to move the long proboscis found in the genus. Therefore the Vermileonidae can be placed in the Tabanomorpha only with uncertainty until more evidence is available to elucidate their phylogenetic position in the Brachycera. The family is apparently an old one. The Upper Jurassic fossil genus *Archirhagio* Rohdendorf is quite possibly a vermileonid (Hennig 1967, 1973; Rohdendorf 1974).

PHYLOGENY OF THE INFRAORDER MUSCOMORPHA

The limits of the infraorder Muscomorpha as I recognize it here are much wider than those used in Volume 1 of this Manual, as I have included nearly all the families of the Asilomorpha, excluding only the Vermileonidae. I prefer to recognize the infraorder broadly because as such it is equivalent to the other recognized infraorders of Brachycera, based on our knowledge of the phylogeny of the suborders at present.

Monophyly of the Muscomorpha. Characters 24–26 support the monophyly of the Muscomorpha.

24. *Antennal flagellum with only four (or fewer) flagellomeres*

All the families in the Muscomorpha have only four antennal flagellomeres, or fewer if further reduction has taken place. This number is contrasted to the plesiomorphic condition of eight flagellomeres for the more primitive Brachycera. All the families of the Xylophagomorpha, Stratiomyomorpha, and Tabanomorpha, with the exception of the Athericidae, have at least some members in which the flagellum has eight flagellomeres, although reduction has occurred independently in several families.

25. *Tibial spurs lost*

No member of the Muscomorpha possesses true tibial spurs. Hennig (1976) stated that they were present in the Phoridae but not in any other family included here in the Muscomorpha. According to Hennig (after Edwards 1938: 6), true tibial spurs could be identified by their pilose vestiture. However, I have examined some phorids in the course of this study and found that although the bristles surrounding the end of the tibia bear pilosity, they are inserted into the tibia itself rather than into the membrane between the end of the tibia and the first tarsomere. Furthermore, although the tibial bristles in phorids are pilose, the larger bristles on the palpi and frons also bear pilosity in a manner identical to that found on the tibial bristles. I have concluded that even though true tibial spurs are pilose, the presence of pilosity on setae can also occur. Therefore, this feature is inconclusive evidence for the homology of these structures. Apparently only the point of insertion may be used to identify true tibial spurs with certainty.

All families in the three other infraorders of Brachycera usually have a majority of members with tibial spurs. Tibial spurs are absent only in most Stratiomyidae, *Litoleptis* (Rhagionidae), and *Trichacantha* (Athericidae) (Nagatomi 1981). In the stratiomyids, tibial spurs are found in *Parhadrestia* (the most primitive extant member of the family) (Woodley 1986) and sporadically in the Chiromyzinae and Beridinae, both primitive subfamilies. They are also found in the Antissinae, a south temperate subfamily that is relatively primitive but more derived than are the previously mentioned subfamilies.

26. *Female cerci one-segmented*

All known Muscomorpha have the cerci composed of one segment in females. Members of the other three infraorders have females with two-segmented cerci. The only exceptions are found in isolated genera of stratiomyids and rhagionids and in all the Athericidae and Tabanidae, where it is assumed to be homoplastic. A one-segmented female cercus is therefore construed as being autapomorphic for the Muscomorpha.

The cladistic relationships of the major subordinate groups within the Muscomorpha are not well resolved. Surveys of characters throughout the infraorder will be necessary before much progress can be made. A few detailed studies within single families are available in which workers have surveyed one character system. Theodor's works on the terminalia of some Asilidae (1976) and Bombyliidae (1983) stand out in this regard. Much more such work is necessary, especially in little-known families such as the Scenopinidae and Acroceridae.

The infraorder Muscomorpha can be separated into four superfamilies, which are of approximately equivalent status cladistically. The Nemestrinoidea, Asiloidea, and Empidoidea are treated below, and their relationship to the Muscoidea is discussed. Details of the relationships within the Muscoidea are treated in Chapter 116.

Phylogeny of the superfamily Nemestrinoidea. Only a single character is known that unites the Nemestrinidae and Acroceridae into a group, and its cladistic value is dubious because of its generality.

27. *Larval stages parasitic, with accompanying hypermetamorphosis*

Hennig (1973) utilized this general character to unite the Nemestrinidae, Acroceridae, and Bombyliidae into his Nemestrinoidea. He noted, however, that the superfamily was one of the most problematic groups of Diptera in terms of its cladistic definition. Based on what I believe are synapomorphic features between bombyliids and families in the Asiloidea and Empidoidea, I cannot accept placing the Bombyliidae in the Nemestrinoidea. I prefer to ascribe their parasitic life history to convergence with the Nemestrinoidea.

Although there is serious question as to the sister-group relationship between the Acroceridae and Nemestrinidae, the two families themselves are distinct and no worker has doubted their individual monophyly.

The Nemestrinidae is a fairly small family composed of 15 genera and about 250 species (Bernardi 1973), which are widespread in distribution and are particularly prevalent in arid areas. Autapomorphic features that characterize the family are not well worked out. The most conspicuous features that have been used to characterize the family are found in the wing venation. The anterior curvature of the radial and anterior medial veins are very characteristic.

One venational feature in the family is unique.

28. *Presence of so-called diagonal vein*

All nemestrinids have a diagonal vein that extends from the radius to the posterodistal margin of the wing. It is apparently composed of elements of Rs, R_{4+5}, r-m, M_1, M_2, M_3, and CuA_1 (see Fig. 44.3; Bernardi 1973: Figs. 18–38). A similar vein is known only in *Exeretoneura* (treated above in Xylophagomorpha), but it is only superficially similar and is not thought to be homologous (Bernardi 1972).

The Acroceridae is another moderately small family, with about 50 genera and 500 species, cosmopolitan in distribution. Characters 29–31 are apparently autapomorphic.

29. *Adult antenna with a single flagellomere (Figs. 43.3–11)*

The antennal flagellomeres of all acrocerids are entirely fused, resulting in an undivided flagellum. Most nemestrinids have four flagellomeres, although a few genera have two flagellomeres or an undivided flagellum, a feature that has apparently evolved convergently. As previously discussed under character 24, most potential outgroups have more than four flagellomeres, although isolated reductions have taken place in some genera. Within the Asiloidea, only the Scenopinidae apparently have the flagellum composed of a single article.

30. *Lower calypter much enlarged*

Acrocerids have a large lower calypter (Figs. 43.1–2). Such a condition has evolved convergently only in the Tabanidae, in calyptrate Muscoidea, and in a very few acalyptrate Muscoidea. Therefore, it is considered autapomorphic for the acrocerids.

31. *Larval stages parasitic on spiders*

The larvae of Acroceridae are true internal parasites of spiders, a mode of life unique in the Diptera. Nemestrinids are parasitic on grasshoppers and beetles, and Bombyliidae are parasites of Hymenoptera, Lepidoptera, Orthoptera, and Diptera. *Petrorossia feti* Zaitsev & Tsharykuliev is the only known bombyliid associated with spiders, but was reared from an egg cocoon rather than from an adult (Zaitsev and Tsharykuliev 1981).

The sister-group relationship between the Nemestrinoidea and remaining Muscomorpha is supported by character 32.

32. *Empodium bristleform*

This character state has been known and utilized in the taxonomy of Diptera for many years. The plesiomorphic character state is found in more primitive Brachycera in which the empodium is pad-like (pulvilliform), appearing much as the pulvilli that are lateral to it. In some classifications the feature has been considered homoplastic, for instance when the bombyliids are included with the Nemestrinoidea (Hennig 1973) or when vermileonids and the Nemestrinoidea, as recognized here, are all included in the Asilomorpha (as in Vol. 1 of this Manual). As I am aware of no evidence that conclusively indicates that Brachycera with a pulvilliform empodium should be placed in a monophyletic group with those in which it is bristleform, I do not believe the character is homoplastic.

Monophyly of the superfamily Asiloidea. The superfamily Asiloidea remains poorly known in a cladistic sense, despite the fact that these flies have attracted the attention of numerous taxonomists over the years because of the general large size and attractive appearances of many of the taxa. Even the most obvious character systems have not received detailed, comparative surveys that would be useful for cladistic analysis.

One character pointed out by Hennig (1972) supports the monophyly of the Asiloidea.

33. *Larval posterior spiracle situated in apparent penultimate abdominal segment*

All larvae of Asiloidea have the posterior spiracle located in the apparent penultimate segment of the abdomen, directed more or less laterally (Figs. 37.23, 38.11, 40.28, 42.76–77). This character state is not found in the Nemestrinoidea or in the other three infraorders of Brachycera. In the Scenopinidae and Therevidae, a more derived state is found, in which the posterior spiracle is apparently in the antepenultimate abdominal segment.

Relationships between families of Asiloidea. The Fig. 115.3 presents a cladogram that shows the putative cladistic relationships of the families included in the Asiloidea. It is based on the few characters that are known well enough to be used to support hypothetical phylogenetic relationships.

The family Bombyliidae is a large one, with slightly more than 200 genera containing 4000 species, found in most areas of the world. Hull (1973) summarized the world genera, and there are several good regional faunal treatments, such as Hall (1975) for the Chilean species. Despite the fact that the taxonomy of the group is moderately well known, the phylogeny of the components of the family is virtually unknown. Hennig (1973) questions the monophyly of the major subdivisions utilized by taxonomists and gives only a provisional classification for the family. Theodor (1983) provides a brief critique of the classification of the family and reviews some important characters. The splitting of some groups now placed in the Bombyliidae and according them family rank is not supported by good evidence and consequently is not recognized here. Rohdendorf (1974) recognized the Cyrtosiidae, Usiidae, and Systropodidae as distinct families separate from the bombyliids. Without explanation, Hall and Evenhuis (1980) and Evenhuis (1982) followed Rohdendorf and recognized the Systropodidae but not his other two segregate families. Rohdendorf himself (1974: 77) stated: "The scheme of division of the bombyliids into four families, as I have proposed, is purely preliminary and must be reasoned out in detail by a special investigation." As these special investigations have not been expedited, recognizing any of these groups is clearly premature. Theodor (1983) also concluded that there was no justification for splitting the Bombyliidae into several families.

I follow Griffiths (1972, citing Bährmann 1960) and Hennig (1972) in relegating the genus *Hilarimorpha* Schiner to the Bombyliidae. The relationship of *Hilarimorpha* to the bombyliids was also inferred by Collin (1961), Saigusa (1973), and Nagatomi (1982*d*), although these authors maintained the genus as a separate family. It was also treated as a separate family by Webb (1974 and in Ch. 46 of this Manual). Webb (1974: 174) concluded that Bährmann (1960: Fig. 49) had examined the terminalia of *Condylostylus singularis* (Dolichopodidae) rather than *Hilarimorpha singularis*, but Bährmann's figure is clearly *Hilarimorpha*, and Webb's supposition is apparently based upon Bährmann's accidental use of Becker as author of the name *singularis* rather than Schiner. Also, Webb (1974: 178, 179) discussed the similarity of *Hilarimorpha* to some rhagionids (following earlier authors), but this similarity is based on plesiomorphic features. Absence of tibial spurs, empodia, and vein M_3, as well as an antennal flagellum with three flagellomeres, indicate placement in the Muscomorpha.

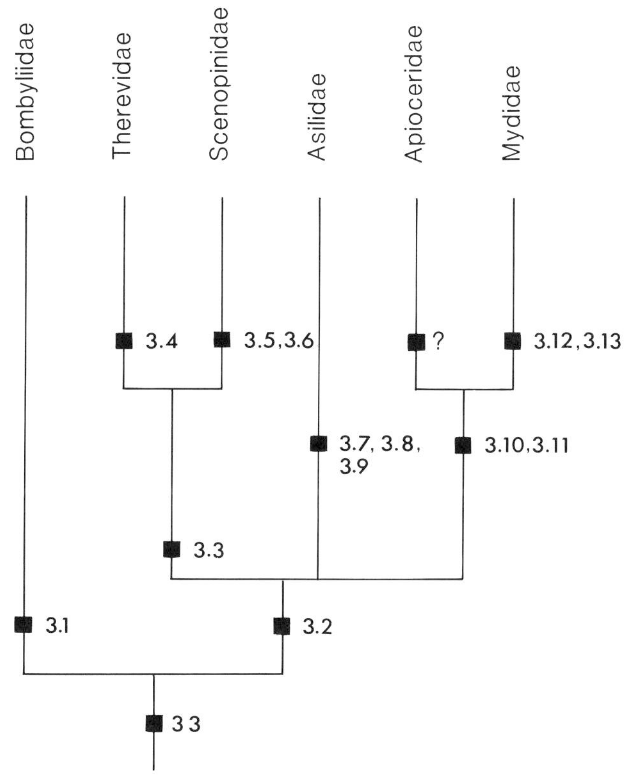

Fig. 115.3. Cladogram showing relationships of families in the superfamily Asiloidea. Numbers indicate points of origin of apomorphic character states and correspond to discussions in the text.

As Webb pointed out, the immature stages, when discovered, will be instrumental in substantiating the placement of *Hilarimorpha*.

Theodor (1983) discusses the peculiar genus *Prorates* Melander, noting the unique structure of its female spermathecae and male aedeagus. Based on the resemblance of the aedeagus to that of one species of Scenopinidae (*Belosta viticolipennis* Kelsey from Arizona) and the presence of so-called sensory areas on abdominal tergite 2 (Theodor 1983: Fig. 23*a*,*b*), Theodor speculates that *Prorates* belongs in the Scenopinidae. As the sensory areas on the abdomen in Theodor's figures of *Prorates* are quite different in appearance from those of the Scenopinidae, and the type of aedeagus in *Prorates* is not more widespread in the Scenopinidae, I feel this relationship requires further evidence before it can be accepted. Additionally, *Prorates* does not have wing venation that is similar to the Scenopinidae. As with other anomalous taxa, the placement of *Prorates* would be greatly facilitated by the discovery of its larva.

The following character supports the monophyly of the Bombyliidae.

3.1. *Larvae parasitic, with hypermetamorphosis*

First-instar bombyliid larvae are active, whereas later instars are grub-like after they have located and infested their host. The only other Brachycera outside of the Muscoidea that are parasitic are the Nemestrinoidea, treated above. The two groups have apparently evolved along these lines convergently, as other evidence suggests the placement of the Bombyliidae in the Asiloidea. The larvae have the posterior spiracles in the penultimate abdominal segment (Malloch 1917: 390), and the first instar larvae bear a resemblance to those of Therevidae and Scenopinidae, especially in having long lateral setae on the thorax and a small, exerted head capsule (Verrall 1909: Fig. 64). Presence of acanthophorites in some adult female bombyliids also indicates placement in the Asiloidea. I therefore conclude that parasitism and hypermetamorphosis are autapomorphic for the Bombyliidae.

That the Bombyliidae and the remaining Asiloidea have a sister-group relationship is indicated by the following character.

3.2. *Larval metacephalic rod hinged*

In the five families of Asiloidea united by this character the internal part of the cranium of the head capsule is separated from the anterior portion by a suture, articulating with that structure (Figs. 37.24; 42.78–79). In the few known bombyliid larvae, the cranium has the more plesiomorphic structure found in Nemestrinoidea and other infraorders of Brachycera, i.e. the cranium is formed of a single sclerite that shows no evidence of subdivision

(Figs. 45.59–60). In the Empidoidea, the metacephalic rods are also articulated but are quite different in structure, being a pair of slender rods rather than a larger median one. Because the larval spiracles are terminal in Empidoidea, I hypothesize that the articulation of the metacephalic rods has evolved independently in the Asiloidea and Empidoidea.

Although their adult stages are quite different in general appearance, the larvae of the Therevidae and Scenopinidae are extremely similar, sharing at least the following character, which indicates that they are sister families.

3.3. *Larval abdominal segments secondarily segmented, the abdomen appearing to have 17 segments (Fig. 38.11)*

Larvae of both therevids and scenopinids, as far as they are known, have this secondary segmentation of the abdomen of their larvae. One result of this segmentation is that the posterior spiracles appear to be in the antepenultimate segment rather than in the penultimate one, as in other Asiloidea. This general form of the larval abdomen is unique in the Brachycera.

An intensive search of the larvae of Asiloidea would probably reveal other characters that support the sister-group relationship of these two families. Reduction of the tentorial rods of the head capsule may be autapomorphic for the group, but *Apiocera maritima* Hardy (Apioceridae, from Australia), according to figures by English (1947: Figs. 7, 8), also has shortened tentorial rods. The small pair of caudal prolegs found in the larvae of both families may also be autapomorphic (Fig. 37.23).

The Therevidae is a moderate-sized family of flies, which is quite poorly known. However, some synthetic work has been done recently, which is beginning to provide a sound taxonomy for the family (e.g. Lyneborg 1976, Irwin and Lyneborg 1981). Some features of the adults are rather generalized and plesiomorphic. For instance, the wing venation is very similar to that found in Rhagionidae (compare Fig. 37.13 with Fig. 33.11). Hennig (1973) noted that defining autapomorphies are not clearly worked out for the Therevidae.

Only one character that I am aware of seems to be autapomorphic.

3.4. *Larval metacephalic rod posteriorly spatulate*

The posterior end of the medial metacephalic rod in the few known therevid larvae is spatulate (Fig. 37.24). Contrastingly, in the Scenopinidae this structure is slender and nearly parallel sided, and in other Asiloidea it is larger and not spatulate (as in Figs. 42.78–79). In the other infraorders the metacephalic rod is also large and not spatulate.

Certain atypical genera are apparently questionably placed in the Therevidae. Irwin and Lyneborg (Ch. 37) doubted the placement of *Apsilocephala* Kröber in the family but declined to give an alternative, although they noted its possible affinity with certain Bombyliidae. The Australian genus *Clesthentia* White, sometimes placed in the Therevidae, is also enigmatic, but it is so poorly known that its status cannot be evaluated at present.

The Scenopinidae is a small family of 18 genera with about 330 species. Because these flies are small and many are apparently quite local in distribution, there are undoubtedly many undescribed species. Their larvae, as already noted, are very similar to those of the Therevidae, and it is not entirely unlikely that they represent a highly autapomorphic group that has evolved from within the Therevidae, and as such they might render the Therevidae paraphyletic. This question can only be resolved when the phylogeny of the Therevidae, at present virtually unstudied except for a few recent papers (e.g. Irwin 1976), is further resolved and characters of importance in that family are surveyed within the Scenopinidae. As mentioned previously in the treatment of the Bombyliidae, Theodor (1983) suggests placement of the aberrant genus *Prorates* in the Scenopinidae, but this cannot be accepted without stronger evidence.

Derived features 3.5 and 3.6 are found in the Scenopinidae.

3.5. *Wing with cell m_1 strongly widened distally*

The wing venation of the Scenopinidae is characteristic, being quite diagnostic for the family. One of the most conspicuous features is the loss of veins M_2 and M_3. This loss, coupled with the apex of vein M_1 being gently curved anteriorly, results in cell m_1 widening rather abruptly from its base to the posteroapical portion of the wing. This conformation of cell m_1 is unique in the Brachycera.

3.6. *Wing with costal vein ending at M_1*

In the scenopinids, the costal vein ends rather abruptly at M_1 (Figs. 38.2–5). Hennig (1973) suggested that this characteristic is an autapomorphic feature of the family. In the Therevidae, most Bombyliidae, the Nemestrinidae, and many Acroceridae the costa is circumambient. I have seen a few bombyliids (such as *Caenotus* Melander), as well as some acrocerids (such as *Ogcodes* Latreille), in which the costa is abbreviated. In these isolated groups the abbreviation of the costal vein is apparently related to small size and has occurred independently several times. Therefore the feature is not synapomorphic.

Three other features of scenopinids may be autapomorphic, but at present their status is ambiguous. All scenopinids lack macrosetae, whereas most therevids have them well developed. Also, some bombyliids have macrosetae, but other groups in the family do not. All Nemestrinoidea lack macrosetae. Macrosetae have probably been lost in independent groups of Asiloidea, but the independent origin of macrosetae in several groups cannot yet be overruled. Another adult feature, mentioned by Hennig (1973) as derived for the family, is the compact antennal flagellum composed of a single flagellomere (Figs. 38.7–

10). Most therevids and bombyliids have three flagellomeres. Some groups of both these families, however, exhibit a reduction in flagellomere number, and in some bombyliids the flagellar structure is rather similar to that of scenopinids (compare Figs. 38.9 and 45.52). Finally, the larvae of scenopinids are said to lack tentorial rods (Hennig 1973), a feature that may be unique in the Asiloidea. Krivosheina (1980: Fig. 1) does not show them for a species of *Thereva* that she compared to scenopinids, although this may be only an omission, because other authors indicate that therevids do possess them (e.g. Fig. 37.24). As so few larvae of therevids are described, and only two or three scenopinids are known from larval stages, the significance of this character is still uncertain.

The Asilidae is a large family of flies, with more than 400 genera and about 5000 species. They are very distinctive, bristly flies and are the only Asiloidea in which the adults are predacious.

Characters 3.7 and 3.8, pertaining to their piercing proboscis, provide the strongest evidence for the monophyly of the family. A third autapomorphic character provides supporting evidence.

3.7. *Adults with labellae of labium strongly reduced, fused with prementum*

Adult asilids have a strongly sclerotized proboscis used for piercing their prey. The labellae and prementum are fused, forming a rigid tube that encloses the hypopharynx (see Ch. 42, discussion; Figs. 42.36–44). Other Asiloidea may have the labellae reduced (e.g. *Rhaphiomidas* Osten Sacken in the Apioceridae, Fig. 41.4; some bombyliids, Figs. 45.3–4, 45.6–7), but they are not as rigidly sclerotized. In these cases the reduced labellae are functionally related to the elongation of the proboscis, which is coupled with nectar feeding.

3.8. *Hypopharynx of adults strongly sclerotized, modified into a hypodermic, needle-like structure*

This character state is unique to the asilids. The hypopharynx has become a piercing organ with which the robber flies pierce their prey, injecting a poisonous salivary secretion that immobilizes the victim and performs extraoral digestion. The hypopharynx is shown extruded in Figs. 42.1 and 42.40.

3.9. *Adult face with vestiture of strong bristles*

The strong patch of bristles on the face of asilids, called the mystax, is unique to the family. It is undoubtedly autapomorphic for the family.

There has been virtually no controversy concerning the limits of the family Asilidae. Martin (1968) attempted to segregate the Leptogastrinae into a separate family, but his reasoning was criticized on practical grounds by Oldroyd (1969). Martin's bases for separating the family were: "(1). Several characters show the Leptogastridae to be primitive to a greater degree than Asilidae. (2). The specialization of the aedeagi of the Leptogastridae and Asilidae has gone in different directions. (3). Several specialized characters are common to all Leptogastridae, but are absent in Asilidae." None of these reasons can be considered a valid basis on which to recognize two families, because the Leptogastrinae share the three apomorphic features delimited above with remaining Asilidae. Oldroyd (1969) criticized the basis on which Martin determined the primitive nature of some of his characters, such as considering the lack of tarsal pulvilli as plesiomorphic. In retrospect Martin clearly did not use sound phylogenetic reasoning. Even if the leptogastrines are found to be the sister group to the remaining asilids, together they still form a monophyletic group; therefore, little is gained by elevating the formal rank of the two groups. This point is discussed briefly in Chapter 42.

The relationship of the Asilidae to other Asiloidea is not clearly resolved at present. Most authors have assumed that the sister group of the Asilidae is the Mydidae, based on the sunken vertex of adults and the overall similarity of the larvae of the two families. However, there is some evidence that the Apioceridae and Mydidae are sister taxa, and the Asilidae are more likely to be the sister group of the Mydidae + Apioceridae. This theory can only be resolved by additional study.

Characters 3.10 and 3.11 support the sister-group relationship of the Mydidae and Apioceridae.

3.10. *Wing with veins R_5 and M_1 strongly curved anteriorly, ending anterior to the wing tip*

The wing venation of the Mydidae (Figs. 40.11–14) and Apioceridae (Figs. 41.2–3) is quite similar, with the most conspicuous feature in common being the anterior curvature of veins R_5 and M_1. Within the Asiloidea, only in the Scenopinidae is M_1 curved forward to a level anterior to the wing tip. In this family, though, the overall conformation of the veins is so different that the similarity has likely resulted from convergence. Within the Muscomorpha, the Nemestrinidae also have a somewhat similar venation (Fig. 44.3), but again there are significant differences in venation to indicate that convergence is likely.

3.11. *Adult with supernumerary rectal papillae*

In the Apioceridae and Mydidae, the adults have been found to have many more rectal papillae than are found in other Diptera. These structures apparently act to reabsorb water as material passes through the rectum (see Richards and Davies 1977). It is not surprising that many additional papillae have evolved in the apiocerids and mydids, because they are predominantly desert-inhabiting flies. Most other Diptera, including the Asilidae, have four rectal papillae, but five have been reported for two asilids (Owsley 1946). *Mydas clavatus* (Drury) has about 33 papillae (Jahn 1930), and other mydids have papillae ranging from 14 to 30 (Fisher, personal commun.); the apiocerids that have been examined have 20–80 (Woodley, personal observation; Fisher, personal commun.). This great increase in number of rectal papillae is construed to be autapomorphic for the Mydidae + Apioceridae.

The Apioceridae is a small family, with five genera: *Apiocera* Westwood (132 species), *Megascelus* Philippi (5 species; southern South America), *Neorhaphiomidas* Norris (7 species; Australia), *Rhaphiomidas* Osten Sacken (13 species), and *Tongamya* Stuckenberg (1 species; southern Africa). Cazier (1941) suggested that the closest relatives of apiocerids were the mydids, whereas Wilcox and Papavero (1971), following Karl (1959), suggested their closest relatives might be found among the asilids. The two characters already discussed support the former author's opinion.

Unfortunately, I have not found one conclusive autapomorphic character that can be used to define the apiocerids as a monophyletic group. In fact, they are possibly no more than plesiomorphic mydids. Some apiocerids (*Megascelus* and *Neorhaphiomidas*) have one-segmented palpi as do mydids, but *Apiocera* has more generalized, two-segmented palpi. Also, the apiocerid genus *Megascelus* has a tendency toward reduction and modification of the ocelli, also found in most mydids. Cazier (1941) suggested that *Apiocera* was the "generalized stock" within the family. A critical morphological survey of both families on a comparative basis is badly needed to delimit clearly cladistically useful characters and to resolve the relationships of the apiocerid genera to the Mydidae.

The Mydidae is another fairly small family of Diptera, with about 54 genera in which are placed about 340 species. They are found worldwide, but are particularly prevalent in arid regions. Again, derived characters unique to the mydids are not well worked out, but two character states appear to be definitive.

3.12. *Hind femora with ventral armature of moderately to very stoutly thickened spine-like bristles*

All mydids have spine-like armature present on the hind femora. This armature is absent in apiocerids, athough a few asilids have a similar armature (compare Figs. 40.15–17 and 42.67). This armature seems most likely to have resulted from convergence, and it would not be surprising if it evolved independently several times in the Asilidae as a modification suiting their predacious habits. Because of lack of data, it cannot be discounted entirely that the Mydidae + Apioceridae evolved from some asilid group with femoral spines, which were subsequently lost in the apiocerid genera.

3.13. *Adult palpus one-segmented*

All adult mydids have a one-segmented palpus. Many asilids and at least the genus *Apiocera* within the Apioceridae have two-segmented palpi. This difference indicates that if these three families are monophyletic as recognized here, one-segmented palpi are autapomorphic for mydids. Again, lack of confidence in the monophyly of the apiocerids leaves this interpretation open to question.

Two other features of the Mydidae may be autapomorphic for the family. The elongate, clavate antennal flagellum is distinctive (Figs. 40.2–10) and was suggested as a possible apomorphy by Hennig (1973). Although the antennae are much shorter in apiocerids, they are strongly thickened apically in *Megascelus* and *Tongamya*, a feature that could be interpreted as derived from the mydid type of antennal structure. The depressed vertex may also be autapomorphic. If the apiocerids, in which the vertex is not depressed, are more closely related to the mydids than are the asilids, then the depressed vertex in these two groups must have evolved independently.

Monophyly of the superfamily Empidoidea. Although the limits of the superfamily Empidoidea have not been questioned by most authors, the relationships of the two included families to other Brachycera have been somewhat controversial. In this section, I discuss the monophyly of the superfamily and its components. In the next section, I consider the varying views concerning the relationships of empidoids to other Brachycera.

Following Volume 1 of this Manual, I prefer to recognize two families, the Empididae and Dolichopodidae, as comprising the superfamily. However, recent work (Chvála 1981, 1983) suggests that the Empididae is a paraphyletic group, as the Dolichopodidae may have evolved from a group within the present Empididae.

Characters 34–36 support the monophyly of the Empidoidea.

34. *Larva with postcranium modified into a pair of slender metacephalic rods (Figs. 47.58; 48.41, 43)*

This feature is unique to the Empidoidea. The only other Brachycera with paired metacephalic rods are discussed previously under character 7, and the division of the postcranium occurs only in the Xylophagidae and a few Asilidae. In these two groups the rods are not as slender and are not nearly parallel sided as in Empidoidea. In general, the larval head capsule in the Empidoidea is much reduced compared with that of other Brachycera, especially the external portion of the cranium. The larvae of empidids and dolichopodids are extremely similar (Dyte 1967, Hennig 1973) and are known only for scattered taxa within both families.

35. *Adult female with a single spermatheca*

Hennig (1973) and Griffiths (1983) pointed out that a reduction in the number of spermathecae to one in females is probably autapomorphic for the Empidoidea. As far as is known, all Empidoidea have this character state, including problematic taxa such as the genera *Atelestus* Walker and *Nemedina* Chandler (Chandler 1981). A further reduction occurs in the empidid subfamilies Ocydromiinae, Hybotinae, and Tachydromiinae, in which the female does not have any spermathecae (Chandler

1981). The primitive number of spermathecae in Diptera is assumed to be three (Downes 1968), and this number is found in most Brachycera more plesiomorphic than Empidoidea. As far as I am aware, only the Xylomyidae (which have two spermathecae) can be characterized at the family level by a reduction in the number of spermathecae. Also, the Muscoidea (= Muscomorpha of Vol. 1) retain the plesiomorphic complement of three spermathecae (Griffiths 1983) in their ground plan. This fact apparently refutes the implication of Chvála's (1983: Fig. 140) hypothetical phylogeny, which shows the Muscoidea as the sister group to his Atelestidae, and the Muscoidea + Atelestidae as the sister group to the Hybotidae + Microphoridae + Dolichopodidae. His phylogeny would relegate the Empidoidea to polyphyletic status, and, as Griffiths (1983) points out, a number of higher categories would be necessary to render this phylogeny into a formal classification. As Griffiths noted, this step has not been done because of conflicting evidence. In my opinion, characters 34 and 35 make it extremely unlikely that the Muscoidea arose from the midst of the Empidoidea.

36. *Adult male with paired apodemes of the genital segment attached to the hypandrium*

Griffiths (1983, 1984) suggested that this character state is synapomorphic for the Empidoidea (which he calls the Orthogenya after Brauer (1883) for reasons explained in his 1972 work), including *Atelestus* and relatives but not the Muscoidea. This reasoning also refutes the evolution of the Muscoidea from some stock within the Empidoidea.

Phylogeny of the Empididae *sensu lato*. As already noted, I prefer to recognize the traditional two families within the Empidoidea, the Empididae and Dolichopodidae, even in light of recent work splitting the former into four families (Chvála 1983). My main reason for this treatment is the lack of sound cladistic evidence for postulating the four groups regarded as family level units. Therefore a short critique of Chvála's work is necessary here.

A major concern is that his figure (1983: Fig. 141) and subsequent characterization of the family Empididae (in which he included the subfamilies Oreogetoninae, Empidinae, Hemerodromiinae, Brachystomatinae, Ceratomerinae, and Clinocerinae) on page 65 fail to give a single conclusive synapomorphy for the included taxa. In his Figure 141, all the apomorphic features shared by the subfamilies included in the Empididae are also found in the Dolichopodidae and other Empidoidea. The Oreogetoninae is indicated as having only plesiomorphic character states for all features discussed. On page 65, the Empididae are characterized by seven character states regarded as apomorphic, even though they are clearly found outside the Empididae in Chvála's sense; most of them are used by Griffiths (1972: 58ff.) to characterize the entire Eremoneura (= Empidoidea + Muscoidea). Chvála then summarized the plesiomorphic character states possessed by his Empididae to show that they may be excluded from his other recognized families.

For the three groups segregated from the traditional Empididae, an inspection of the section outlining the supposedly defining characters for each group reveals that some of the characters are clearly not unique to a single group; some of these discrepancies are actually illustrated as such in Figure 141. As far as I can tell, the Atelestidae are characterized most definitely by Chvála's characters 2 and 9, and possibly by 3, 8, and 11 (1983: 67). The Microphoridae are more poorly characterized, as most of the characters are also present in the Dolichopodidae (1983: 70ff.). For the microphorids, only character 3 is possibly definitive. The Hybotidae is the best-characterized segregate, as was previously noted by Tuomikoski (1966); its monophyly is supported by Chvála's characters 4–7, 11, and 12, the last two of which were proposed by Tuomikoski and were not reviewed by Chvála.

A final point I wish to make concerning this work by Chvála is its preliminary nature. It is clearly based primarily on Palaearctic material and only secondarily on Nearctic material. The vast and poorly collected faunas of the south temperate regions are not considered to any extent. Some of the more primitive members of the Empididae are expected to be found in these regions. For instance, many *Oreogeton*-like taxa are known from Chile (Collin 1933). These primitive taxa will be important in the resolution of the relationships of Empidoidea.

In summary, because the analysis of the characters currently known within the Empidoidea is incomplete and thus results in tentative hypothetical phylogenies, and because the knowledge of many important exotic faunas is meager, I prefer not to recognize segregate families split off from the traditional Empididae. Contributions such as that made by Chvála (1983) are very important, in that they provide some of the first studies of the Empidoidea utilizing a cladistic theoretical framework. This approach is necessary and such works initiate a systematic study of characters rather than the haphazard methodology of the past. However, I am not a proponent of changing formal nomenclature every time a new hypothesis is proposed. When these hypotheses are tentative, the frequent nomenclatural changes they engender become incomprehensible to all but specialists. I think such nomenclatural changes are best saved until the phylogenies are rather better understood than they are currently in the Empidoidea.

Despite my recognition of only the single family Empididae, it must still be considered likely to be paraphyletic or possibly even polyphyletic, depending on the relationships of the Microphorinae to the rest of the Empididae. Evidence presented by Chvála (1983: Fig. 142) indicates that the microphorines are the most likely candidate for sister-group relationship to the Dolichopodidae. Ulrich (1984) has recently presented additional support for this relationship, based upon characters of the thoracic skeletal and muscular structure.

The Empididae *sensu lato* is a large family, with about 3500 described species in many genera. In the south temperate regions and mountainous tropics they are poorly collected, but works based on available material (Collin 1928, 1933; Smith 1969) suggest that the faunas are vast and probably rich in undescribed forms. I am aware of no conclusive apomorphic character state that can be used to define the Empididae exclusive of the Dolichopodidae. None was given by Hennig (1973) or Chvála (1983). The group is essentially recognized on a practical basis only, comprising Empidoidea with more plesiomorphic wing venation and other structures, in comparison with dolichopodids, which is not particularly surprising in a probable paraphyletic group.

Monophyly of the Dolichopodidae. The Dolichopodidae is another quite large family of Diptera, with more than 150 genera and about 6000 species widespread throughout the world. Members of the family are very distinctive and rather uniform in appearance, such that no recent author has questioned the status of the family or inclusion of any of its members within it.

I am aware of two characters that support the monophyly of the Dolichopodidae.

37. *Adult mouthparts with sharp, strongly sclerotized, epipharyngeal armature*

Dolichopodid flies have strongly developed, tooth-like processes termed the epipharyngeal armature (Fig. 2.51) at the sides of the labrum, which are strongly sclerotized and sharply pointed, sometimes apically bifurcated to form a pair of teeth (see Cregan 1941: Figs. 65–96 for a good survey of the variation found in these structures within the family). These teeth are presumably utilized in cutting and tearing soft tissue of small insects upon which dolichopodids prey. It is apparently a unique development in the dolichopodids and thus autapomorphic for the family.

38. *Adult with base of wing strongly reduced*

The wing venation of the dolichopodids is very distinctive and easily allows for practical recognition of the family. The most conspicuous feature of the venation is the extreme reduction of the base of the wing, which is best indicated by the origination of the radial sector near the humeral crossvein and the extreme reduction of cells br and cu*p*, the latter sometimes being absent (Figs. 48.25, 48.35). Cell bm is also reduced, and the crossvein separating it from the discal cell is absent. Sometimes some of these features have been regarded as separate apomorphic features (Hennig 1973), but I suspect that they are all correlated with reduction of the wing base. The basal area of the wing is not reduced as much in any of the Empididae *sensu lato*.

Hennig (1973) mentioned the reduction of the costa beyond vein M_1 and the simple vein R_{4+5} as being derived for the dolichopodids, but these character states are also found in various Empididae. Therefore, although they are apomorphic, their level of synapomorphy remains ambiguous until the exact relationship of Dolichopodidae to empidid subgroups is known. Chvála (1983) also deemed the simple vein M_1 as being apomorphic for the family. However, he mentioned that a stub of M_2 is found in some Sciapodinae within the dolichopodids and notes that a simple M_1 is found in some Empididae. Therefore this character is also ambiguous. Its autapomorphic nature in the dolichopodids also presumes the sister-group relationship of Dolichopodidae with the Parathalassiini, as proposed by Chvála.

One additional feature noted by Chvála (1983), which is possibly autapomorphic for the Dolichopodidae, is the reduction of the subcostal vein. In most members of the family, it is quite short and fuses apically with R_1. However, in some Hydrophorinae it is somewhat longer and its end is free (Fig. 48.32). This state approaches the condition found in many Empididae. Thus the autapomorphic nature of this character state is uncertain. Further resolution of the relationships within dolichopodids are needed to determine whether or not the state found in Hydrophorinae is an independent development within dolichopodids relative to empidids.

Relationships of Empidoidea to other Brachycera. The sister-group relationship of the Empidoidea to other brachycerous Diptera has been the subject of controversy in recent years. Some authors have maintained that the Empidoidea is the sister group to the Asiloidea (Hennig 1972, 1973; Vol. 1 of this Manual); others have considered the Empidoidea to be the sister group to the Muscoidea (Griffiths 1972, 1983, 1984; Hennig 1976; Chvála 1983).

The relationship of the Empidoidea to the Asiloidea is supported by two characters deemed important:

- presence of only three flagellomeres in the adult antennal flagellum
- presence of acanthophorites (= hemitergites)

Three flagellomeres in the adult. This feature was reviewed in detail by Hennig (1972, 1976). A flagellum composed of three flagellomeres is found virtually throughout the Asiloidea and Empidoidea, except where it has become even further fused. A very few Empididae that have four flagellomeres are known, but according to Chvála's (1983) classification of the Empididae *sensu lato*, these are in derived lineages. They include *Dryodromia* Rondani (Chvála 1983: Fig. 13) and *Meghyperus* Loew (from California, Chvála 1983: Fig. 560), but in the latter genus, four flagellomeres are not found in all species. Other genera of empidids supposedly with four flagellomeres include *Hesperempis* Melander, *Ragas* Walker, and *Hilara* Meigen. However, Chvála's figures (1983: Figs. 2, 3, and 8) are somewhat vague and seem to indicate a closer similarity to the form of flagellum found in the Asiloidea, in that the last "flagellomere" appears similar to the hyaline style found in many asiloid flies.

Reports of four flagellomeres in the genus *Ruppellia* Wiedemann (Therevidae; Hennig 1972) have recently been found erroneous (Lyneborg 1983). In the Muscoidea, the antennal flagellum has four flagellomeres, which Hennig believed to be more plesiomorphic than the state with three flagellomeres. This condition is somewhat similar to that found in many Nemestrinidae. The presence of only three flagellomeres in *Opetia* Meigen (Platypezidae) has been considered of interest in this argument (Hennig 1976, Chvála 1981). Its significance remains unclear, though, as reduction could have occurred independently in *Opetia*; the loss of flagellomeres by fusion has without doubt taken place many times in the Brachycera.

The cladistic implications of the number of flagellomeres in these groups need support from corroborating characters before we can determine whether or not the four flagellomeres found in Muscoidea are plesiomorphic relative to the Empidoidea or whether the number merely represents a secondary addition of another flagellomere.

Acanthophorites (= hemitergites). This feature has been regarded as synapomorphic for the two superfamilies Asiloidea and Empidoidea (Hennig 1972, 1976). Acanthophorites are the divided lobes of the female abdominal tergite 10; they bear large spine-like setae that are used for oviposition, especially in sandy substrates (Figs. 41.8–10). They are found in at least some members of every family of Asiloidea. Their scattered occurrence led Adisoemarto and Wood (1975) to regard them as a derived feature of the Asiloidea, rather than as structures that had evolved independently several times within the superfamily. Irwin (1976) considered the enlarged setae found on the acanthophorites of some Therevidae to be apomorphic within the family, but this idea was discounted by Lyneborg (1983). As large spines are present in most Asiloidea with acanthophorites, large spines seem to represent the most reasonable plesiomorphic state, with reduction in size of these structures being relatively more apomorphic.

Acanthophorites are also found in the Empididae and Dolichopodidae (Chvála 1983). In addition to the few genera Chvála mentioned (1983: 38), I have observed them in *Oreogeton*-like empidids from Chile, although acanthophorites are apparently not found in Holarctic members of *Oreogeton* Schiner. These Chilean Oreogetoninae may be among the most primitive empidids, if Chvála's phylogeny of his restricted Empididae is correct. I cannot agree with Chvála (1983: 38) and Griffiths (1983: 484) that the scattered distribution of acanthophorites within the Empidoidea indicates that the character has no cladistic significance. The fact remains that if the structures are homologous and synapomorphic for the Asiloidea and Empidoidea, only two alternatives are suggested: either the Asiloidea and Empidoidea are sister groups and thus are more closely related to each other than either one is to the Muscoidea; or the Asiloidea, Empidoidea, and Muscoidea shared synapomorphic acanthophorites, which were subsequently lost in the Muscoidea. Hennig's (1976) assertion that acanthophorites were not present in the ground plan of Empididae seems to be supported by little or no data. We really do not know what the most plesiomorphic empidids are, because of our poor knowledge of the south temperate faunas.

Two other characters were mentioned by Hennig (1976) in support of the hypothesis of the monophyly of the Asiloidea + Empidoidea. One was the absence of tibial spurs in the Asiloidea and Empidoidea and their supposed presence in the Phoridae in the Muscoidea. I have noted above that the structures found in phorids are probably not true tibial spurs. The second character Hennig discussed is the hinged metacephalic rods found in larvae of the Asiloidea and Empidoidea. The validity of this character as support for this hypothesis seems questionable in view of the absence of articulation in the larval cranium in known Bombyliidae (see character 3.2). Bombyliidae have the larval posterior spiracle in the penultimate abdominal segment, and some are known to have acanthophorites. They should, therefore, probably be placed in the Asiloidea, a move that makes the hinged metacephalic rods suspect as a synapomorphy between the Asiloidea and Empidoidea.

The alternative hypothesis, that the Empidoidea is the sister group to the Muscoidea, is supported primarily by characters of the male terminalia (Griffiths 1972, 1984). These characters are rather complicated and were criticized by Hennig (1976), whose interpretations of homologies of the genital structures involved were considerably different from those of Griffiths. As I cannot personally evaluate these characters and their distributions within the Brachycera, I shall simply list those characters given by Griffiths (1984) as supporting the monophyly of the Empidoidea + Muscoidea and refer the reader to discussions found in the works by Griffiths and Hennig.

The characters deemed synapomorphic for the Empidoidea + Muscoidea by Griffiths are (verbatim from Griffiths 1984) as follows:

- gonocoxites expanded dorsally
- gonocoxites not closed distally (with separate bacilliform sclerites on inner surfaces of gonocoxae)—this partial desclerotization of the gonocoxae is the origin of the processus longi of Calyptratae, etc., and equals the Skleritleiste of Empididae
- separation of the paired (gonocoxal) apodemes from the gonocoxites (precise ground plan condition unclear, but in neither of the subgroups are the apodemes attached to the gonocoxites as in other Diptera—probably separation of these apodemes was a result of the partial desclerotization of the gonocoxae already stated)
- hypandrial arms fused above base of aedeagus (i.e. Hennig's bridge, which is well developed in Platypezidae as well as in Orthogenya (= Empididae + Dolichopodidae), so probably belongs to the ground plan although lost in many Cyclorrhapha)

- hypandrium with a pair of posterior processes
- tenth tergite of male lost (not a very useful character, as the same has occurred independently in other groups)
- female with complete loss of the sclerites of the ninth abdominal segment

Although Hennig (1976) criticized some of these characters proposed by Griffiths (in the 1972 version), he still felt that the most likely sister group of the Empidoidea was the Muscoidea.

CONCLUSIONS

This review of the phylogeny of the Brachycera exclusive of the Muscoidea is intended to provide a framework from which future studies can progress. Some of the characters discussed are based on very limited surveys. Only supplementation of this limited knowledge can provide a firmer foundation for their cladistic use. Many lineages are supported by only one or a few characters, and thus searches for additional characters of cladistic use are badly needed for most major taxa.

The phylogenies of intrafamilial components of most families are poorly known or not known at all. Therefore this field is fertile for investigation. Knowledge of relationships within families will prove useful as an end to itself. In addition, knowledge of the distribution of character states within families will also be helpful in establishing a ground plan or most primitive taxon. This framework can then be used to estimate what ancestral members of each family may have been like.

Certain areas require special attention. We need character information that will help to resolve the relationships of the infraorders of Brachycera, a subject that has received little attention. Also, the relationships within the Asiloidea are especially in need of investigation; our fairly good understanding of the taxonomy of the families that constitute the major components should facilitate the sampling of the major taxa within each for character surveys. Studies to resolve the placement of the Nemestrinoidea are also needed.

Finally, the need for general surveys of characters cannot be overemphasized. Far too often, cladistic studies are too limited taxonomically and fail to provide an adequate survey of character-state distributions in taxa that are potential outgroups. Such studies are not useful when another worker attempts to utilize their results in a study of broader scale. Many character systems, such as internal anatomy, have received little study. They may well provide a wealth of additional character suites that can be used to test existing hypotheses from an independent character base. With the broadening awareness of the importance of cladistic methodology among entomologists in many fields, much exciting information useful in resolving the phylogeny of the Brachycera should be forthcoming.

References

Adisoemarto, S., and D. M. Wood. 1975. The Nearctic species of *Dioctria* and six related genera (Diptera, Asilidae). Quaest. Ent. 11: 505–576.

Austen, E. E. 1899. On the preliminary stages and mode of escape of the imago in the dipterous genus *Xylomyia*, Rond. (*Subula*, Mg. et auct.), with especial reference to *Xylomyia maculata*, F., and on the systematic position of the genus. Ann. Mag. nat. Hist. (7)3: 181–190.

Bährmann, R. 1960. Vergleichend-morphologische Untersuchungen der männlichen Kopulationsorgane bei Empididen (Diptera). Beitr. Ent. 10 (5 / 6): 485–540.

Bernardi, N. 1972. Notes on the genus *Exeretoneura* Macquart, and its removal from the family Nemestrinidae (Diptera, Brachycera). Papéis Dep. Zool. S Paulo 26 (3): 47–50.

Bernardi, N. 1973. The genera of the family Nemestrinidae (Diptera: Brachycera). Archos Zool. Est. S Paulo 24 (4): 211–318.

Brauer, F. 1883. Die Zweiflügler des kaiserlichen Museums zu Wien. III. Systematische Studien auf Grundlage der Dipteren-Larven nebst einer Zusammenstellung von Beispeilen aus der Literatur über dieselben und Beschreibung neuer Formen. Denkschr. Math.-naturw. Classe kaiser. Akad. Wiss., Wien 47: 1–100.

Brunetti, E. 1920. Diptera Brachycera. Vol. I. Pages 1–401 *in* A. E. Shipley, ed. The fauna of British India, including Ceylon and Burma. Taylor and Francis, London.

Cazier, M. A. 1941. A generic review of the family Apioceratidae with a revision of the North American species (Diptera-Brachycera). Am. Midl. Nat. 25: 589–631.

Chandler, P. 1981. *Nemedina alamirabilis* sp. n., a new genus and species of Diptera Eremoneura, of uncertain affinities, from Hungary. Acta zool. hung. 27 (1–2): 103–113.

Chvála, M. 1981. Classification and phylogeny of Empididae, with a presumed origin of Dolichopodidae (Diptera). Ent. scand. Suppl. 15: 225–236.

Chvála, M. 1983. The Empidoidea (Diptera) of Fennoscandia and Denmark. II. General Part. The families Hybotidae, Atelestidae and Microphoridae. Fauna ent. scand. 12: 1–279.

Colless, D. H., and D. K. McAlpine. 1970. Diptera (flies). Ch. 34, pages 656–740 *in* The insects of Australia. Melbourne University Press, Carlton, Victoria.

Collin, J. E. 1928. New Zealand Empididae based on material in the British Museum (Natural History). British Museum, London.

Collin, J. E. 1933. Empididae. Vol. 4, Pages 1–334 *in* Diptera of Patagonia and South Chile. British Museum (Natural History), London.

Collin, J. E. 1961. Empididae. Vol. 6, Pages 1–782 *in* British flies. Cambridge University Press, Cambridge.

Cregan, M. B. 1941. Generic relationships of the Dolichopodidae (Diptera) based on a study of the mouth parts. Illinois biol. Monogr. 18(1): 1–68.

Daniels, G. 1976. The Xylomyidae (Diptera) of Australia and Papua New Guinea.J. Aust. ent. Soc. 15: 453–460.

Downes, J. A. 1968. Notes on the organs and processes of sperm-transfer in the lower Diptera. Can. Ent. 100: 608–617.

Dyte, C. E. 1967. Some distinctions between the larvae and pupae of the Empididae and Dolichopodidae (Diptera). Proc. R. Ent. Soc. Lond. Ser. A, 42 (7–9): 119–128.

Edwards, F. W. 1938. British short-palped craneflies. Taxonomy of adults. Trans. Soc. Br. Ent. 5 (1): 1–168.

English, K. M. I. 1947. Notes on the morphology and biology of *Apiocera maritima* Hardy (Diptera, Apioceridae). Proc. Linn. Soc. N.S.W. 71:296–302.

Evenhuis, N. L. 1982. New east Asian *Systropus* (Diptera: Systropodidae). Pacif. Insects 24 (1): 31–38.

Greene, C. T., and F. W. Urich. 1931. The immature stages of *Pantophthalmus tabaninus*, Thunberg. Trans. R. ent. Soc. Lond. 79 (2): 277–282: plates vii–ix.

Griffiths, G. C. D. 1972. The phylogenetic classification of Diptera Cyclorrhapha, with special reference to the structure of the male postabdomen. Series entomologica, 8. Dr. W. Junk, N.V., The Hague.

Griffiths, G. C. D. 1983. Book review of Chvála, M. 1983. The Empidoidea of Fennoscandia and Denmark. II. The families Hybotidae, Atelestidae and Microphoridae. Fauna Entomologica Scandinavica, Volume 12. Quaest. Ent. 19 (3 / 4): 483–485.

Griffiths, G. C. D. 1984. Note on characterization of Eremoneura, Orthogenya and Cyclorrhapha. Privately published, 1 page.

Hackman, W., and R. Väisänen. 1982. Different classification systems in the Diptera. Annls zool. Fenn. 19 (3): 209–220.

Hall, J. C. 1975. The Bombyliidae of Chile (Diptera: Bombyliidae). Univ. Calif. Publ. Ent. 76: 1–278; 6 plates.

Hall, J. C., and N. L. Evenhuis. 1980. Homoeodactyla and Asilomorpha. Part 13. Bombyliidae. Number 1, Pages 1–96, Vol. 5 *in* G. C. D. Griffiths, ed. Flies of the Nearctic region. Schweizerbart, Stuttgart.

Hardy, G. H. 1927. On the phylogeny of some Diptera Brachycera. Proc. Linn. Soc. N.S.W. 42: 380–386.

Hennig, W. 1950. Grundzüge einer Theorie der Phylogenetischen Systematik. Deutscher Zentralverlag, Berlin.

Hennig, W. 1954. Flügelgeäder und System der Dipteren unter Berücksichtigung der aus dem Mesozoikum beschriebenen Fossilien. Beitr. Ent. 4(3 / 4): 245–388.

Hennig, W. 1967. Die sogenannten "niederen Brachycera" im Baltischen Bernstein (Diptera: Fam. Xylophagidae, Xylomyidae, Rhagionidae, Tabanidae). Stuttg. Beitr. Naturk. 174: 1–51.

Hennig, W. 1968. Kritische Bemerkungen über den Bau der Flügelwurzel bei den Dipteren und die Frage nach der Monophylie der Nematocera. Stuttg. Beitr. Naturk. 193: 1–23.

Hennig, W. 1972. Eine neue Art der Rhagionidengattung *Litoleptis* aus Chile, mit Bemerkungen über Fühlerbildung und Verwandtschaftsbeziehungen einiger Brachycerenfamilien (Diptera: Brachycera). Stuttg. Beitr. Naturk. 242: 1–18.

Hennig, W. 1973. Ordnung Diptera (Zweiflügler). Handb. Zool. 4 (2) 2 / 31(Lfg. 20): 1–337.

Hennig, W. 1976. Das Hypopygium von *Lochoptera lutea* Panzer und die phylogenetischen Verwandtschaftsbeziehungen der Cyclorrhapha (Diptera). Stuttg. Beitr. Naturk. Ser. A, 283: 1–63.

Hennig, W. 1981. Insect phylogeny. *Translated and edited by* A. C. Pont. John Wiley & Sons, New York.

Hull, F. M. 1973. Bee flies of the world. The genera of the family Bombyliidae. Smithsonian Institution Press, Washington.

Irwin, M. E. 1976. Morphology of the terminalia and known ovipositing behaviour of female Therevidae (Diptera: Asiloidea), with an account of correlated adaptations and comments on phylogenetic relationships. Ann. Natal Mus. 22 (3): 913–935.

Irwin, M. E., and L. Lyneborg. 1981. The genera of Nearctic Therevidae. Bull. Ill. St. nat. Hist. Surv. 32 (3): 191–277.

Jahn, L. A. 1930. The internal anatomy of the mydas fly. Ohio J. Sci. 30: 85–94.

Karl, E. 1959. Vergleichend-morphologische Untersuchungen der männlichen Kopulationsorgane bei Asiliden (Diptera). Beitr. Ent. 9: 619–680.

Kovalev, V. G. 1981. The oldest representatives of the Diptera with short antennae from the Jurassic in Siberia. Paleont. J. 15 (3): 84–100. [*Translated from* Paleont. Zh. 15 (3): 85–101.]

Kovalev, V. G. 1982. Some Jurassic Diptera-rhagionids (Muscida, Rhagionidae). Paleont. J. 16 (3): 87–99. [*Translated from* Paleont. Zh. 16 (3): 88–100.]

Krivosheina, N. P. 1967. Comparative characteristics of the larva of *Arthropeas sibirica* Loew (Diptera, Xylophagidae). Zool. Zh. 46: 954–956.

Krivosheina, N. P. 1971. The family Glutopidae, Fam. N., and its position in the system of the order Diptera (Diptera Brachycera Orthorrhapha). Ent. Rev. 50: 387–395. [*Translated from* Ént. Obozr. 50: 681–694.]

Krivosheina, N. P. 1980. New Palearctic Scenopinidae (Diptera). Ent. Rev. 59 (1): 157–164. [*Translated from* Ént. Obozr. 59 (1): 197–205.]

Kröber, O. 1930. Tabanidae. Vol. 5, Pages 106–161 *in* Diptera of Patagonia and South Chile. British Museum (Natural History), London.

Lindner, E. 1949. Handbuch. Bd. 1, Pages 1–422 *in* Die Fliegen der palaearktischen Region. Schweizerbart, Stuttgart.

Lyneborg, L. 1976. A revision of the therevine stiletto-flies (Diptera: Therevidae) of the Ethiopian region. Bull. Br. Mus. Nat. Hist. Ent. 33 (3): 189–346.

Lyneborg, L. 1983. A review of the Palaearctic genera of Phycinae (Insecta, Diptera, Therevidae). Steenstrupia 9 (8): 181–205.
Mackerras, I. M., and M. E. Fuller. 1942. The genus *Pelecorhynchus* (Diptera, Tabanoidea). Proc. Linn. Soc. N.S.W. 67 (1–2): 9–76; 1 plate.
Macquart, J. 1846. Diptères exotiques nouveaux ou peu connus. Suppl. Mém. Soc. Sci. Agric. Lille 1844: 133–364.
Malloch, J. R. 1917. A preliminary classification of Diptera, exclusive of Pupipara, based upon larval and pupal characters, with keys to imagines in certain families. Part I. Bull. Ill. St. Lab. nat. Hist. 12 (3): i–v, 161–409; plates 28–57.
Malloch, J. R. 1932. Rhagionidae, Therevidae. Vol. 5, Pages 199–257 *in* Diptera of Patagonia and South Chile. British Museum (Natural History), London.
Martin, C. H. 1968. The new family Leptogastridae (the grass flies) compared with the Asilidae (robber flies) (Diptera). J. Kans. ent. Soc. 41 (1): 70–100.
McFadden, M. W. 1967. Soldier fly larvae in America north of Mexico. Proc. U.S. Nat. Mus. 121 (3569): 1–72.
Moucha, J. 1976. Horse-flies (Diptera: Tabanidae) of the World Synoptic Catalogue. Acta ent. Mus. nat. Prague Suppl. 7: 1–319.
Nagatomi, A. 1961. Studies in the aquatic snipe flies of Japan. Part III. Descriptions of the larvae (Diptera, Rhagionidae). Mushi 35 (3): 11–27.
Nagatomi, A. 1970. Rachiceridae (Diptera) from the Oriental and Palaearctic regions. Pacif. Insects 12 (2): 417–466.
Nagatomi, A. 1977. Classification of lower Brachycera (Diptera). J. Nat. Hist. 11 (3): 321–335.
Nagatomi, A. 1981. Some characters of the lower Brachycera (Diptera) and their plesiomorphy and apomorphy. Kontyû 49 (3): 397–407.
Nagatomi, A. 1982*a*. The genera of Rhagionidae (Diptera). J. Nat. Hist. 16: 31–70.
Nagatomi, A. 1982*b*. The genus *Pseudoerinna* (= *Bequaertomyia*) (Diptera, Rhagionidae). Kontyû 50 (1): 97–99.
Nagatomi, A. 1982*c*. Geographical distribution of the lower Brachycera (Diptera). Pacif. Insects 24 (2): 139–150.
Nagatomi, A. 1982*d*. Genitalia of *Hilarimorpha* (Diptera, Hilarimorphidae). Kontyû 50 (4): 544–548.
Nagatomi, A., and K. Iwata. 1976. Female terminalia of lower Brachycera—I. Beitr. Ent. 26 (1): 5–47.
Nagatomi, A., and K. Iwata. 1978. Female terminalia of lower Brachycera—II.Beitr. Ent. 28 (2): 263–293.
Nagatomi, A., and A. Tanaka. 1971. The Solvidae of Japan (Diptera). Mushi 45 (6): 101–146.
Oldroyd, H. 1969. The family Leptogastridae (Diptera). Proc. R. ent. Soc. Lond., Ser. B, 38 (1–2): 27–31.
Oldroyd, H. 1977. The suborders of Diptera. Proc. ent. Soc. Wash. 79 (1): 3–10.
Owsley, W. B. 1946. The comparative morphology of internal structures of the Asilidae (Diptera). Ann. ent. Soc. Am. 39: 33–68.
Paramonov, S. J. 1953. A review of Australian Nemestrinidae (Diptera). Aust. J. Zool. 1 (2): 242–290.
Richards, O. W., and R. G. Davies. 1977. Imms' general textbook of entomology. Tenth edition. Volume 1: Structure, physiology and development. Chapman and Hall, London.
Rohdendorf, B. 1974. The historical development of Diptera. B. Hocking, H. Oldroyd, and G. E. Bull, eds. University of Alberta Press, Edmonton, Alberta. [*Translated from* Russian.]
Rozkošný, R. 1973. The Stratiomyioidea (Diptera) of Fennoscandia and Denmark. Fauna ent. scand. 1: 1–140.
Rozkošný, R. 1982. A biosystematic study of the European Stratiomyidae (Diptera). Vol. 1. Introduction, Beridinae, Sarginae and Stratiomyinae. Series entomologica, 21. Dr. W. Junk, N.V., The Hague, 1–401.
Rozkošný, R. 1983. A biosystematic study of the European Stratiomyidae (Diptera). Volume 2. Clitelariinae, Hermetiinae, Pachygasterinae and Bibliography. Series entomologica, 25. Dr. W. Junk, N.V., The Hague.
Saigusa, T. 1973. The brachycerous family Hilarimorphidae newly recorded from Japan (Diptera). Sieboldia 4 (3): 161–166.
Smith, K. G. V. 1969. The Empididae of southern Africa (Diptera). Ann.Natal Mus. 19: 1–347.
Steyskal, G. C. 1953. A suggested classification of the lower brachycerous Diptera. Ann. Ent. Soc. Am. 46: 237–242.
Steyskal, G. C. 1974. Recent advances in the primary classification of the Diptera. Ann. Ent. Soc. Am. 67 (3): 513–517.
Stuckenberg, B. R. 1960. Diptera (Brachycera): Rhagionidae. S. Afr. Anim. Life 7: 216–308.
Stuckenberg, B. R. 1973. The Athericidae, a new family in the lower Brachycera (Diptera). Ann. Natal Mus. 21 (3): 649–673.
Teskey, H. J. 1970. The immature stages and phyletic position of *Glutops rossi* (Diptera: Pelecorhynchidae). Can. Ent. 102 (9): 1130–1135.
Theodor, O. 1976. On the structure of the spermathecae and aedeagus in the Asilidae and their importance in the systematics of the family. The Israel Academy of Sciences and Humanities, Jerusalem.
Theodor, O. 1983. The genitalia of Bombyliidae (Diptera). The Israel Academy of Sciences and Humanities, Jerusalem.
Tuomikoski, R. 1966. The Ocydromiinae group of subfamilies (Diptera, Empididae). Suom. Hyönt. Aikak. 32 (4): 282–294.
Ulrich, H. 1984. Skelett und Muskulatur des Thorax von *Microphor holosericeus* (Meigen) (Diptera, Empidoidea). Bonn. zool. Beitr. 35 (4): 351–398.
Val, F. C. 1976. Systematics and evolution of the Pantophthalmidae (Diptera, Brachycera). Archos Zool. Est. S Paulo 27 (2): 51–164; 2 plates.
Verrall, G. H. 1909. Stratiomyidae and succeeding families of the Diptera Brachycera of Great Britain. Volume 5, Pages 1–780 *in* G. H. Verrall, ed. British flies. Gurney and Jackson, London.

Webb, D. W. 1974. A revision of the genus *Hilarimorpha* (Diptera: Hilarimorphidae). J. Kans. ent. Soc. 47 (2): 172–222.

Webb, D. W. 1977. The Nearctic Athericidae (Insecta: Diptera). J. Kans. Ent. Soc. 50 (4): 473–495.

Webb, D. W. 1979. A revision of the Nearctic species of *Xylophagus* (Diptera: Xylophagidae). J. Kans. ent. Soc. 52 (3): 489–523.

Webb, D. W. 1983*a*. The genus *Coenomyia* (Diptera: Coenomyiidae) in the Nearctic region and notes on generic placement. Proc. ent. Soc. Wash. 85 (4): 653–664.

Webb, D. W. 1983*b*. A new genus and species of Nearctic coenomyiid (Diptera: Coenomyiidae). Proc. ent. Soc. Wash. 85 (4): 822–825.

Webb, D. W. 1984. A revision of the Nearctic species of *Rachicerus* (Diptera: Rachiceridae). J. Kans. ent. Soc. 57 (2): 298–315.

Webb, D. W., and E. A. Lisowski. 1983. The immature stages of *Dialysis fasciventris* (Loew) (Diptera: Coenomyiidae). Proc. ent. Soc. Wash. 85 (4): 691–697.

Westwood, J. O. 1840. An introduction to the modern classification of insects; founded on the natural habits and corresponding organisation of the different families. Longman, Orme, Brown, Greene, and Longmans, London.

Wilcox, J., and N. Papavero. 1971. The American genera of Mydidae (Diptera), with the description of three new genera and two new species. Archos Zool. Est. S Paulo 21 (2): 41–119.

Williston, S. W. 1908. Manual of North American Diptera. 3rd Ed. James T. Hathaway, New Haven, Connecticut.

Woodley, N. E. 1983. The world genera of Beridinae with a discussion of cladistic relationships within Stratiomyioidea and Stratiomyidae (Diptera). Ph.D. Thesis. Harvard University. Available on demand from University Microfilms International, Ann Arbor, Mich.

Woodley, N. E. 1986. Parhadrestiinae, a new subfamily for *Parhadrestia* James and *Cretaceogaster* Teskey (Diptera: Stratiomyidae). Syst. Ent. 11: 377–387.

Zaitsev, V. F., and D. Tsharykuliev. 1981. On the biology of the genus *Petrorossia* Bezzi (Diptera, Bombyliidae) with the description of a new species from Turkmenia. Ent. Rev. 60: 158–160. [*Translated from* Ént. Obozr. 60: 914–916.]

PHYLOGENY AND CLASSIFICATION OF THE MUSCOMORPHA 116

J. F. McAlpine

INTRODUCTION

Definition of the Muscomorpha

The infraorder Muscomorpha is defined here to include only the cyclorrhaphous Brachycera (Cyclorrhapha of most authors), as opposed to the orthorrhaphous Brachycera, which includes several infraorders, treated in Chapter 115. The name Muscomorpha (Crampton 1944*a*, p. 24) is used in preference to Cyclorrhapha (Brauer 1880) to conform with the names applied to the other infraorders of Diptera. Because rules for ordinal group names were rejected as part of the Code at the London Congress, suitability is as good a basis as any for choosing them. When Crampton proposed the name Muscomorpha (spelled Muscamorpha and Muscomorpha on the same page), he used it as a replacement for Calyptratae. Hennig (1948*a*, 1952, 1954) extended its application to a broader taxon, the Eremoneura of Lameere (1906), which includes all of the cyclorrhaphous Brachycera and the Empidoidea (Empididae + Dolichopodidae) (= Empidiformia Hennig 1948*a*, = Orthogenya Brauer 1883). In those papers Hennig regarded the cyclorrhaphous Brachycera as "subsection" Musciformia of the "section" Muscomorpha. Later, however, Hennig (1972*b*, 1973) removed the Empidoidea to his infraorder Asilomorpha, and recognized the Muscomorpha (as Cyclorrhapha) as an infraorder coordinate with the Asilomorpha, Tabanomorpha, Bibionomorpha, and so on. This concept of the group is adopted here, notwithstanding the fact that Hennig (1976*b*) again entertained the idea that the Eremoneura may be a monophyletic taxon. As shown by Krivosheina (1969), by Hennig (1972*b*), and by the following discussion, a sister-group relationship between the Muscomorpha and the Empidoidea cannot be convincingly demonstrated.

Griffiths (1972), who also treated the Eremoneura (= Muscomorpha *sensu* Hennig 1948*a*, 1952, 1954) as a monophyletic taxon, rejected for unassigned reasons the long-standing use of a subordinal category for the Brachycera (orthorrhaphous Brachycera + cyclorrhaphous Brachycera) and used the infraordinal category in its place. Thus, he called the suborder Brachycera the infraorder Brachycera. Then he substituted the category "superphalanx" (Kéler 1963) for the infraorder Eremoneura, i.e. he called the infraorder Eremoneura the superphalanx Eremoneura. Such changes in relative rank associated with well-known taxonomic categories are confusing and should, if possible, be avoided. Fortunately, neither Steyskal (1974) nor Chvála (1983) accepted Griffiths' proposals in these respects.

On the other hand, Hackman and Väisänen (1982) used the infraordinal name Muscomorpha as a replacement name for the suborder Brachycera, which proposal is likewise unnecessary and confusing. It arbitrarily repudiates the long-standing application of the name Brachycera, and inappropriately associates a new, inflated rank with the name Muscomorpha.

In Chapter 115, Woodley proposed yet another infraordinal concept for the Muscomorpha by assigning to it the Nemestrinoidea, the Asiloidea, the Empidoidea, and his "Muscoidea" (= Muscomorpha *sensu* J.F. McAlpine *et al.* 1981). Under this proposal the rank of the taxon containing all of the cyclorrhaphous Diptera, the members of which comprise fully half the order, is reduced to a single superfamily (Muscoidea) coordinate with the Empidoidea. I find these proposals unsatisfactory on both practical and cladistic grounds.

Rohdendorf (1964, 1974, 1977) frequently accorded high formal ranks to highly modified groups. For example, he ranked the phorid subfamily Termitoxeniinae as "infraorder Termitoxenomorpha," the family Braulidae as "infraorder Braulomorpha," and the families of the Hippoboscoidea as "infraorder Nycteribiomorpha," "infraorder Streblomorpha," "superfamily Glossinidea," and "superfamily Hippoboscidea." As pointed out by Griffiths (1972), many of Rohdendorf's higher groupings cannot be rationalized with the rankings of coordinate units recognized in a cladistic classification of the order. Consequently, they have not been accepted.

The superfamily names employed in the Muscomorpha are the same as those proposed in the outline of classification for the order, given in Volume 1 of the Manual (McAlpine *et al.* 1981, Ch. 1). In choosing these names, we followed the International Code of Zoological Nomenclature (Principle of Coordination), which specifies that a superfamily name should be based on the oldest family-group name among its components. I am very grateful to Dr. Curtis Sabrosky for assistance and expert advice in this respect. Most of the systematic concepts and terms employed throughout this chapter are explained and defined by Holmes (1980); the term monophyletic (*sensu* Hennig 1966*a*) is used in preference to holophyletic (*sensu* Ashlock 1971).

Most of the characters mentioned throughout this chapter are illustrated in the introductory and respective family chapters; to conserve space such figures are usually not cited in the text.

Ground-plan Characters of the Muscomorpha

Enumeration of the ground-plan characters of the Muscomorpha is a prerequisite for any discussion on the cladistics of the group. Such a list, including both plesiomorphic (primitive, diagnostic) and apomorphic (derived, constitutive) characters (Hennig 1981), is essential for defining the group, for documenting how it resembles its progenitors, and for showing how it has changed from them. The following list of characters was compiled from the literature and from observations by my colleagues and myself. I have relied heavily on the results of the investigations published in Volume 1 of the Manual (especially J. F. McAlpine, Chs. 2 and 4; Teskey, Chs. 3 and 5) for selecting and interpreting these characters. The most important papers from the standpoints of selecting and analyzing characters of the Muscomorpha and of synthesizing earlier literature are by Hennig (especially Hennig 1958, 1973, 1976*a*,*b*) and Griffiths (1972).

My primary goals are to adduce, first, as much as possible about the sister group of the Muscomorpha, presumably in the orthorrhaphous Brachycera, and, second, the sister-group relationships (cladistics) of the taxa that comprise the Muscomorpha. In Hennigian cladistics, rigorous arguments for sister-group relationships must be based on shared apomorphic character states (synapomorphies) that can be accepted as homologous. Consequently, to achieve the goals mentioned it is necessary to analyze the ground-plan characters in terms of plesiomorphy and apomorphy.

The designations plesiomorphic and apomorphic are relative terms, and the labeling of character states as one or the other involves judgements based on knowledge of characters as they occur both inside (in-group character analysis) and outside (out-group character analysis) the taxon concerned. For example, an orbicular flagellum comprised of four discrete flagellomeres (composite first flagellomere and a three-articled arista) is a plesiomorphic character with respect to the ground plan of the Muscomorpha, whereas it is clearly an apomorphic one with respect to the ground plan of the orthorrhaphous Brachycera (flagellum consisting of eight discrete flagellomeres). Criteria that can be employed in forming such judgements, or in testing them, were discussed in general by Hennig (1966*a*, 1981), Stevens (1980), and Wiley (1981) and are not repeated here. The out-group comparison method of character analysis was fully dealt with by Watrous and Wheeler (1981).

It is assumed from the outset that the Muscomorpha stemmed from the lower (= orthorrhaphous) Brachycera, and that synapomorphies between the muscomorphan ground plan and certain members of the orthorrhaphous Brachycera comprise the main evidence for adducing the primordial sister group of the Muscomorpha. It is also assumed that the sister group of the Muscomorpha is a subgroup, rather than the whole, of the orthorrhaphous Brachycera, from which it follows that the latter is a paraphyletic group. But that problem relates primarily to classificatory ranking and does not affect the significance of the characters discussed. Thus, for present purposes, the general out-group that I have utilized for the Muscomorpha is the orthorrhaphous Brachycera as a whole, and the main thrust of my efforts is to reach conclusions as to which subgroup of the orthorrhaphous Brachycera is cladistically most closely related to the Muscomorpha.

The main subgroups of the orthorrhaphous Brachycera have been discussed in Chapter 115. That chapter was prepared after this one was written, and so I have not been able to deal with all the data and ideas presented there as fully as I would have liked; limitations in this respect are candidly recognized. However, I wish to specially acknowledge authors D. M. Wood and A. Borkent (Ch. 114) and N. E. Woodley (Ch. 115), for their many helpful suggestions and criticisms in connection with the introductory sections of this chapter.

The ground-plan characters of the Muscomorpha that I have selected as being useful are categorized as follows:

1. Character states that are judged to be *plesiomorphic* with respect to the ground plan of the Brachycera.

There can be no doubt that these conditions did not first arise within the Muscomorpha and, therefore, that they are primitive characters of the group. They cannot be used as indicators of close cladistic relationship, but because they are diagnostic of the Muscomorpha they should not be ignored. They do not contradict propinquity of descent and sometimes can play a supportive role in discussions on cladistic relationships. For example, it seems likely that the primitive sister group of the Muscomorpha would possess such plesiomorphic characters as a holoptic head in the male, ocelli present in both sexes, and three spermathecae in the female. Thus, a taxon without these characters in its ground plan can probably be excluded as a likely candidate for the primitive sister group of the Muscomorpha.

2. Character states that are judged to be *autapomorphic* for the Muscomorpha (synapomorphies of the included taxa), with respect to the ground plan of the Brachycera.

These conditions arose within the Muscomorpha. Because they are present in its ground plan and are not shared with any other group, they comprise the best evidence for the monophyly of the Muscomorpha but reveal little or nothing about its sister group.

3. Character states that are judged to be *apomorphic* with respect to the ground plan of the Brachycera, but which are known also to occur in certain taxa within the orthorrhaphous Brachycera.

These characters provide the best evidence available for adducing the primitive sister group of the Muscomorpha because they represent possibilities for synapomorphies between the Muscomorpha and its potential sister group. It is frequently difficult to determine whether these similarities are, in fact, synapomorphies or whether they are convergences that have arisen independently through parallel evolution or homoplasy.

This kind of character analysis serves also as a basis for recognizing the polarity of changes that have occurred within the Muscomorpha, a prerequisite for working out the cladistics of the component taxa.

Plesiomorphic characters

- *Larva. Mandibles parallel to each other, operating in a vertical plane (Teskey Ch. 3)*

Vertical orientation of the mandibles is a ground-plan condition of the Brachycera as a whole, and, although it is an important diagnostic character, its occurrence in the Muscomorpha reveals nothing about the precise ancestry of that group within the orthorrhaphous Brachycera.

- *Metacephalic rods undeveloped*

Metacephalic rods originate from the mid-posterodorsal extremities of the head capsule and serve as struts for muscle attachment. Their presence and development coincide with the retraction of the head capsule, and two evolutionary processes appear to have led to their occurrence. First, they probably represent remnants of the head capsule that were retained during the desclerotization of the posterodorsal part of the cranium, as in Xylophagidae, and second, they became secondarily enlarged as internal apodemes that are hinged flexibly to the cranium, as in many asiloid families. The beginning of these rods possibly occurs in the larvae of Xylophagidae, and they reach full development in such families as the Therevidae, Scenopinidae, Mydidae, Apioceridae, Asilidae, Empididae, and Dolichopodidae (Teskey, Ch. 3). Because metacephalic rods are absent in the Stratiomyoidea and Tabanoidea, it is assumed that their absence is a plesiomorphic ground-plan character of the Brachycera and also of the Muscomorpha. In these groups the anterior tentorial arms and tentorial phragmata replaced the external head capsule (and metacephalic rods) as points for muscle attachment. There is no evidence that metacephalic rods were secondarily lost in the Muscomorpha, and their absence speaks against a sister-group relationship with asilomorphan groups, including the Empidoidea, in which they are so well developed.

- *Respiratory system amphipneustic, i.e. anterior and posterior spiracles, only, present*

Although apomorphic with respect to the ground plan of the Diptera, this type of respiratory system occurs most commonly throughout the order. It is characteristic of the Tanyderidae, Axymyiidae, most Psychodidae, Trichoceridae, Anisopodidae, Thaumaleidae (Nematocera), most orthorrhaphous Brachycera, and all Muscomorpha (Teskey, Ch. 3).

- *Adults. Tracheal system with air sacs in head, thorax, and abdomen (Faucheux 1971)*

Air sacs, especially in the thorax and abdomen, are best developed in groups that are strong fliers. Their distribution and relative development vary in both the Nematocera and orthorrhaphous Brachycera, but the presence of all three is very constant in the Muscomorpha. The air sacs in the head, thorax, and especially the abdomen of certain Stratiomyidae are more similar to those in the Muscomorpha than are those of any other orthorrhaphous Brachycera. It is also noteworthy that the Empididae and Dolichopodidae are the only families of the Brachycera in which thoracic air sacs are absent (Faucheux 1971).

- *Some hairs and bristles, especially of the head and thorax, pilose, i.e. palynophilic (Holloway 1976*b*)*

This character state is most evident in more densely haired members of such families as Stratiomyidae, Tabanidae, Nemestrinidae, Bombyliidae, and Asilidae, but it seems to be absent in the Empididae and Dolichopodidae. It occurs in all families of the Aschiza, but is very rare or absent in the Schizophora (genus *Telothyria* Wulp, family Tachinidae).

- *Head holoptic in male (J. F. McAlpine and Munroe 1968)*

- *Frons uniformly sclerotized, i.e. aschizometopous*

- *Bristles of frons and vertex weakly differentiated*

The propensity for certain differentiated bristles, such as ocellar and vertical bristles, was probably already present in the ground plan of the Muscomorpha.

- *Face broad, flat, and bare*

- *Antenna short and with nearly straight horizontal axis (Hennig 1971*a*)*

- *Scape, pedicel, and flagellum setulose*

- *Scape and pedicel more or less equal in length*

- *Arista separated from first flagellomere by a distinct articulation*

In the basic plan of the Brachycera, the flagellum consists of eight flagellomeres (Hennig 1967*a*, 1972*b*, 1973). Each successive flagellomere is clearly separated from the preceding one, and all of them are more or less equal in length and are gradually more slender distally. This type of flagellum still occurs in the Xylophagidae, Pantophthalmidae, nearly all Xylomyidae, some Stratiomyidae, some Pelecorhnchidae, some Tabanidae, some Rhagionidae, and some Vermileonidae. Within most of these families and in all other brachycerous families the number of flagellomeres is reduced and there is a sharp distinction between the larger basal portion (first flagellomere) and the more slender apical portion (stylus or arista). There is good evidence that these reductions (certainly apomorphic with respect to the ground plan of the

Brachycera) occurred repeatedly and independently, and in different ways, even within families. In many Stratiomyidae, for instance (Jorgensen and James 1968), the swollen basal part consists of up to five flagellomeres and the slender apical part consists of up to three flagellomeres. Within that family there is a gradual increase in the degree to which the basal flagellomeres are fused and a gradual reduction in the number of annulations present in both parts of the flagellum. The Sarginae and Pachygastrinae have the most advanced types of the family, i.e. the basal part is fused into a homogeneous mass (four flagellomeres in Sarginae, four of five flagellomeres in Pachygastrinae) and the slender apical part (one-segmented in Sarginae, two- or three-segmented in Pachygastrinae) forms an arista resembling that of the Muscomorpha. In the Tabanidae and Rhagionidae, also, there is a sharp distinction between the more swollen basal part of the flagellum and the more slender apical part. However, in both these families the basal part consists of flagellomere 1, only, and the slender apical part consists of up to seven flagellomeres. Thus, the composition of both parts of the flagellum differs from those mentioned in the Stratiomyidae, and they simply converge in overall form. Regardless of these basic differences, however, it is clear that a two-part condition of the flagellum, similar to the one retained in the ground plan of the Muscomorpha, arose more than once within the lower Brachycera. In the Muscomorpha, the swollen basal part probably consists of four fused flagellomeres as in some Pachygastrinae and Sarginae. (For further discussions of the first flagellomere and the arista, see "Apomorphic characters.")

- *Thoracic bristles weakly differentiated*

A propensity for certain thoracic hairs to become differentiated as bristles, such as notopleural bristles, supra-alar bristles, dorsocentral bristles, and scutellar bristles (as in the Anisopodidae), was probably carried over from the lower Diptera into the ground plan of the Muscomorpha.

- *C without humeral or subcostal costal breaks*
- *Sc complete, free from R_1*
- *M_2 present as a separate vein*
- *A_1 complete, reaching wing margin at least as a fold*
- *A_2 incomplete, not attaining wing margin even as a fold*
- *Crossvein sc-r present*
- *Cell dm complete*
- *Pulvilli well developed and pad-like*

Pad-like pulvilli are present in most Nematocera and are, therefore, considered to be a plesiomorphic ground-plan condition in both the orthorrhaphous and cyclorrhaphous Brachycera.

- *Pilose tibial spurs present (Hennig 1976*b*)*

A single ventro-apical bristle is present on all tibiae in many Phoridae (in addition to other pilose apical tibial bristles). From this, Hennig (1976*b*) concluded that the presence of true tibial spurs belongs to the ground plan of the Muscomorpha. Whether this bristle is, in fact, a true tibial spur (Edwards 1938, p. 6) or is simply a palynophilic bristle (as discussed) is uncertain. Woodley (Ch. 115, character 25) considers that true tibial spurs were lost in all Muscomorpha, including the Phoridae.

- *Coxopleural streak present*

This feature derives from the suture between the mesopleural katepimeron and the meron (J. F. McAlpine, Ch. 2), and its presence is a plesiomorphic ground-plan feature of the Muscomorpha.

- *Abdominal tergites and sternites 1–8 present and free in both sexes*

This statement is not meant to deny the existence of a certain amount of torsion and asymmetry in the male tergites and sternites of segments 7 and 8 (associated with circumversion of the terminalia) in the ground plan of the Muscomorpha.

- *Abdominal bristles absent or weakly differentiated*
- *Abdominal spiracles 1–5 in pleural membrane*
- *Male with tergite 8 and sternite 8 present*
- *Tergite 9 forming a dorsal epandrium (Hennig 1976*b*, J. F. McAlpine, Ch. 2)*

I regard Griffiths' (1972, 1981, 1983, 1984) views that tergite 9 (epandrium) is absent in all Muscomorpha as incorrect. The relative position and linkage of this sclerite, with tergite 8 anteriorly, with gonopodal portion of hypandrium laterally, and with elements of tergite 10 (surstyli) and dorsal elements of the proctiger (epiproct and cerci) posteriorly in primitive Aschiza (e.g. Platypezidae, Fig. 50.21) and in primitive Schizophora (e.g. Coelopidae, Fig. 82.5), identifies it as tergite 9. Thus, the epandrium of the Muscomorpha is, in fact, homologous with that of the orthorrhaphous Brachycera (e.g. Xylomyidae, Figs. 2.121–123 and Empididae, Figs. 2.124–126). I quite agree with Hennig (1976*b*) that, apart from tergite 9, no other tergal (tergite 10 or 11) or sternal (sternite 9) or gonocoxal elements are involved in the formation of the dorsal wall of the epandrium in the ground plan of the Muscomorpha.

- *Sternite 9 forming the anterior portion of a ventral hypandrium*

Discussion is given in conjunction with the following character.

- *Gonopods (one pair) joined proximally with hypandrium and laterally with the epandrium, free from each other medially, and each consisting of a gonocoxite and a gonostylus (J. F. McAlpine, Ch. 2)*

In the ground plan of the Muscomorpha, even though the bases of the gonocoxites are more or less fused with the hypandrium and each gonostylus is reduced and fused with the gonocoxite, the lateral articulation between the gonocoxite and the epandrium is maintained. These conditions are illustrated in certain Platypezidae (J. F. McAlpine, Ch. 2, Figs. 2.132–133; Griffiths 1972, Fig. 10). The reduced gonostylus is shown in my figures of a species of *Callomyia* Meigen; it is also present at the apex of the elongate gonocoxite ("pregonite" of Griffiths 1972) of *Plesioclythia agarici* (Willard), but it was omitted in Griffiths' figure. In most members of the Muscomorpha, however, a discrete gonostylus is not evident. I regard as wrong Griffiths' (1972, 1984) views that the originally biarticled processes, which I call gonopods, (his "pregonites") are "additional paraphyses" or "sensory structures," and that the gonocoxites migrated dorsally and became fused along their lateral margins to form a "periandrium." Much evidence throughout the orthorrhaphous Brachycera indicates that the gonopods are repeatedly reduced and incorporated with the hypandrium as in the Platypezidae. This same fusion and reduction occurs in the Empididae (J. F. McAlpine, Ch. 2, Figs. 2.124–126), notwithstanding Griffiths' (1983, 1984) statements to the contrary. In my opinion, the biarticled gonopods that occur in the ground plan of the Muscomorpha are homologous with those of the Nematocera and orthorrhaphous Brachycera, but, with circumversion and folding under of the terminalia, they have come to serve a different function in the Muscomorpha than they do in the lower Diptera.

- *Basal gonocoxal rims and apodemes extended posteromedially and fused with each other to form a dorsal bridge behind the aedeagus, thereby turning the hypandrial complex into a closed ring (Hennig 1976*b*)*

This condition occurs in many Nematocera, especially in the Bibionomorpha, and may be a ground-plan feature of the Diptera (Hennig 1976*b*); it is the usual condition throughout the orthorrhaphous Brachycera, whence it probably has been passed on into the ground plan of the Muscomorpha. In *Platypezina connexa* Boheman the dorsal bridge is very similar to that found in relatively primitive Brachycera, e.g. a species of *Rhagio* Fabricius (Hennig 1976*b*). Usually in the Muscomorpha, there is a line of weakening (hinge-like in many Lonchaeidae, for example) between the anterior, more plate-like portion of the hypandrium (sternite 9 in the strict sense) and the posterior part. The posterior part bears the gonopods and forms the so-called hypandrial arms and the dorsal bridge, which enclose the aedeagus and the parameres. This posterior part, exclusive of the aedeagus and parameres, is probably derived from the gonocoxites and their apodemes. In keeping with the structure of the male terminalia of the lower Diptera, the point of articulation between the epandrium (tergite 9) and its ventral counterpart is always with the posterior portion of the hypandrium, derived from the gonocoxite, rather than with the anterior portion derived from sternite 9.

- *Aedeagal guide present (J. F. McAlpine, Ch. 2)*

This external process of the hypandrium appears to be a fundamental part of the male terminalia of Diptera. The exogenous, internal, aedeagal apodeme of the Muscomorpha develops as an infolding from the base of the distal end of the aedeagal guide.

- *Parameres (one pair) free, articulating primarily with posterolateral base of aedeagus and inner posterior base of gonocoxite (J. F. McAlpine, Ch. 2)*

By definition, parameres are unsegmented processes attached to the basiphallus (Verhoeff 1893). Ontogenetically they are paraphallic lobes (paraphyses) arising as a secondary division of the primary phallic lobe of segment 10 (Christophers and Barraud 1926, Snodgrass 1957) and should not be confused with the gonopods, which arise on segment 9. True parameres occur regularly in most endopterygote orders and are clearly distinguishable in most families of the Diptera. Although long recognized and called parameres in the Nematocera and certain orthorrhaphous Brachycera, the neutral term, postgonites, is the name most commonly applied to them in the Muscomorpha. In my opinion, ample evidence is presented in this Manual to show that these processes are homologous throughout the Diptera. Plesiomorphically they are free from each other. Apomorphically they are fused with each other and the gonocoxal apodemes, forming a sheath around the aedeagus. This condition occurs in many Nematocera, especially in the Bibionomorpha, and in many orthorrhaphous Brachycera. In the ground plan of the Muscomorpha, they are separate and never form a sheath around the aedeagus.

- *Aedeagus comprised of a basiphallus and distiphallus*

In the ground plan of the Muscomorpha, these parts were probably rather similar to those in the platypezid, *Grossoseta pacifica* (Kessel) (Fig. 50.21), in which the distiphallus is relatively slender and flexible.

- *Sternite 10 (ventral epandrial sclerite) forming a simple ventral plate behind the hypandrium and aedeagus; articulating anteriorly with posterior margins of hypandrial arms and dorsal bridge, laterally with inner bases of surstyli, and posteriorly with hypoproct (J. F. McAlpine, Ch. 2)*

- *Proctiger comprised of a simple epiproct, a hypoproct, and a pair of cerci (J. F. McAlpine, Ch. 2)*

- *Female with tergite 8 and sternite 8 undivided*

- *Tergite 10 simple, without strong spines (acanthophorites)*

- *Three sclerotized spermathecae present*

It seems virtually certain that the immediate muscomorphan ancestor possessed three spermathecae, for this is the basic number present in the orthorrhaphous Brachycera and in all the major sections of the Muscomorpha, i.e. the Aschiza, Schizophora, Acalyptratae, and Calyptratae. This condition argues against, but does not refute the possibility of, a sister-group relationship between the Muscomorpha and Empidoidea, which has only one spermatheca in its ground plan.

- *Three separate spermathecal ducts opening independently into anterodorsal wall of genital chamber*

I disagree with Griffiths' (1984) statement that two of the three spermathecae share a common branching duct in the ground plan of the Muscomorpha. Three ducts with independent openings occur in a number of the most primitive families of the Nematocera, and this could well be the primitive pattern in the Diptera (J. A. Downes 1968, Hennig 1973). The same pattern, as exemplified in the Tabanidae (Mackerras 1955), in the Asilidae (Theodor 1976), and in the Bombyliidae (Theodor 1983), apparently occurs in the ground plan of the Brachycera. Similarly, it occurs in both the Platypezidae and the Syrphidae of the Aschiza (Sturtevant 1925–1926) and in the Calyptratae of the Schizophora. Fusion of these ducts at their bases to form one or two common ducts has occurred independently many times throughout the order, producing various branched conditions, including the one described by Griffiths that certainly applies to the ground plan of the Acalyptratae. However, it seems probable that three separate spermathecal ducts with independent openings into the oviduct is the ground-plan state for the Muscomorpha as a whole, and that the branched condition characteristic of the Acalyptratae is an apomorphic condition.

- *Two accessory glands opening into dorsal wall of uterus just posterior to openings of spermathecal ducts*

- *Ventral receptacle absent*

Autapomorphic characters

- *Larva. Subesophageal, thoracic, and abdominal ganglia fused to form a single ventral ganglion (Fraser 1959, M. J. Roberts 1969*a*, Krivosheina 1969)*

In the ground plan of the Brachycera the three pairs of cerebral ganglia are fused to form a single pair of cephalic ganglia, and the three pairs of subesophageal ganglia are also fused. All these head ganglia are displaced into the thorax in the lower Diptera, but, in the ground plan at least, fusion between the thoracic and abdominal ganglia as occurs in the Muscomorpha, has not occurred. The ground-plan state still exists in the Rhagionidae (M. J. Roberts 1969*a*), but the larvae of many orthorrhaphous Brachycera, especially in the Tabanidae, have a central nervous system that somewhat resembles that of the Muscomorpha. They differ, however, in showing a pronounced external segmentation, which does not occur in the Muscomorpha (M. J. Roberts 1969*a*, Krivosheina 1969). According to Krivosheina (1969) and Rozkošný (1982), the nervous system of the Stratiomyidae is remarkably concentrated, perhaps resembling that of the Muscomorpha more than the Tabanoidea. Nevertheless, the consolidation of all these ganglia into a unified, ventral, ganglial mass is considered to be an autapomorphic condition of the Muscomorpha.

- *Corpora cardiaca, corpora allata, and peritracheal glands (= prothoracic glands) coalesced to form a ring gland (= Weismann's ring) (Hennig 1973)*

In the ground plan of the Brachycera, the corpora allata, which are paired and separate in various Nematocera, are coalesced. In *Tabanus atratus* Fabricius, for example, they are joined to form a complex at the aorta, but the corpora cardiaca still remain separate (Hennig 1973). Knowledge of these brain-associated glands throughout the Diptera is still fragmentary, but it is assumed that their coalescence into a ring gland (= Weismann's gland) is an autapomorphic condition of the Muscomorpha. However, because this gland is the source of the hormone that inititates pupariation, a study of comparative conditions in the Stratiomyoidea, whose members also form puparia, would be interesting.

- *Three larval instars instead of four or more (Hennig 1973)*

Precise knowledge about the numbers of larval stages is fragmentary for most families of Diptera. Usually there are from four to six in both the Nematocera and orthorrhaphous Brachycera, but the number varies in both groups (Krivosheina 1969, Hennig 1973). There are seven to nine in Tabanidae (Pechuman and Teskey, Ch. 31), six to ten in the Stratiomyidae (Rozkošný 1982), six in some Asilidae (Krivosheina 1969), and five in Therevidae (Irwin and Lyneborg, Ch. 37). It seems very doubtful if the six larval stages reported for a species of *Lonchoptera* Meigen (Baud 1973) actually relate to six true larval instars (Hennig 1976*b*). The alleged fourth larval skin that is found in the muscomorphan puparium (Hinton 1949) is in reality only an inner layer of the endocuticle of the third-instar larva (Whitten 1957). Consequently, it is assumed that reduction to three larval instars, as occurs in all known members of the Muscomorpha, is an autapomorphic ground-plan condition of the infraorder.

- *External head capsule desclerotized to form a mostly membranous cephalic segment (Teskey, Ch. 3)*

Transitions from a relatively fully developed prognathous head capsule to a completely reduced acephalic state can be found in the larvae of orthorrhaphous Brachycera, depending largely on the type of feeding involved. The tendency toward reduction can be traced in various clades, leading to repeated membranization of the external elements of the head capsule. Transition from the massive head capsule of the Stratiomyidae, Tabanidae, and Rhagionidae, in which all the constitutive elements are expressed, is traceable through the abbreviated head capsule of the Xylophagidae, Asilidae, and Therevidae, in which frontal and lateral sclerites are still distinct, to the greatly reduced head capsule of the final instar of certain Stratiomyidae, e.g. all Sarginae and Hermetiinae (Rozkošný 1982, p. 17) and all stages of the Empidoidea. But no clear-cut transition is evident between any of these types and the completely reduced type in the ground plan of the Muscomorpha. Although it is now agreed that the muscomorphan head represents a logical further development of the features of the head capsule of larvae of the orthorrhaphous Brachycera, as pointed out by Cook (1949) and Ludwig (1949) and adopted by Teskey (Ch. 3), it is also generally agreed that the acephalic condition of muscomorphan larvae is an autapomorphic character state with respect to that found in all families of orthorrhaphous Brachycera.

- *Desclerotized anterior portion of larval head capsule invaginated to form an atrium (oral pocket) (Hennig 1973)*

The atrium is present in most Muscomorpha (said to be absent in the Lonchopteridae) between the functional mouth and the cibarium. It was created by the additional retraction of the head, including the mandibles within the cephalic lobe and the thorax, and it is considered to be a unique ground-plan character of the Muscomorpha (Teskey, Ch. 3).

- *Labrum forming a median labral tooth, retained as an independent structure in first instar only and adapted for breaking the egg (Hennig 1973)*

In the orthorrhaphous Brachycera, the labrum is rather slender, laterally compressed, and wedge-shaped in all instars and is used during feeding (Teskey, Ch. 3). Its peculiar form and occurrence as a sclerotized tooth in the first instar only (in both the Aschiza and Schizophora) is considered to be an autapomorphic condition in the ground plan of the Muscomorpha. In the Syrphidae it is used for rupturing the egg (Hartley 1963), and in the Tachinidae it is used for penetrating the cuticle of the host (Hennig 1973; Wood, Ch. 110), as well as for rupturing the egg.

- *Internal structures of head forming a peculiar cephalopharyngeal skeleton consisting of a tentopharyngeal sclerite, a hypopharyngeal sclerite, and a pair of modified mandibles*

The muscomorphan cephalopharyngeal skeleton differs radically from the larval mouth parts of more primitive Diptera. Just as in the case of retraction of the head capsule, there is no clear-cut series of connectant clades showing the gradual transformation of the cephalopharyngeal skeleton from homologous structures in families of the orthorrhaphous Brachycera. Consequently, most workers agree that the unique cephalopharyngeal skeleton, as an integrated assemblage, is an autapomorphic ground-plan feature of the Muscomorpha.

At the same time, however, it is generally agreed that the cephalopharyngeal skeleton of the Muscomorpha represents a logical further development of pre-existing structures present in the larvae of orthorrhaphous Brachycera (Hennig 1973; Teskey, Ch. 3), not a complex of completely new structures. Cook (1949), Ludwig (1949), and Krivosheina (1969) outlined the major evolutionary sequences in its development and interpreted them in terms of trends expressed in various more primitive representatives. Teskey (Ch. 3) summarized the main trends as follows: further reduction and complete desclerotization of all external elements of the maxillae and head capsule, together with the loss of all direct connection of the tentorial phragmata with the external body cuticle; complete fusion of the tentorial arms with the pharynx; and additional phragmatal growth posteriorly to the tentorial phragmata and to the fused tentorial arms and pharynx. Cook (1949) noted that the tentorial development in *Odontomyia alticola* James (Stratiomyidae) presented a different picture than in other orthorrhaphous Brachycera and found there an indication of a method by which the inner framework of the muscomorphan larval head may have developed. It is perhaps specially noteworthy that the process, whereby the tentorium becomes fused to the pharynx, begins in the Stratiomyidae and occurs only there and in the Muscomorpha. Krivosheina (1969) emphasized that the apparent similarity between the cephalic segment of the Empidoidea and Muscomorpha is not confirmed by the basic organizational features or the morphological details. She concluded on the basis of the entire complex of cephalic features of empidoid larvae that the Empidoidea is a specialized group the larvae of which bear only a purely superficial similarity to those of the Muscomorpha.

- *Antennal lobes largely fused with the maxillary lobes (Teskey, Ch. 3)*

In the orthorrhaphous Brachycera the antennae and the maxillary palpi are more closely approximated than in the Nematocera, but they still arise from separate sclerites on the head capsule. In the Muscomorpha, where the external head capsule is atrophied and the maxillae arise on a pair of membranous antennomaxillary lobes (comprising a bilobate cephalic segment), each antennal papilla is situated dorsally to each maxillary papilla (Teskey, Ch. 3). This condition has no close counterpart outside the Muscomorpha and is considered to be autapomorphic with respect to the ground plan of the orthorrhaphous Brachycera.

- *Mandible with a single condyle articulating with the hypopharyngeal sclerite*

In the Nematocera and most orthorrhaphous Brachycera there are two mandibular condyles, an inner epicondyle articulating with the tentorial phragma, and an outer hypocondyle articulating with the external head capsule (Cook 1949). It is assumed that fusion or loss of the hypocondyle has occurred in the Muscomorpha (Roberts 1971), and the socket joint has become shifted to the hypopharyngeal sclerite (which is a fusion product of the tentorial phragmata and the pharynx). The same condyle is lost in the Asiloidea and Empidoidea and is probably associated with the heavily sclerotized maxilla present in these larvae (Teskey, Ch. 3). Loss of the hypocondyle is apomorphic with respect to the ground plan of the Brachycera, and the more or less similar conditions in the Asilomorpha and Muscomorpha are probably the result of convergence (Krivosheina 1969).

- *Tentorial phragmata produced posteriorly and greatly expanded, forming dorsal and ventral cornua*

The bases of the U-shaped tentoropharyngeal sclerites are fused with the walls of the pharynx and represent, at least in part, the tentorial phragmata (Teskey, Ch. 3). The two dorsal cornua probably represent additional posterior expansions of these same elements, but the two ventral cornua apparently also include elements of the anterior tentorial arms. These conditions are unique to the Muscomorpha.

- *Hypopharyngeal sclerite forming a heavily sclerotized, H-shaped supporting structure for the cibarium (Teskey, Ch. 3)*

This sclerite is apparently a fusion product of several structures of which the hypopharynx is almost certainly a part (Teskey, Ch. 3). The lateral bars of the hypopharyngeal sclerite may be partially derived from anterior extensions of the points at which the tentorial phragmata articulate with the mandibles in the orthorrhaphous Brachycera (Hartley 1963). However, the heavily sclerotized crossbar between the lateral bars, which crosses the floor of the cibarium, appears to be at least partially derived from the labium (Ludwig 1949). In the lower Diptera, the salivary duct always opens into the cibarium between the base of the labium and the hypopharynx, and, because this duct enters just behind the transverse bar in muscomorphan larvae, it is assumed that the transverse bar is at least partially derived from the labium or hypopharynx or both (Teskey, Ch. 3). In any event, the peculiar, H-shaped form of the hypopharyngeal sclerite is an autapomorphic feature with respect to the ground plan of the orthorrhaphous Brachycera.

- *Anterior spiracles prominently situated on a spiracular stalk (Teskey, Ch. 3)*

Throughout the Diptera the anterior spiracles differ from the posterior spiracles, not only in structure, but probably also in function; in general, the posterior ones are more strongly developed in all forms except those with hemipneustic or propneustic systems (Keilin 1944). The strongly developed, distinctly stalked condition of the anterior spiracle in the Muscomorpha seems to be a more or less unique ground-plan feature. By way of contrast, the anterior spiracles of the Empidoidea are very small or vestigial (Peterson 1951).

- *Posterior spiracular plate with branching spiracular hairs associated with hypodermic spiracular glands that secrete a hydrofuge substance (Hennig 1973; Teskey, Ch. 3)*

The occurrence and structure of these hairs is rather fragmentary. According to Teskey (Chs. 3, 5), they occur in at least some members of the Aschiza, e.g. Syrphidae (Fig. 5.20), and in most of the Schizophora (Figs. 5.39, 5.70, 5.78, 5.85–87, 5.95, and 5.104). Normally four are present on each spiracular plate. Their development is evidently related to the life of the larvae in moist environments (Hennig 1973). Although spiracular glands are present on functional spiracles of all dipterous larvae (Keilin 1944), the occurrence of these peculiar, branched hairs appears to be unique to the Muscomorpha.

- *Head without an ecdysial suture*

A distinct ecdysial suture is present in all dipterous larvae with well-developed head capsules. Its absence in the Muscomorpha is an autapomorpic condition associated with the desclerotization and invagination of the head.

- *Pupa. Anterior thoracic spiracle of pupa forming a prothoracic, pupal respiratory horn that penetrates the puparium in the first abdominal tergite*

When a puparium is formed from the last larval skin, it retains the more or less completely evaginated anterior larval spiracles, but these probably become nonfunctional. The pupa is smaller than the puparium and the space between the two cuticles, especially in the anterior region, is filled with air (Keilin 1944). The pupa, itself, develops a pair of functional anterior pupal spiracles, which, for at least the initial stages of pupation, are enclosed within the puparium. In all the Aschiza except the Platypezidae, in practically all the Calyptratae, and in a few Acalyptratae the outer walls of the anterior pupal spiracles and their felt chambers are produced into a pair of sclerotized horns (prothoracic pupal respiratory horns) covered with spiracular papillae. Both these horns are then thrust out through a pair of small apertures in abdominal tergite 1 of the puparium (for review see Keilen 1944). This phenomenon appears to be unique to the Muscomorpha, and, because it is present in the ground plans of all three main divisions of the infraorder, it is assumed to be an autapomorphic condition with respect to the ground plan of the Brachycera. Excluding the prothoracic pupal respiratory horns that occur in many aquatic Nematocera, the closest approach to the condition found in the Muscomorpha occurs in the Stratiomyidae. Similar protrusive, pupal respiratory horns, which originate from the abdominal spiracles, are protruded through abdominal tergites 1–6 of the puparium (James 1957; Rozkošný 1982, 1983).

- *Adult. Male with segment 9 and following parts of terminalia circumverted, i.e. rotated clockwise through 360° around the long axis of the abdomen; main internal ducts of the genital, nervous, and tracheal systems twisted around the hind gut*

This phenomenon (for review see Griffiths 1972) is one of the most outstanding autapomorphies in the ground plan of the Muscomorpha. It enables the males to store their terminalia in a protected position beneath the abdomen and permits both sexes to mate in an upright, undirectional position (J. F. McAlpine, Ch. 2). Although rotation of up to 180° (inversion of the terminalia) has occurred repeatedly in the Nematocera and the orthorrhaphous Brachycera (for review see J. F. McAlpine, Ch. 2), true circumversion is unique to the Muscomorpha. In the ground plan, only the first 180° rotation takes place within the puparium, as in the Platypezidae, and the additional 180° takes place during the teneral state immediately following emergence (for review see Griffiths 1972). In such cases, circumversion is reversible to an inverse position, thereby permitting males and females of some families, including Platypezidae and Lonchaeidae, to assume an upright, tail-to-tail orientation during the final stage of mating. In most cases, however, especially in the Schizophora, it appears that the ontogenetic process of circumversion is completed at an earlier stage in relation to the overall ontogeny of the animal, and a fixed 360° rotation occurs within the puparium (Griffiths 1972).

- *Terminalia folded ventrally and anteriorly (ventroflexed) when at rest, so that their ventral surface is in close contact with the ventral surface of the pregenital segments*

Hennig (1976*b*) identified this jack-knife-like folding of the male terminalia as a modified ground-plan feature of the Muscomorpha, along with circumversion. How closely it is linked with circumversion is uncertain. In the basic pattern of the Diptera, the male terminalia were probably relatively unflexed when at rest (J. F. McAlpine, Ch. 2). In many Nematocera and in most orthorrhaphous Brachycera, including the Empididae, they are usually carried in a more or less dorsoflexed position. Except for an isolated case in the Mycetophilidae (Vockeroth, Ch. 14, p. 226), the closest approach to the ventroflexed condition of the Muscomorpha that I have found among the orthorrhaphous Brachycera is in the Xylomyidae and some Stratiomyidae. According to Woodley (personal commun.), it evolved independently in these two families for no flexion of the male terminalia has been observed in any members of the three most primitive subfamilies of the Stratiomyidae. Where flexion does occur in both families, tergite 8 is greatly reduced and the terminalia are strongly downwardly and forwardly directed, but they are not so tightly folded against the ventral surface of the much-reduced pregenital sternites (Ch. 2, Figs. 2.121–122). Probably the tightness of the folding in the Muscomorpha comes as a result of physical tension created by circumversion, for any flexible cylindrical article such as a tube or rope that is twisted through 360° will always fold tightly forward under itself. This jack-knife condition is certainly autapomorphic for the Muscomorpha.

- *Segment 8 inverted, i.e. rotated clockwise 180° around the long axis of the abdomen; both tergite and sternite asymmetric; sternite relatively large, mainly in a dorsal position, but connected dexteroventrally to hypandrium; tergite reduced to a narrow mainly ventral band, but connected sinistrodorsally with epandrium*

This peculiarly twisted ground-plan condition is clearly evident in the platypezid, *Grossoseta pacifica* (J. F. McAlpine 1967, Fig. 16; Kessel, Ch. 50, Fig. 50.21). No truly transitional patterns are known between this condition and those found in members of the orthorrhaphous Brachycera that show partial rotation. Superficial resemblances of these sclerites to homologous ones in some Empididae where some twisting occurs, and in the Dolichopodidae, where sternite 8 occupies a left lateral position, probably result from independent evolution.

- *Segment 7 with tergite and sternite asymmetric and rotated clockwise about 90° around the long axis of the abdomen*

This condition, as in the case of the previous character, is also unusually clearly evident in *G. pacifica* (see same figures). Tergite 7, which is connected to tergite 6 anterodorsally and to tergite 8 posteroventrally, is slender and corkscrew-like. Sternite 7, which is connected to sternite 6 anterolaterally (in left side) and diagonally to sternite 8 dorsolaterally, is relatively large and plate-like. This conformation and arrangement is not found in the orthorrhaphous Brachycera and is considered to be an autapomorphic ground-plan feature of the Muscomorpha.

- *Epandrial and hypandrial complexes in two, instead of three, horizontal planes (Hennig 1976*b*)*

In the ground plan of the Diptera and throughout the Nematocera and orthorrhaphous Brachycera, the male genital complex is arranged in three horizontal planes, i.e. a primarily tergal plane, a pleural plane, and a sternal plane. The dorsal plane is composed mainly of tergites 9, segment 10, and the proctiger. The median plane is composed of the biarticled gonopods (which are fundamentally pleural in origin and serve as genital claspers), the aedeagus, and the parameres. The ventral plane is composed mainly of sternite 9. In the Muscomorpha, one of the most obvious changes, apparently associated with circumversion and a change in the function of the gonopods, is the reduction in the size of the gonopods and the change in their position in relation to sternite 9 (hypandrium). Whereas they occupy a pleural position above sternite 9 in the lower Diptera, they have come to lie posteriorly to sternite 9 and in almost the same horizontal plane in the Muscomorpha. In the Muscomorpha, the gonocoxal apodemes and adjoining rims of the gonocoxae are fused to form a connecting bridge behind the aedeagus and the parameres, homologous with the dorsal bridge of the Bibionomorpha and Brachycera (see discussion of gonopods under "Plesiomorphic characters"), but it lies in almost the same horizontal plane as the anterior part of the hypandrium (sternite 9), rather than dorsal to it.

Hennig (1976*b*) was of the opinion that only two planes could be distinguished in the Empidoidea (his Empidiformia), as in the Muscomorpha, but my investigations (J. F. McAlpine 1967; J. F. McAlpine, Ch. 2) indicate that there is no real difference between the ground-plan condition of this character in the Empidoidea and that of the remainder of the orthorrhaphous Brachycera. The three-plane arrangement is evident, for example, in *Gloma luctuosa* Melander (Ch. 2, Figs. 2.124–126), an empidid species in which both the gonocoxites and the gonostyli are more strongly developed than in most members of that family. The basal arms of the dorsal bridge (formed primarily from the gonocoxal apodemes) are at right-angles to the plane of sternite 9, and the dorsal bridge lies in a plane well above that of sternite 9 (see J. F. McAlpine 1967, Fig. 22). Therefore, I consider the two-plane condition, as exemplified by the Platypezidae, to be an autapomorphic ground-plan feature of the Muscomorpha.

- *Aedeagus slender with its apex directed anteroventrally*

The basic condition in the Diptera is for the aedeagus to be directed posterodorsally, which appears also to be the ground plan and prevalent condition in the orthorrhaphous Brachycera including the Empididae. The Xylomyidae, in which the terminalia are ventroflexed and usually twisted about 90°, is an exception. Also in the Dolichopodidae, in which the terminalia are folded anterolaterally against sternite 8 (J. F. McAlpine, Ch. 2), the aedeagus is often, if not always, anteroventrally directed. As both these latter cases were attained by means of different processes than in the Muscomorpha, the muscomorphan condition is considered to be autapomorphic. I agree with Griffiths (1972, p. 59) that the aedeagus was probably slender in the ground plan of the Muscomorpha, but I disagree with his assumption that it was probably upcurved, i.e. recurved so as to be directed posteriorly. As evidence for his belief he cited figures of conditions in *Plesioclythia agarici* (Platypezidae), *Lamprolonchaea smaragdi* (Walker) (= *aurea* (Macquart)) (Lonchaeidae), and *Cryptochetum nipponense* Tokunaga (Cryptochetidae). For the last mentioned species, he mistook the female terminalia for the male terminalia, so that evidence must be rejected. The other two species probably reflect relatively derived conditions in both families. The ground-plan condition for the Platypezidae, and probably for the Muscomorpha as a whole, is probably more as in the primitive platypezid, *Grossoseta pacifica* (Kessel, Ch. 50, Fig. 50.21), and for the Lonchaeidae as in the primitive lonchaeid, *Protearomyia nigra* (Meigen), (J. F. McAlpine 1983, Fig. 14). In both cases the aedeagus is slender and directed anteroventrally.

- *Aedeagus more or less flexible and capable of swinging through a wide arc*

Griffiths (1972) was probably the first to draw attention to this peculiar feature. When the male terminalia are circumverted and folded anteroventrally, such an ability is a physical necessity for copulation, especially in those primitive forms, such as the Platypezidae and Lonchaeidae, that may assume an upright, tail-to-tail position during coitus. Griffiths included as a contingent factor, the articulation between the aedeagal apodeme (see next character) and the basiphallus, but my investigations indicate that an articulation exists at this point wherever there is such an aedeagal apodeme, i.e. throughout the Muscomorpha, even in those members that have a completely rigid aedeagus, e.g. many Lonchaeidae and Chamaemyiidae. Probably the capability to swing the aedeagus through a wide arc rests mainly on the flexibility of the distiphallus. In the more primitive state, the entire distiphallus is slender and flexible as in *G. pacifica*; in more advanced states where the distiphallus becomes at least partially secondarily sclerotized, as in most members of the Muscomorpha including the Syrphoidea, the Ephydroidea, and the Calyptratae, there remains a flexible hinge as it were, between the basiphallus and the distiphallus. Where flexibility of the distiphallus has been entirely lost, as in *P. agarica* (Platypezidae), many Lonchaeidae, and most Chamaemyiidae, the distiphallus is either reduced to a virtual pore, as in species of *Dasiops* Rondani (Lonchaeidae), or it is secondarily recurved posteriorly. It seems likely that a wholly flexible distiphallus, as exemplified in *G. pacifica*, is probably the ground-plan state for the Muscomorpha and is an autapomorphic condition of the clade.

- *Aedeagal apodeme developed as an ingrowth from the posterior base of the aedeagal guide*

It is perhaps unfortunate that the term aedeagal apodeme (phallopodeme) was applied to this unique ground-plan feature of the Muscomorpha, because the structure involved is not an apodeme of the aedeagus and it is not homologous in its origin or in its main function with the structures to which the same name is applied in the Nematocera and orthorrhaphous Brachycera. Much speculation exists concerning its origins and homologies in the Muscomorpha (for detailed analysis see Hennig 1976*b*), but no completely satisfactory interpretation has yet been proposed.

In the Nematocera, the term aedeagal apodeme, or its equivalent, is usually applied to the ejaculatory apodeme which is usually solidly united with the internal walls of the basiphallus. Similarly, in the orthorrhaphous Brachycera, it is often applied to the ejaculatory apodeme, which, as in the Nematocera, is usually fused with the internal wall of the basiphallus. Sometimes, however, the same name is applied to an extension of the external walls of the basiphallus (Figs. 32.6, 33.14, 37.17; see also Rozkošný 1982, Fig. 3). In the ground plans of the Nematocera and orthorrhaphous Brachycera, the apodemes in question arise from, and are solidly interconnected with, the aedeagus. The muscles inserted on them serve mainly as motor muscles for the sperm pump, but, in some cases, they may also serve for certain movements of the aedeagus (Hennig 1976*b*).

In the Muscomorpha, the so-called aedeagal apodeme develops as an ingrowth from the posterior base of a strap-like sclerital connection or infolding of the integument that extends from the median line of the hypandrium to the base of the aedeagal apodeme, i.e. from the posterior base of the aedeagal guide (see J. F. McAlpine 1967, Fig. 18). Hennig (1958, p. 539; 1973, p. 218) was the first to note this peculiar connection between the hypandrium and the aedeagal apodeme in the Muscomorpha. He emphasized that it had nothing to do with the gonopods (pregonites), and that the sole function of the aedeagal apodeme in the Muscomorpha is to position the aedeagus (as opposed to having an ejaculatory function in the lower Diptera). Later, Hennig (1976*b*) listed the same feature as one of two important ways in which the Muscomorpha differs in its ground plan from all other Diptera, including the Empidoidea. Here he emphasized that, in most cases, there is a distinct joint between the aedeagal apodeme and the basiphallus. He also pointed out correctly that the axis of the aedeagal apodeme in the Muscomorpha forms a sharp dorsoventral angle with that of the aedeagus.

Hennig (1976*b*) concluded that the possession of a separate sperm pump independent of the aedeagal apodeme, and the presence of a striking, at least partially sclerotized, connection between the aedeagal apodeme and the hypandrium were probably interrelated features. He proposed three possible alternatives for interpreting these features:

1. The ejaculatory apodeme in association with the sperm pump of the Muscomorpha is homologous with the aedeagal apodeme of the Nematocera, the orthorrhaphous Brachycera, including the Empidoidea, and the so-called aedeagal apodeme of the Muscomorpha represents a neomorphic structure of the group.
2. The aedeagal apodeme of the Muscomorpha is homologous with that of the remaining Diptera, and the ejaculatory apodeme in association with the sperm pump of the Muscomorpha is a neomorphic structure in that group.
3. The aedeagal apodeme and the ejaculatory apodeme in association with the sperm pump have arisen through the splitting of a uniform structure, which belongs to the ground plan of the Diptera and still to that of the Eromoneura, and each of the products of that splitting has taken over one of the originally combined functions of the former aedeagal apodeme.

I believe that the first of these alternatives is most likely correct, but Hennig favored the third alternative.

Griffths (1972), citing Schräder (1927) and Black (1966), was probably correct in accepting that ontogenetically the aedeagal apodeme of the Muscomorpha develops as an ingrowth of the integument at (not from) the base of the aedeagus, but his conclusion that the presence of any sclerotized link between the aedeagal apodeme and the hypandrium or body wall is an apomorphic condition with respect to the ground plan of the Muscomorpha, is probably wrong. I agree with three other statements made by the same author (Griffiths 1981): first, "the aedeagal apodeme of the Cyclorrhapha is an exogenous apodeme, and so cannot be homologous with the ejaculatory apodeme of the Orthogenya and other (lower) Diptera, since the latter is endophallic in origin"; second, "it could not have arisen by the splitting process suggested by Hennig"; and third, "the unpaired apodemes (of the Nematocera and orthorrhaphous Brachycera) that are involved in the ejaculatory function are homologous with one another and with the ejaculatory apodeme of the Cyclorrhapha (= Muscomorpha)." However, Griffiths' indication here and his definite statement later (Griffiths 1984) that the unpaired aedeagal apodeme of the Muscomorpha is a fusion product of the paired gonocoxal apodemes can scarcely be taken seriously within the context of his periandrial theory. How can one rationalize a concept, which holds on the one hand, that the gonocoxae migrate in opposite directions to the dorsum and fuse along their outer margins to form a periandrium, and on the other hand, that the endogenous apodemes of these gonocoxae fuse along their midlines to form a median ventral apodeme attached to the anterior margin of the basiphallus?

All the data amassed during my studies on the families of the Muscomorpha support the opinion that the aedeagal apodeme in this infraorder develops as an ingrowth of the hypandrial integument at the distal end of the aedeagal guide.

- *Aedeagal apodeme rod-like (cuneiform), attached to the body wall only at the point where it arises (see above)*

- *Longitudinal axis of aedeagal apodeme forming a sharp dorsoventral angle with the longitudinal axis of the aedeagus (see above)*

- *A membranous articulation present between aedeagal apodeme and basiphallus (see above)*

- *Female with sternite 9 mostly membranized, furca atrophied*

The formation of a genital chamber, which in the Diptera opens externally between sternites 8 and 9, involves the invagination of the ventral abdominal region immediately behind sternite 8. In the process of invagination sternite 9 is carried forward and inward above sternite 8, so that the external ventral surface forms the roof of the genital chamber (Bonhag 1951, Irwin 1976). In the ground plans of both the Nematocera and the orthorrhaphous Brachycera, the posterolateral margins of sternite 9 have internal foldings that usually are fused anteromedially to form a more heavily sclerotized, usually Y-shaped furca (genital fork) for muscle attachments. With the progressive membranization of the main plate of sternite 9, it is often reduced to little else than the furca itself. Nagatomi and Iwata (1976, 1978) showed that a well-developed furca is present in most of the families of the orthorrhaphous Brachycera, but they did not recognize it as a derivative of sternite 9. Nagatomi's (1982) conclusions, that sternite 9 is fused with sternite 8 and that the furca is an independent structure, seem incorrect. The evidence presented by both Bonhag (1951) and Irwin (1976) leaves little room for doubt about its origins from sternite 9.

In the Muscomorpha, reduction of sternite 9 appears to have proceeded to the extent that the furca is no longer present as a discrete structure, which would seem to be an autapomorphic feature of the infraorder.

Apomorphic characters

- *Larva. Saprophagous, feeding on "particles" such as bacteria, fungi, yeasts, and algae, in decaying organic matter; mouthparts adapted for filter-feeding (Dowding 1967, Roberts 1971, Ferrar 1987)*

Amongst endopterygote larvae in general, and probably in the ground plan of the orthorrhaphous Brachycera, larvae are primarily predacious on invertebrates as is usually exemplified by the Tabanoidea, Asiloidea, and Empidoidea (M. J. Roberts 1969*a*). In general, the larval mouthparts are adapted for carnivorous feeding. The Stratiomyoidea and Muscomorpha stand apart from the rest of the Brachycera not only in being primarily particle- or detritus-feeders but also in having mouthparts that are adapted for filter-feeding (Dowding 1967, Roberts 1971). Such adaptations are considered to be apomorphic with respect to the ground plan of the orthorrhaphous Brachycera, but it is uncertain whether their occurrences in the Stratiomyoidea and Muscomorpha represent synapomorphies, parallelisms, or convergences.

- *Head retracted into thorax*

Retraction of the larval head into the thoracic segments occurred repeatedly to varying degrees in the Nematocera and in the orthorrhaphous Brachycera. For example, it is often retracted for at least half its length in the Stratiomyidae; in the Tabanidae and Rhagionidae it may be fully retracted as in the Muscomorpha. Although the fully retracted condition in the Muscomorpha is extreme, it can scarcely be considered unique. At the same time, none of the retracted states in the orthorrhaphous Brachycera stands out clearly as a synapomorphy with that of the Muscomorpha.

- *Maxillae lost through reduction, desclerotization, and fusion with mandibles*

Fusion of the maxillae with the mandible is almost exclusively a brachyceran character, but the beginning of the process is evident in certain Culicidae (Cook 1949). In most orthorrhaphous Brachycera, the mandibular part of each maxillo-mandibular complex consists of a basal and an apical part (M. J. Roberts 1969*a*), and fusion that has occurred between the maxilla and the apical part of the mandible has not proceeded to the extent that questions arise about the actual homologies. In many families, distinct elements of the maxillae are modified to form a kind of lateral sheath for the apical part of the mandible. This modification is readily apparent in the Tabanidae, Mydidae, Asilidae, and so on (Teskey, Ch. 3). In the Stratiomyidae, however, fusion and membranization have proceeded to such a degree that, except for the maxillary palpi, the elements of the maxillae and the apical part of the mandible can scarcely be distinguished from each other (Cook 1949, M. J. Roberts 1969*a*, Rozkošný 1982). A progressively skeletonized appearance of the maxillae is also evident in larvae of the Vermileonidae, Empididae, and Dolichopodidae (Teskey Ch. 3). In these families, sclerotized parts of the maxillae can usually, if not always, be distinguished. In any event, whereas the complete loss of the maxillae through fusion with the mandible and progressive membranization is an apomorphic condition with respect to the ground plan of the orthorrhaphous Brachycera, it is not entirely unique to the Muscomorpha. Perhaps the closest counterpart of the muscomorphan condition is that found in the Stratiomyidae (see especially the penultimate instar of *Sargus bipunctatus* (Scopoli), Fig. 60.5 in Rozkošný 1982). Whether the similarity results from parallel or convergent evolution is uncertain.

- *Mandibles reduced, basal and apical parts solidly fused*

A two-part mandible is prevalent throughout the orthorrhaphous Brachycera and is not uncommon in the Nematocera (Krivosheina 1969, M. J. Roberts 1969*a*, Hennig 1973). The undivided condition in the Muscomorpha probably came about through fusion and reduction of one or both of these parts, and it certainly represents an apomorphic condition with respect to the ground plan of the Brachycera. Hennig (1973) suggested that the basal part was lost as a result of the reduction of the head capsule, but Teskey (Ch. 3) drew attention to a lateral pore leading to the lumen that is present on the basal mandibular sclerite of tabanid larvae, and which is also present on the enlarged triangular base of the mandible of the Muscomorpha. On the basis of this feature, plus the general similarity of the form of the base of the muscomorphan mandible to that of certain orthorrhaphous Brachycera, he concluded that the structures are homologous, and that the two parts of the mandible have become solidly fused in the Muscomorpha. There is perhaps an indication in the structure of the maxillo-mandibular complex of the penultimate larval instar of certain Stratiomyidae as to how this development might have occurred (see especially that of *S. bipunctatus*, Fig. 60.5 in Rozkošný 1982).

- *Mandibular muscles shifted from outer head wall to inner head framework (tentorial skeleton)*

In general, each mandible in larval Diptera is operated by several bundles of both dorsal abductor muscles and ventral adductor muscles. In the Nematocera and some orthorrhaphous Brachycera these muscles are inserted on the outer wall of the head, but concurrent with the reduction of the outer head wall in some orthorrhaphous Brachycera and in all the Muscomorpha, these muscles became shifted to the inner tentorial skeleton. In the Muscomorpha, the two bundles of abductor muscles and the three bundles of adductor muscles join the lateral surfaces of the ventral cornua, which are believed to be derived mainly from the anterior tentorial arms (Teskey, Ch. 3). This final state is clearly apomorphic with respect to the ground plan of the orthorrhaphous Brachycera, but the beginnings of the muscle shift are evident especially in the Stratiomyidae and Therevidae in which one of the three adductor muscle bundles originates on the apex of the anterior tentorial arm (Cook 1949; Teskey, Ch. 3).

- *Posterior arms of tentorium reduced, and anterior arms fused with the cibaro-pharyngeal wall*

There is a general tendency throughout the orthorrhaphous Brachycera toward reduction of the posterior tentorial arms (Cook 1949), but fusion of the anterior arms with the walls of the cibaro-pharyngeal complex occurs only in the Stratiomyidae and Muscomorpha. In both groups these conditions are apomorphic with respect to the ground plan of the orthorrhaphous Brachycera, but the significance of the structural similarities in these two groups is uncertain. Cook (1949) observed that the stratiomyid structure is one method by which the inner framework of the muscomorphan head may have developed, and Hartley (1963) noted that the stratiomyid structure resembles that of the Syrphidae more than that of the orthorrhaphous Brachycera.

- *Pharyngeal filter present, composed of longitudinal ridges and troughs in the ventral floor of the pharynx bridged by numerous convergent microfilaments*

Pharyngeal filters are biological sieves for concentrating suspensions of particulate matter such as bacteria, protozoa, algae, fungal spores, and yeasts from liquid or semiliquid media (Dowding 1967). In the Brachycera they occur only in the Stratiomyoidea and Muscomorpha, where they are special adaptations for the filter-feeding habits that occur in the ground plans of both these groups. It is generally agreed that the presence of a pharyngeal filter is an apomorphic character with respect to the ground plan of the orthorrhaphous Brachycera (M. J. Roberts 1969*a*). It is also generally accepted that it is a plesiomorphic character in the ground plan of the Muscomorpha, and that within the Muscomorpha it has undergone subsequent loss or radiative specializations in connection with carnivorous, phytophagous, parasitic, and highly evolved filter-feeding types (Roberts 1971, Hennig 1973). The cladistic significance of the occurrence and the similarities of filter-feeding apparati in the Stratiomyoidea and Muscomorpha is uncertain. They are usually interpreted as having resulted from convergent evolution. This interpretation may be correct for the details of the pestle-and-mortar-type pharyngeal mills that have developed in specialized branches of both groups (M. J. Roberts 1969*b*) and even in the nematocerous family, Ptychopteridae, but perhaps it is less applicable when extended to the basic design of pharyngeal filters in the Brachycera as a whole.

- *Anterior and posterior spiracles of type III structure (Keilin 1944)*

In the ground plan of the orthorrhaphous Brachycera, both anterior and posterior spiracles in mature larvae appear to be of type II structure. In this type the felt chamber of each successive instar is developed *around* the felt chamber of the previous instar, and the trachea of the previous instar is withdrawn through an opening in the middle of the new felt chamber and the new spiracular plate. In the basic plan the spiracular plate is symmetric with numerous peripheral spiracular openings and with a central ecdysial scar (Teskey, Ch. 3, Fig. 3.18D). In the ground plan of the Muscomorpha, both these spiracles are of type III structure in which the felt chamber of each successive instar is developed *beside* the felt chamber of the previous instar, and the trachea of the previous instar is withdrawn through a laterally placed hole in the base of the new felt chamber and the rim of the new spiracular plate; in the basic plan the spiracular plate is asymmetric with only three radially arranged openings and a laterally placed ecdysial scar (Teskey, Ch. 3, Fig. 3.19E and F).

Type III spiracles are probably apomorphic with respect to type II spiracles and are therefore apomorphic with respect to the ground plan of the Brachycera, but they are not unique to the Muscomorpha. Keilin (1944) illustrated their occurrence, especially with reference to the anterior spiracles in several nematocerous families, e.g. the Anisopodidae, Mycetophilidae, and Scatopsidae, and to the posterior spiracles of the Therevidae. Peterson (1951) shows that a type III condition also occurs both in the anterior spiracles of the Xylomyidae (*Solva* Walker) and several genera of the Stratiomyidae, and in the posterior spiracles of the Sciaridae and Scenopinidae. Further documentation of this condition throughout the orthorrhaphous Brachycera might provide interesting clues concerning the origin of the Muscomorpha.

- *Pupa. Pupation occurring within the hardened cuticle of the last larval instar, i.e. third larval skin forming a tough cocoon (puparium) for the delicate pupa. Pupation within the last larval skin, i.e. in a kind of cocoon called a puparium, which is unique to certain Diptera*

By this means an exceptional protective structure is produced by the larva to shield the pupa against adverse environmental elements. D. M. Wood (personal commun.) observed that almost all the Nematocera and the orthorrhaphous Brachycera (which do not form puparia) overwinter as larvae and have a pupal stage of short duration, whereas almost all the Muscomorpha (which do form puparia) overwinter as pupae and may spend many months and even years in the pupal stage. He suggested that perhaps the main adaptive advantage of pupariation to the Muscomorpha is to enable individuals to pass long periods safely, especially during seasons that are unsuitable for larval or adult activities such as winters, dry seasons, and wet seasons, in the quiescent, energy-conserving pupal stage.

In the Nematocera, pupariation in the last larval skin is found in all Perissommatidae, some Bibionidae, and all Scatopsidae (Colless 1962). An analogous condition occurs in some Cecidomyiidae, in which pupation occurs within the skin of the penultimate larval instar (Hennig 1973, Colless 1962). As is usual in the Nematocera, the pupae in all these instances are obtect, i.e. the appendages are fastened down to the body wall—a ground-plan feature of the Diptera (Hinton 1949). This same condition also applies in all the orthorrhaphous Brachycera with the possible exception of some Stratiomyidae (Hinton 1949). In all the Muscomorpha and possibly in some Stratiomyidae, the pupae are exarate, i.e. the appendages are free from body wall, a state which, in Diptera, is clearly derived from the obtect condition (Hinton 1949). A somewhat intermediate state is evident in the Xylomyidae. Another feature that is associated with these conditions is that, in the Muscomorpha and Stratiomyidae, the pupal cuticle is shed within the puparium during the process of emergence. This character also shows an intermediate condition in the Xylomyidae, in which the pupal exuviae is wedged in the opening of the puparium in a partially extruded condition (Rozkošný 1982).

If the cladistics of the Nematocera, as presented previously (Wood and Borkent, Ch. 114) are correct, the phenomenon of pupariation must have arisen independently several times in the Diptera. It seems to have occurred at least four times in the Nematocera, i.e. in the Perissommatidae, Bibionidae, Cecidomyiidae, and Scatopsidae. It also occurred at least once, if not twice, in the Brachycera, i.e. in the Stratiomyoidea and Muscomorpha. More work is needed to establish whether these last two cases represent evolutionary convergence or whether they were inherited from a common ancestor. Although certain details about pupariation in the Stratiomyoidea differ from those in the Muscomorpha (more than three larval instars involved, pupation probably occurs in spring rather than autumn, less radical change in form of puparium, and so on), there are distinct similarities, as discussed, and the whole process of pupariation in the Stratiomyoidea may reflect to some extent the evolution of the same phenomenon in the Muscomorpha (Jusbaschjanz 1916).

- *Pupa not motile, i.e. pupae without posteriorly directed abdominal spines (Hinton 1949)*

Primitively the pupae of the Nematocera and orthorrhaphous Brachycera are well armed with stiff, posteriorly directed spines. These spines enable them to wriggle out of their surrounding cocoon and substrate to an exposed site for emergence of the adult. Such spines are almost always lost in species that shed the pupal cuticle while still in their cocoon. Thus, they are absent on the pupa of many aquatic Nematocera, some Bibionidae, some Mycetophilidae, many Stratiomyidae, and all Muscomorpha. Loss of these spines is to be regarded as an apomorphic condition (Hinton 1949), and it is interesting to note that in the Brachycera it is shared by some Stratiomyidae and the Muscomorpha, but not by the Empidoidea and most other members of the Asilomorpha.

- *Body wall (epidermis) of adult renewed from imaginal discs (Hennig 1973)*

In many Nematocera, e.g. the Simuliidae and Culicidae, the body wall grows by cell division during the larval period and is passed on from the larva to the pupa to the adult (Hinton 1961). In the Muscomorpha, the larval epidermis is partially or entirely replaced in a process which spreads from imaginal discs (Hinton 1961, Bautz 1971, Evans 1967*a–c*). The entire adult integument, except for the abdomen, arises from 10 pairs of imaginal discs and a genital disc (Postlethwait and Schneiderman 1974). According to Jusbaschjanz (1916), a similar pattern exists in the Stratiomyidae, but comparable data on other families of the orthorrhaphous Brachycera are unknown.

- *Cleavage lines of puparium H-shaped, restricted to abdominal tergites 1 to 3*

Puparial cleavage lines are lines of weakness in the puparium that facilitate the escape of the adult from its special cocoon. They occur in definite preformed positions and show distinctive patterns throughout the Muscomorpha (see Brauns 1954). The relative lengths and positions of these lines depend to a large extent on the source of the pressure and the points at which pressure is applied on the interior surface of the puparium by the emerging adult.

On the frons of adults of most Schizophora is a well-developed, protrusible, membranous bladder or ptilinum. When the imago is ready to emerge from its puparium, the ptilinum is inflated by blood forced into it by strong contractions of the abdomen and thorax. Alternate inflations and deflations of the ptilinum eventually force the anterior end of the puparium to open along the preformed cleavage lines. Because the force is localized in the ptilinal area in these flies, the cleavage lines are shorter and more restricted than in more primitive forms. Also, the operculum lifts from the front and opens posterodorsally rather than vice versa. When the ptilinum is absent or weakly developed, as in the Aschiza, the internal pressure for splitting the puparium is less localized and a different opening pattern is evident.

Probably the pattern of cleavage lines and the method of opening in the aschizous family, Lonchopteridae, is the one that has remained closest to the ground plan of the Muscomorpha. The H-shaped cleavage lines consist of a median longitudinal seam through abdominal tergites 1–3 inclusive; at each end of this seam is a transverse seam but no horizontal lateral seams (see Brauns 1954, Fig. 45A). The basic pattern of the Phoridae is similar to that of the Lonchopteridae, except that the posterior transverse seam (on abdominal segment 3) is extended laterally, thence forward horizontally along the pleural wall on each side to a point just in front of the anterior transverse seam; at this latter point it suddenly bends ventrally. As a result, a dorsal lid or operculum, which includes abdominal segments 1–3 and the prothoracic pupal respiratory horns, may be pushed up when the adult emerges (see Brauns 1954, Figs. 53A, 53G). Neither of these families has a ptilinum. The emerging adult probably forces the puparium open by expanding the thorax and possibly, to a limited extent, by expanding the head.

In the next evolutionary advance, the median longitudinal cleavage line is lost and the two horizontal lateral lines are extended to the anterior end of the puparium. Various degrees of this condition are evident in the Syrphidae (see Brauns 1954, Figs. 46–50). This family has a rudimentary ptilinum (Hinton 1946), and the puparium, which is distinctly spheroid anteriorly, is forced open from internal pressure exerted primarily from the adult head. In all cases, however, the upper operculum includes portions of several abdominal tergites and the prothoracic, pupal, respiratory horns. The Platypezidae show a rather similar pattern, but the anterodorsal area of the puparium is distinctly flattened as far back as the posterior margin of abdominal tergite 3, and the prothoracic pupal respiratory horns are not protruded through the puparium.

In the Schizophora, the pattern of cleavage lines and the way the puparium is opened are more or less uniform. The transverse cleavage line, which appears to be derived from the primitive anterior transverse line, encircles the puparium in the vicinity of the anterior margin of abdominal segment 1; also, the two lateral horizontal seams extend anteriorly until they meet at the extreme anterior end of the puparium. A dorsal and a ventral operculum are thus formed, which are opened from the anterior end when the adult emerges. The upper operculum does not include any abdominal tergites, and the prothoracic, pupal, respiratory horns, when present, remain intact on abdominal tergite 3, which remains attached to the main body of the puparium. As already indicated, in the Schizophora the pressure for opening the puparium is provided by a well-developed ptilinum.

The pattern of cleavage lines and the way the puparium is forced open in the Stratiomyoidea is similar to that deduced for the ground plan of the Muscomorpha. Although it is sometimes stated that adults of the Stratiomyidae escape through a T-shaped slit (e.g. Rozkošný 1982), the cleavage lines are, in fact, H-shaped (as in the Lonchopteridae) in all examples I have seen (Beridinae, Sarginae, Stratiomyinae, Clitellariinae, and Pachygastrinae). The same is true for the Xylomyidae. The H-shaped cleavage lines shown for *Chiromyza rubriceps* (Macquart) (Hinton 1946, Fig. 36) are typical of both families. In the Stratiomyoidea, however, the prothoracic, pupal, respiratory horns do not penetrate the puparium; instead, abdominal, pupal, respiratory horns penetrate the puparium on abdominal tergites l-6 (Rozkošný 1982, 1983). Whether the similarities noted are synapomorphies, parallelism, or convergences is uncertain.

- *Adult. Clypeus reduced, U-shaped, separated from face by a relatively broad, frontoclypeal membrane*

A detailed comparative study to elucidate the contradictory and confusing array of views on the evolution of the clypeal region of the Diptera is needed (Hennig 1973). In mandibulate Nematocera and orthorrhaphous Brachycera, which presumably manifest the primitive state in the Diptera, the clypeus is large and occupies most of the mid-facial area. Even in most non-mandibulate forms in the lower Diptera, it is relatively large and closely united with the lower margin of the face. Sometimes, e.g. the Blephariceridae, Trichoceridae, and Xylophagidae, both its components, the proximal postclypeus and distal anteclypeus, are evident. In the Muscomorpha, however, it is usually reduced to a relatively narrow, stirrup-shaped sclerite, which is sharply separated from the anterior margin of the face by a fairly broad frontoclypeal membrane. This feature is probably an adaptive modification associated with the general mobility of the proboscis, and with the ability to withdraw it deep into the subcranial cavity. The most nearly equivalent condition that I have found is in the Stratiomyidae (certain Pachygastrinae and Sarginae), but, at least superficially, similar conditions also exist in some Bombyliidae, some Empididae, and probably in other families. There is little question that the condition in the Muscomorpha is apomorphic with respect to the ground plan of the Brachycera, but more work is needed to establish whether it is autapomorphic or synapomorphic with somewhat similar conditions in members of the orthorrhaphous Brachycera.

- *Occiput with a well-developed, median, occipital sclerite (J. F. McAlpine, Ch. 2)*

According to Hendel (1928) and Hennig (1958), this sclerite, which is marked off by internal dorsolateral thickenings, is indistinct or absent in the Nematocera, and its presence in many Brachycera constitutes an important diagnostic character. In the orthorrhaphous Brachycera it is well developed in the Xylophagidae, Stratiomyoidea, and Tabanoidea, but it is poorly developed or absent in the Nemestrinoidea, Asiloidea, and Empidoidea. The strong similarity of its structure in the Muscomorpha with that of the stratiomyoid–tabanoid condition would seem to indicate a closer cladistic relationship with one or other of these groups or their ancestor(s) than with the Nemestrinoidea, Asiloidea, Empidoidea, or their ancestors.

Certainly the presence of a well-developed, median, occipital sclerite is an apomorphic character with respect to the ground plan of the Nematocera. However, it is still uncertain whether it first appeared in the ground plan of the Brachycera as a whole, or only in certain clades of the Brachycera, i.e. the Xylophagidae, Tabanoidea, Stratiomyoidea, and Muscomorpha. If the former, its presence in the Muscomorpha would represent a symplesiomorphic condition carried over from the ground plan of the Brachycera. If the latter, it may represent a synapomorphy between a clade of the orthorrhaphous Brachycera in which it occurs and the Muscomorpha. More research is required on all aspects of the character.

- *Apex of pedicel dorsolaterally notched to receive the base of the first flagellomere, and with a membranous, cone-shaped, distal condyle that is deeply inserted into the base of the first flagellomere (Hennig 1971*a*)

That this suite of conditions is a ground-plan feature of the Muscomorpha is proven by its occurrence in primitive members of both the Aschiza and Schizophora. It is well documented in the Ironomyiidae (J. F. McAlpine 1967), the Sciadoceridae (Tonnoir 1916), the Phoridae (Peterson, Ch. 51), the Syrphidae (Hennig 1971*a*), and the Pipunculidae (personal observation). In the Aschiza, the antennae of the Platypezidae (including *Opetia* Meigen) and the Lonchopteridae differ from the basic muscomorphan plan in that the cone-shaped condyle is not deeply inserted into the base of the first flagellomere. Nevertheless, the membranous cone is present and readily seen in macerated specimens. In the light of the more typical conditions that prevail in all the other families of the Aschiza and in all the Muscomorpha, the somewhat different conditions in the Platypezidae and Lonchopteridae may well be the result of secondary modifications.

The connection between the pedicel and the first flagellomere in the empidid subfamily, Ceratomerinae, is superficially similar to that in the Muscomorpha (see Chvála 1983, Fig. 17), but maceration and dissection reveals that they are quite different in basic structure. The invaginated portion of the pedicel arises from a projection on the inner apex of the pedicel and, in macerated specimens, is readily eversible through a relatively large opening on the inner base of the first flagellomere. Also, the antennae of the dolichopodid genus *Syntormon* Loew are somewhat similar in external appearance to conditions in the Ironomyiidae and Sciadoceridae, but maceration revealed them to be similar to that in the Ceratomerinae, except that no membranous pouch is inserted into the base of the first flagellomere. Chandler (1981), in comparing the conditions of the Platypezidae with typical conditions in the Empididae, referred to the platypezid type of antennae as "empididform antennae." In view of other unrelated evidence that mitigates against a close cladistic relationship between the Empidoidea and Muscomorpha, it seems likely that these similarities in the antennae are the result of homoplasy. Among the orthorrhaphous Brachycera, the Stratiomyidae show more transitional stages between the filiform antennal types of the Nematocera and the aristate types of the Muscomorpha than any other family (Jorgensen and James 1968). Outside the Muscomorpha, it is in the Stratiomyidae that the peculiar muscomorphan type of attachment between the pedicel and the first flagellomere is first encountered. It is a common feature in the subfamily Pachygastrinae, in which the basal part of the flagellomere is, in many cases, reduced to three or four fused flagellomeres, and the arista is commonly composed of three aristomeres (Jorgensen and James 1968). Occurrence of these conditions in the Stratiomyidae shows that they did not arise first in the Muscomorpha, and their striking similarity to conditions in the Muscomorpha indicates they are probably significant clues to the origins of the Muscomorpha.

- *Flagellum consisting of a composite first flagellomere and a three-segmented, pubescent, dorsoapical arista*

In the basic plan of the Brachycera, the flagellum probably consists of eight flagellomeres, which are all relatively similar to each other (Hennig 1967*a*, 1972*b*). This type of flagellum is found in representatives of only a few families, i.e. the Xylophagidae, Pantophthalmidae, Tabanidae, Vermileonidae, Xylomyidae, Stratiomyidae, Rhagionidae, and Pelecorhynchidae, all of which are often combined under the name Homoeodactyla (Hennig 1973). All the other brachycerous families, sometimes referred to as the Heterodactyla, have a reduced number of flagellomeres, i.e. never more than four and often even fewer. Similar reductions occur in most families of the Homoeodactyla, and there is firm evidence, especially in the Stratiomyidae (Jorgensen and James 1968), that these reductions have occurred independently and in different manners. For example, the fused basal part of the flagellum in the Pachygastrinae consists of three to five flagellomeres and the arista consists of two or three aristomeres, whereas in the Sarginae the basal part consists of four flagellomeres and the arista consists of a single aristomere (Jorgensen and James 1968).

Hennig (1972*b*) contributed much to our understanding of the origin and evolution of the arista throughout the Brachycera, but he said little or nothing about the composition of the first flagellomere (referred to by him as the funiculus). He believed that the sharp separation between the first flagellomere and the reduction of the arista to, at most, three aristomeres might be synapomorphic attributes that suggest the monophyly of the Heterodactyla. I believe that this assumption rests on weak grounds until the basic composition of the first flagellomere, as well as the arista, is established. Perhaps it consists of a different number in the Asilomorpha than in the Muscomorpha, as is the case with the arista.

I know of no study purporting to show the number of original flagellomeres that are embodied in the first flagellomere of the Asilomorpha or Muscomorpha. That it does, in fact, contain elements of at least three or four flagellomeres is perhaps evidenced by the distribution of microscopic sensory pits (Jorgensen and James 1968), which can be detected on most forms. In groups of the Stratiomyidae in which fusion of these flagellomeres has occurred, there is a strong tendency for these pits to become concentrated on the distal edges of flagellomeres 1–4. Thus, the alignment of the sensory pits reflects the original segmentation (Jorgensen and James 1968, Figs. 3D, E, F, H, I, K, L). Possibly the arrangement of sensory pits and other microstructures in the first flagellomeres of certain members of the Asilomorpha and Muscomorpha reflect the original segmentation involved. Unfortunately, insufficient comparative data are available to resolve this problem. Despite this, however, it is evident that the antennae of the Muscomorpha resemble those of certain members of the Stratiomyidae, especially in the subfamilies Pachygastrinae and Sarginae, and it seems plausible to assume that the stratiomyid antennae illustrate the kind of transition that occurred between the filiform type of the Nematocera and the aristate type of the Muscomorpha (Jorgensen and James 1968). Hennig's (1972*b*) findings, that in the ground plan of the Asilomorpha the number of aristomeres is two and that the apical segment is a relatively short, blunt, nonpubescent, hyaline stylus, are very significant. These conditions are usual in the Therevidae, Scenopinidae, Mydidae, Apioceridae, Asilidae, Empididae, and Dolichopodidae. In certain cases within a number of these families, the arista is reduced to one aristomere, but the clear terminal stylus is usually retained. In the Muscomorpha, the basic number of aristomeres is three. The terminal segment, which is the longest, is gradually tapered to a fine tip, and it is pubescent and of uniform texture throughout. On the basis of the three-segmented condition, the Muscomorpha are more primitive than the Asilomorpha, including the Empidoidea (Hennig 1970, 1973). The pubescent condition and the absence of a hyaline terminal stylus are probably also primitive conditions. It was mainly these differences between the aristae of the Empidoidea and Muscomorpha that caused Hennig (1972*b*) to renounce the idea that the Empidoidea are most closely related to the Muscomorpha. The closest conditions to those of the Muscomorpha are probably found in the Stratiomyidae, especially in the Pachygastrinae (Jorgensen and James 1968).

Hennig (1972*b*) noted that in some Nemestrinidae the arista consists of three separate aristomeres of which the last is the longest as in the Muscomorpha. However, in this family the last segment is usually endowed with a nonpubescent, hyaline, terminal stylus as in the Asilomorpha. Also, in some representatives, the arista is unsegmented and the terminal stylus is lost. According to Woodley (personal commun.), Greathead's (1967) statement that the arista of *Trichopsidea* Westwood consists of six segments is in error. Because the peculiar terminal stylus is present on the arista in the ground plan of the Nemestrinidae, it seems probable that the three-segmented condition here and in the Muscomorpha is the result of homoplasy.

- *Proboscis highly mobile, capable of being retracted far into the head capsule, adapted for lapping and sucking, and mandibles absent in both sexes*

In the basic plan of the Diptera, and in the ground plan of the Brachycera, the mouthparts are adapted for piercing and sucking (J. A. Downes 1958, Hennig 1973). This applies to the females only, most of which require a blood meal to provide the proteins needed to develop their eggs. In such forms the proboscis is relatively closely attached to the head capsule and fully exposed as, for example, in females of most Blephariceridae, Simuliidae, and Tabanidae. The proximal border of the proboscis lies between the postmentum and the prementum on the ventral side, and between the face and the clypeus on the dorsal side. But this basic design was retained in relatively few instances in both the Nematocera and the Brachycera. In the Nematocera it was retained in the ground plans of the Tanyderidae, Blephariceridae, Psychodidae, Dixidae, Chaoboridae, Culicidae, Simuliidae, Thaumaleidae, Ceratopogonidae, and Chironomidae, but it was lost within each of these on various occasions. In the Brachycera it was retained only in the Tabanoidea, and, within that group, it was lost repeatedly as the necessity for blood-feeding disappeared.

Proboscis forms that deviate from this basic pattern occur frequently throughout the order. Usually in groups in which the proboscis is greatly elongated, e.g. the Culicidae, some Mycetophilidae, some Tabanidae, the Acroceridae, the Nemestrinidae, the Bombyliidae, some Empididae, some Conopidae, and also in the Stomoxinae and Glossinidae, it is mainly the haustellum that is elongated. In other cases, however, the labella are also involved, e.g. in certain Conopidae, Tephritidae, Milichiidae, and Chloropidae, in which the proboscis is strongly geniculate. In still other cases, e.g. the Nymphomyiidae, several Acroceridae, and many Oestridae, the proboscis is entirely reduced. Placing these and other obvious deviations aside, however, it is evident that, throughout the Diptera, mouthparts became independently adapted for lapping and sucking on many different occasions. The important difference is that in this type of proboscis the mandibles are lost in both sexes, and that both the carbohydrate diet for normal energy needs and the protein-rich diet for ovarian development are obtained by a nonmandibulate type of feeding. Mandibles are missing in the ground plan of the Tipulomorpha, Bibionomorpha, Xylophagoidea, Stratiomyoidea, Nemestrinoidea, Asiloidea, Empidoidea, and Muscomorpha. This adaptation opened the way to a much broader range of protein sources, other than living animals, and is probably one of the main factors that led to the great diversification of the Diptera.

Although the derived form is almost certainly apomorphic with respect to the ground plan of the Brachycera, it could be autapomorphic for many lineages within the Brachycera, including the Muscomorpha. If it could be shown that the Muscomorpha have the same commonly evolved character state as another clade within the Brachycera it would be an important synapomorphy.

In the ground plan of the Muscomorpha, the prementum is separated from the head capsule wall (formed on the ventral side by a continuous hypostomal bridge) by a wide membranous area. The same membranous area extends over the dorsal side of the proboscis between the clypeus and the face, and between the clypeus and the labrum. Thus, the proboscis is divided into three, more or less separated parts, i.e. a large, membranous basiproboscis (rostrum), an intermediate basiproboscis (haustellum), and a distal distiproboscis (labella) (Hennig 1973). These modifications, especially in the rostral area, render the proboscis very flexible and enable it to be withdrawn deeply into an enlarged subcranial cavity. According to Hennig (1973), rostral structures, which in some respect resemble those of the Muscomorpha, are found in the Stratiomyidae, Bombyliidae, and Empidoidea. With regard to the Empididae, however, he stated that the proximal margin of the clypeus is firmly fused with the face, and that the clypeal membrane is present only between the distal margin of the clypeus and the labrum. More research is needed to establish whether the similarities between the mouthparts of the Stratiomyidae, Bombyliidae, and other brachycerous groups are truly homologous with those of the Muscomorpha.

- *Maxilla reduced to a slender lacinia and a one-segmented maxillary palpus*

In the basic pattern of the Diptera the maxilla consists of a cardo, a stipes, one endite comprising a slender, blade-like lacinia, and a five-segmented palpus (J. F. McAlpine, Ch. 2). As shown by Bonhag (1951) for the Tabanidae, the border between the cardo and the stipes is still in evidence in the ground plan of the Brachycera, but it is missing in most of the Diptera, including all the Muscomorpha (Hennig 1973). Unfortunately, the detailed morphology of the maxilla is still unclear for most groups so it is difficult to draw any meaningful comparisons. According to Hennig (1973) the paired stipites have shifted inwards in the Schizophora and in the Empidoidea, but in the Schizophora they are in contact with the labrum, whereas in the Empididae they are in contact with the labium.

A two-segmented palpus has been reported for most of the orthorrhaphous Brachycera, but a one-segmented condition similar to that present in the ground plan of the Muscomorpha is known in some members of many families, including the Stratiomyidae (Pachygastrinae) (Rozkošný 1982), the Therevidae (Irwin and Lyneborg, Ch. 37), the Scenopinidae (Kelsey, Ch. 38), the Apioceridae (Peterson, Ch. 41), the Asilidae (G. C. Wood, Ch. 42), the Acroceridae (Schlinger, Ch. 43), and the Bombyliidae (Hall, Ch. 45). It is one-segmented in all Mydidae (Wilcox, Ch. 40), Empididae (Chvála 1983), and Dolichopodidae (Robinson and Vockeroth, Ch. 48).

- *C ending at M*

C surrounds the wing in the ground plans of both the Nematocera and the Brachycera, but it fades out and becomes absent beyond the costal insertions of R and M in the Deuterophlebiidae, in some Blephariceridae, in all Chironomidae, in all Bibionomorpha except the Cecidomyiidae, in all Stratiomyoidea, in some Empididae, in all Dolichopodidae, and in all the Muscomorpha except Lonchopteridae and some Platypezidae (*Opetia* spp.) (J. F. McAlpine, Ch. 2). It seems likely that it was developed secondarily in the Cecidomyiidae, Lonchopteridae, and *Opetia*. It is perhaps significant that in the major clades of the Brachycera it ends near the wing apex only in the ground plan of the Stratiomyoidea and Muscomorpha, but it is uncertain whether the reduction results from evolutionary parallelism or convergence. However, the fact that C surrounds the wing in the ground plan of the Empidoidea (Chvála 1983) and in the Asilomorpha in general, probably argues against the cladistic validity of the Eremoneura.

- *R_s two-branched, i.e. neither R_{2+3} or R_{4+5} forked*

In the ground plan of the Diptera, R_s is twice branched, but the four free branches are retained only in the Tanyderidae and Psychodidae. In the ground plan of the Brachycera, R_{2+3} is unbranched, but R_4 and R_5 are separate. The free tips of R_4 and R_5 are retained in the ground plans of most of the major clades of the Brachycera except the Muscomorpha, but R_{4+5} also occurs as a single vein within many families of the orthorrhaphous Brachycera. Therefore, the unforked condition of R_{4+5}, although apomorphous with respect to the ground plan of the Brachycera, cannot be considered unique to the Muscomorpha. On the other hand, because the same condition occurs so commonly outside the Muscomorpha it is of little value for establishing cladistic relationships.

- *M two-branched, i.e. M_3 absent*

A free M_3 is present in the ground plan of the Brachycera and is retained in the ground plans of all the major subordinate clades except the Empidoidea and Muscomorpha. However, just as in the case of the previous character, it has been lost independently within many groups of the orthorrhaphous Brachycera, and it seems likely that its loss in the Empidoidea and Muscomorpha was likewise realized through independent processes. But there is no direct evidence for or against this hypothesis, and, as with the reduced branching of R_s, it also appears to be of little use in establishing the origins of the Muscomorpha.

- *Cell cup acutely closed near the wing margin, i.e. CuA_2 not far removed from the wing margin and not recurved on A_1*

A closed cell cu*p* is formed by the fusion of CuA_2 and A_1. This fusion never occurs in the Nematocera, but it occurs or nearly occurs in the ground plans of virtually all families of the Brachycera. Cell *cup* is relatively widely open to the wing margin in many Bombyliidae and in a few Asilidae (*Leptogaster* Meigen), only. In the ground plan of this condition, CuA_2 joins A_1 at an internally acute angle, but with the accompanying trend for cell *cup* to become progressively shorter, the internal angle at which CuA_2 joins A_1 becomes progressively greater. In extreme cases, such as occurs in the ground plan of the Empidoidea (Chvála 1983) and in some specialized members of the Schizophora, CuA_2 becomes recurved, thus forming an obtuse angle at its juncture with A_1. These trends are well illustrated in the series of wing illustrations provided for the "Key to Families" (J. F. McAlpine, Ch. 4). The most generalized states of this character occurring in the Muscomorpha are probably shown by members of the Syrphidae and Pipunculidae in the Aschiza. Because relatively similar conditions occur so commonly throughout the orthorrhaphous Brachycera, this character state can tell us little or nothing about the origins of the Muscomorpha. On the other hand, its peculiarly specialized state in the ground plan of the Empidoidea speaks strongly against a sister-group relationship between the Empidoidea and Muscomorpha.

- *Empodia setiform*

The question of whether a setiform empodium arose from a pulvilliform one, or vice versa, is still unsettled (Hennig 1981). Ross (1965) thought that the lobe-like empodium of the Homoeodactyla is a derived character, but there is some evidence that it is, in fact, a plesiomorphic condition of the Diptera that was carried over into the Brachycera. The presence of a pulvilliform empodium in the Mecopteroidea, in most Bibionomorpha, and in the most primitive orthorrhaphous Brachycera is usually taken as evidence that the pad-like form is more generalized than the seta-like form. In the Nemestrinidae, most members have pulvilliform empodia, but in some members it is reduced to a relatively slender process. In many Bombyliidae it is reduced to a short, triangular process, and it is absent in the Mydidae and Apioceridae. These reduced conditions are almost certainly derived states. The setiform empodium that occurs in many orthorrhaphous Brachycera and in all Muscomorpha is certainly homologous with the pulvilliform empodium of the Bibionomorpha and the so-called Homoeodactyla. It arises as an unpaired median process from the plantar region of the aroleum (J. F. McAlpine, Ch. 2) and I believe that the setiform condition is apomorphic with respect to the ground plan of the Brachycera. A cursory analysis of its condition in the Muscomorpha shows that it always has a rather broad base, is frequently relatively coarsely pubescent throughout, and is sometimes relatively broad and blunt-tipped. It varies from family to family in size, length, shape, curvature, and vestiture. For instance, in the Platypezidae and Lonchopteridae it is very reduced or absent, whereas in the Nothybidae and Strongylophthalmyiidae it is relatively broad and flat; in the Coelopidae it has a plume of long hairs at the base, and in the Dryomyzidae it is unusually long and recurved. Comparative study of this structure throughout the Diptera would undoubtedly provide many interesting cladistic insights, but in our present state of knowledge, it cannot be effectively used as an indicator of the ancestry of the Muscomorpha.

- *Mid coxal prong present*

This peculiar process arises from the upper posterior margin of the excised outer surface of the mid coxa and projects anteroventrally over a cluster of sensory setulae on the mid trochanter (J. F. McAlpine, Ch. 2). It was first recorded by Malloch (1923), who observed that its absence in the Pyrgotidae and Conopidae separated those two families from nearly all other Muscomorpha. It is clearly a ground-plan feature of the Muscomorpha, for it occurs almost universally in both the Aschiza and Schizophora. With regard to the Aschiza, I reported earlier (J. F. McAlpine, Ch. 2) that it was absent in the Platypezidae, but I have since found that it occurs in a number of genera including *Opetia*, *Melanderomyia* Kessel, *Platypezina* Wahlgren, *Grossoseta* Kessel & Kirby, *Callomyia*, and others. It is absent, however, in the Ironomyiidae and seems to be weakly developed or absent in many Pipunculidae. Nevertheless, it is sufficiently widespread throughout the Aschiza to indicate that it was present in the ground plan of that clade. In the Schizophora it is absent only in reduced or peculiarly adapted forms. It is not absent in all Pyrgotidae as reported by Malloch, for I have found it in the South American genera *Pyrgotosoma* Malloch and *Tetrura* Bigot, and in an unidentified member of the family from Lake Albert, Uganda. However, I have not found it in any Conopidae, and it also seems to be absent in such bizarre families as the Braulidae, Mormotomyiidae, Mystacinobiidae, Hippoboscidae, Nycteribiidae, and Streblidae. Again however, it is so widespread throughout the Schizophora, that it must have been present in the ground plan. In an extensive survey of representatives of almost all families of the Nematocera and the orthorrhaphous Brachycera for the presence of this feature, I found it in two families, the Stratiomyidae and Tabanidae. In the Stratiomyidae it is weakly to strongly developed in seven of the nine subfamilies that are currently recognized (Rozkošný 1983) (absent in all Beridinae and Chiromyzinae examined). In the Tabanidae, it is well developed in the Chrysopinae and Tabaninae, but I did not find it in any Pangoniinae. I found no indication of it, whatever, in any of the Nematocera, nor in any of the remaining orthorrhaphous Brachycera examined.

It is rather difficult to rationalize the occurrence of this little-known character within the cladistic framework of the Brachycera as perceived by most workers, including Woodley (Ch. 115). How can an apparently homologous feature of this nature be expressed only in the Stratiomyidae, Tabanidae, and Muscomorpha? If it were a ground-plan feature of the Brachycera that was carried over into these groups, one would expect to see traces of it in related clades such as the Xylophagidae, the Xylomyidae, and additional members of the Tabanoidea as well as in the supposedly most primitive clades of the Stratiomyidae (Chiromyzinae and Beridinae) and the Tabanidae (Pangoniinae). On the basis of its presently known occurrence it perhaps seems most parsimonious that it evolved independently three times rather than being lost in the more primitive Stratiomyidae. However, it may yet be found elsewhere, for absence of proof is not always proof of absence. Although the possibility that its occurrence in these three groups only is the result of homoplasy cannot be entirely ruled out, it is difficult to accept that such a peculiar feature would evolve spontaneously in exactly the same position, and in such similar form, on three separate occasions. Certainly these processes are homologous within each of the three groups involved, and when one studies examples of them as they occur in, e.g. *Sargus* Fabricius, *Tabanus* Linnaeus, and *Syrphus* Fabricius, they give a definite impression of being homologous structures that were inherited from a common ancestor. There can be no doubt that the mid coxal prong did not first arise in the Muscomorpha, and the possibility that it is a synapomorphic feature of its bearers should not be dismissed lightly.

- *Male terminalia. Gonocoxites fused with hypandrium, and gonostyli reduced and fused with gonocoxites (J. F. McAlpine, Ch. 2)*

These character states were discussed to some extent, under the plesiomorphic character "Gonopods joined (not fused) with hypandrium, ..." and also under the autapomorphic character "Epandrium and hypandrial complexes in two instead of three planes...." Here, it is necessary to emphasize that the fusion of the gonocoxites with the hypandrium and the fusion and reduction of the gonostyli are apomorphic states with respect to the ground plan of the Brachycera. However, similar fusion and reduction of these parts occur so commonly in various clades of the orthorrhaphous Brachycera that, by themselves, these states are of little help in establishing the ancestry of the Muscomorpha.

As explained earlier, I reject Griffiths' (1972, 1981, 1983, 1984) statements to the effect that in the Eremoneura, not only have the gonocoxites migrated dorsally in opposite directions and become fused along their previously outer margins to form a periandrium that replaces the epandrium, but also that the gonostyli have somehow become linked up posteriorly with the dorsal sclerites of the proctiger and have replaced the surstyli.

- *Sperm pump free from aedeagus and aedeagal apodeme, and with a separate ejaculatory apodeme and sperm duct*

In the ground plans of both the Nematocera and orthorrhaphous Brachycera, the sperm pump is enclosed within the base of the aedeagus and is closely associated with what is often referred to in these groups as the aedeagal apodeme (phallapodeme). That apodeme arises from, and is firmly attached to, the inner walls of the aedeagus, and the muscles that operate the sperm sac are attached to it (see discussion under the autapomorphic character "Aedeagal apodeme developed as an ingrowth from the posterior base of the aedeagal guide"). In this way, the sperm pump is only loosely joined to the base of the aedeagus by means of the ejaculatory duct (which continues through the lumen of the aedeagus or endophallus to the gonopore), and the ejaculatory apodeme is completely separate from the body wall.

Possession of a separate sperm pump with its own independent ejaculatory apodeme for attachment of muscles that operate the sperm sac evidently belongs to the modified ground plan of the Muscomorpha, for it is present in all but a few members of both the Aschiza and Schizophora. In the Aschiza it is possibly absent in the Phoridae (Hennig 1976*b*), but it is present in all other families including the Ironomyiidae (J. F. McAlpine 1967, Fig. 11) and the Sciadoceridae (personal observation). Its absence in some Platypezidae and possibly in all Phoridae is almost certainly as a result of secondary reduction. It is probably also secondarily reduced in certain members of the Schizophora (Griffiths 1972).

Hennig (1936, 1976*b*) suggested that both the ejaculatory apodeme and the aedeagal apodeme of the Muscomorpha may have arisen by splitting of a previously uniform structure, i.e. the so-called aedeagal apodeme (phallopodeme) of the lower Diptera. I agree with Griffiths (1972) that no convincing evidence supports that hypothesis. I believe, however, that the muscomorphan ejaculatory apodeme, by itself, may be derived from the so-called aedeagal apodeme of the lower Diptera. Perhaps the key to its origins in the Muscomorpha is revealed in the Stratiomyidae. In the subfamily Beridinae, for example, the ejaculatory apodeme resembles the generalized muscomorphan type, and, although it is still partially inserted into the base of the aedeagus, it is free from the walls of the aedeagus (see Rozkošný 1973, Fig. 96; 1982, Fig. 11.8). In most other members of this family, however, the same structure is attached to the inner walls of the aedeagus. These varying conditions within the Stratiomyidae reflect how a separate ejaculatory apodeme could arise by gradual detachment from the inner walls of the aedeagus, and, in this way, how the withdrawal of the entire sperm pump from the base of the aedeagus would be possible. Perhaps these conditions indicate the method by which the muscomorphan conditions arose. Certainly, the separate conditions that exist in some Stratiomyidae approach the basic conditions in the Muscomorpha more closely than do the conditions in any other group that I have examined.

- *Tergite 10 atrophied dorsally, and with its lateral parts adapted as a pair of hinged, clasper-like surstyli (J. F. McAlpine, Ch. 2)*

In the ground plans of both the Nematocera and Brachycera, a well-developed tergite 10 follows tergite 9. In the Nematocera, e.g. in the Trichoceridae, it may occur as a simple transverse strip; in other cases it may be either divided medially, or fused with tergite 9, or atrophied. In some nematocerous families, e.g. Ptychopteridae, Blephariceridae, Tipulidae, Mycetophilidae, and Synneuridae, it may bear lateral processes called surstyli. Similarly, tergite 10 varies in form in the orthorrhaphous Brachycera. In the Tabanoidea, for instance, either it may be present as a free, transverse sclerite, or it may be divided on the midline into two variously reduced sclerites, or it may be completely atrophied. Clasper-like surstyli occur in some Xylomyidae, some Stratiomyidae, some Asilidae, and in most Empididae and Dolichopodidae. Their widespread occurrence in empidids and dolichopodids indicates that they are probably a ground-plan feature of the Empidoidea. In the Muscomorpha the presence of articulated surstyli is a basic feature of practically all families, and this character is a well-recognized, ground-plan feature of the infraorder. Absence of surstyli in a few isolated taxa within the Muscomorpha is the result of secondary reduction.

Hennig (1976*b*) showed convincingly that the surstyli of the Muscomorpha probably arose by the splitting of tergite 10 (similar to the splitting of tergite 9 that occurs in many Asiloidea and Empidoidea) into two parts, the outer walls of which remain attached to the posterolateral margins of the epandrium (tergite 9) and the inner walls of which join with sternite 10 (ventral epandrial plate). In this way the main elements of segments 9 and 10, together with the proctiger are integrated into a uniform structure, in which the clasping function formerly performed by the gonopods and their gonostyli are now performed, at least in part, by the lateral margins of the epandrium and the surstyli. Loss of flexibility at the outer junctures of the surstyli with the epandrium, and at their inner junctures with sternite 10 are interpreted as secondary modifications. Similarities in the musculature of the Rhagionidae, Empididae, and Muscomorpha led Hennig to two inescapable conclusions: first, that in the Empidoidea and Muscomorpha the gonocoxites and hypandrium are fused with each other; and second, that the assumptions by Griffiths (1972) and others, that the muscomorphan surstyli are homologous with the gonostyli (telomeres), cannot be defended. The question of whether the splitting of tergite 10 into two parts (and the development of these parts as surstyli) belongs to the modified ground-plan characteristics of a superordinate group that embraces the Muscomorpha and one or more of these groups cannot be ascertained without further research.

Possibly the presence of surstyli in the Xylomyidae and its sister group, the Stratiomyidae, is a synapomorphy, but if so, they must have been lost secondarily in virtually all primitive Stratiomyidae (Chironomyzinae and Beridinae). Probably the presence of surstyli is a synapomorphy of the Empididae and its sister group the Dolichopodidae. Similarly, their occurrence in certain primitive clades of the Asilidae and in the ground plan of the Empidoidea is synapomorphic. But the possibility that the presence of surstyli in all these groups, including the Muscomorpha, is the result of evolutionary parallelism or even homoplasy cannot be excluded. Nevertheless, it now seems clear that true surstyli are derived from tergite 10 throughout the Diptera, that they are more or less homologous in all the groups in which they occur, and that they did not first arise in the Muscomorpha.

- *Female with tergite 10 combined with tergite 9 and proctiger*

Tergite 10 is separate from tergite 9 in the ground plans of both the Nematocera and Brachycera (J. F. McAlpine, Ch. 2, Figs. 2.79, 2.94), but usually the basal tergite of the proctiger (tergite 11) is incorporated with it or is otherwise lost. Thus, the cerci give the appearance of being attached directly to tergite 10. Commonly tergite 10 is divided into two lateral hemitergites. These features are especially apparent throughout the homoeodactyline series of the orthorrhaphous Brachycera (Xylophagidae, Tabanoidea, and Stratiomyoidea) as illustrated by Nagatomi and Iwata (1976, 1978). In numerous members of the heterodactyline series of the orthorrhaphous Brachycera (Asilidae, Mydidae, Apioceridae, Therevidae, Scenopinidae, Bombyliidae, Empididae, and Dolichopodidae) tergite 10 is divided into two heavily spinose plates (acanthophorites). In many, the modified plates are united with tergite 9 (J. F. McAlpine, Ch. 2, Figs. 2.98, 2.102).

In the Muscomorpha tergite 10 is always more or less united with tergite 9; it may be divided as in the homoeodactyline series of the orthorrhaphous Brachycera, but it is never modified as in the heterodactyline series. Leher (1971) indicated that the morphological conditions in the Syrphidae permit some interesting generalizations about the basic structure of the abdomen in the Muscomorpha, which may well be so. However, he misinterpreted a secondary median sclerotization at the posterior margin of sternite 8 as being sternite 9 and concluded wrongly that the female genital opening in the Syrphidae (and by implication in the Muscomorpha) lies between sternite 9 and 10. In fact it is immediately behind sternite 8 as is usual throughout the order.

The occurrence of tergite 10 as heavily spinose plates in the ground plan of the Empidoidea (Chvála 1983, wrongly interpreted by him as tergite 9) probably indicates a relatively close cladistic relationship between the Empidoidea and Asiloidea, but, unless it can be shown to have been lost secondarily in the Muscomorpha, it speaks rather definitely against a sister-group relationship between the Empidoidea and Muscomorpha (Hennig 1973, p. 218). I have been unable to establish which, if any, of the conditions of tergite 10 occurring in the remaining orthorrhaphous Brachycera resemble most closely those of the Muscomorpha.

- *Cerci one-segmented, free from each other*

The female cerci are two-segmented in the ground plans of the Nematocera and Brachycera, but commonly they are independently reduced to a one-segmented condition, especially in the orthorrhaphous Brachycera (see

Nagatomi and Iwata 1976, 1978). Consequently, from this aspect at least, they can be of little use for establishing the ancestry of the Muscomorpha.

Monophyly of the Muscomorpha

The monophyly of the Muscomorpha is one of the best substantiated and most universally accepted assumptions in the phylogeny of the Diptera. This assumption is based on the fact that its ground plan includes numerous autapomorphic characters that set it apart from its sister group (see "Autapomorphic characters"). It is noteworthy that these characters occur in the larvae, the pupae, and in both sexes of the adults. Moreover, most are distinctive and appear regularly in both main subdivisions of the Muscomorpha.

Ancestral sister group of the Muscomorpha

The infraorder Muscomorpha falls readily into the suborder Brachycera. It follows that its immediate ancestor would also be a member of the Brachycera. But, because the Muscomorpha is such a discrete and easily definable clade, it is certain that there is a relatively broad evolutionary gap between it and its nearest existing relatives. Fully intermediate forms probably no longer exist.

The oldest fossil representative is an unnamed fly from a Lower Cretaceous (Neokom) deposit in Lebanon (Hennig 1971*b*). This fossil consists of a head and fragments of one leg preserved in amber. Hennig assigned it with certainty to the Muscomorpha, but was unable to place it any more precisely within the infraorder. Four other fossil species from Cretaceous formations in western Canada have been named, described, and assigned to modern families within the Muscomorpha, i.e. *Cretonomyia pristina* McAlpine (1973) (Aschiza: Ironomyiidae), *Sciadophora bostoni* and *Prioriphora canadambra* McAlpine & Martin (1966) (Aschiza: Sciadoceridae), and *Cretaphormia fowleri* McAlpine (1970) (Schizophora: Calliphoridae). These fossils are late Cretaceous in age (70–73 million years old), but none reveals anything more about the muscomorphan archetype than do extant members of the same families. They simply establish that within the Muscomorpha both the Aschiza and Schizophora were already well differentiated by late Cretaceous times. Consequently, the only means for ascertaining the primitive sister group of the Muscomorpha is by comparing its taxonomic characters with homologous characters in representatives of the orthorrhaphous Brachycera.

As already indicated, the best clues about the primitive ancestor of the Muscomorpha are those characters that are plesiomorphic within the infraorder, but which are, at the same time, apomorphic with respect to the ground plan of the Brachycera (see 28 apomorphic characters already discussed). In reviewing these characters it is striking that in so many cases the conditions most similar to those in the Muscomorpha are found in the Stratiomyoidea (Stratiomyomorpha of Woodley, Ch. 115), a group which, like the Muscomorpha, has long been considered a specialized offshoot of the brachyceran stock (Oldroyd 1964, M. J. Roberts 1969*a*). Perhaps the following 12 characters are the most significant apomorphic conditions shared by these two clades:

1. Larvae saprophagous feeding on particles, such as bacteria, yeasts, algae, and fungi;
2. Larval mouthparts adapted for filter feeding;
3. Larvae with a well-developed internal pharyngeal filter;
4. Larval maxillae lost;
5. Larvae with anterior arms of tentorium fused with cibaropharyngeal walls;
6. Pupation occurring within a puparium formed from the last larval skin;
7. Imagines with a well-developed occipital sclerite;
8. Apex of pedicel notched, and with a cone-shaped condyle deeply inserted into base of first flagellomere;
9. Flagellum consisting of a composite first flagellomere and a three-articled arista;
10. C ending at M;
11. Mid coxal prong present (also shared with some Tabanidae); and
12. Male with ejaculatory apodeme free from aedeagus and body wall.

In these 12 characters and many of the others discussed, one is faced with deciding whether their similarities in the Stratiomyoidea and the Muscomorpha result from propinquity of descent or from independent evolution, i.e. whether they represent synapomorphies or convergences. This problem is especially perplexing with respect to characters 8, 9, 11, and 12, because the plesiomorphic conditions of all three obtain in the more primitive clades of the Stratiomyoidea. If it is assumed that they are synapomorphies of the Muscomorpha and Stratiomyoidea, only two cladistic possibilities can be hypothesized; either the Muscomorpha arose from within the Stratiomyoidea, or the Muscomorpha and Stratiomyoidea arose from a common ancestor in which these characters occurred in a weakly developed or potential state. In subsequent independent clades either they became intensified and consolidated as in the more specialized Stratiomyidae and in all Muscomorpha, or they reverted to the more plesiomorphic conditions as in the primitive Stratiomyoidea.

I reject the first possibility because the Muscomorpha do not belong to the Stratiomyoidea any more than do those Tabanidae that possess a mid coxal prong. I agree with Hinton (1946, p. 301) that such similarities between the specialized Stratiomyidae and the Muscomorpha must have been attained through parallel or convergent evolution after they separated from a common ancestor.

As indicated in the discussion of the mid coxal prong, however, it is difficult to accept that such peculiar modifications would evolve by completely independent processes in completely unrelated systems. If each character is considered separately, it is perhaps easy to attribute certain similarities shown by both groups, e.g. C ending at M, to evolutionary convergence. But when all these peculiarities, which do in fact occur together, are considered together, the possibility that they evolved completely spontaneously appears less likely. In my opinion, they are probably indicators of true cladistic relationship.

On this basis it seems reasonable to assume that the original ancestor of the Muscomorpha was a stratiomyid-like brachyceran rather than a tabanoid, nemestrinoid, asiloid, or empidoid type. There is little or no convincing evidence that the Muscomorpha is more closely related to any or all of the Asilomorpha than to the stem that gave rise to the Stratiomyoidea + Tabanoidea. All the evidence I have been able to find indicates that the ancestry of the Muscomorpha bypasses the Asilomorpha and traces farther back in the cladistics of the orthorrhaphous Brachycera. It seems inescapably clear that there are more significant similarities between the Muscomorpha and Stratiomyoidea than there are between the Muscomorpha and any other out-group. Again, this is not to say that the Stratiomyoidea (or any of its subgroups), as known today, is *the* sister group of the Muscomorpha, but simply to indicate that the ancestor of the Muscomorpha was probably fairly similar to the ancestor that gave rise to the Stratiomyoidea.

As stated previously, no form is known that embodies all the characters necessary to qualify as the immediate ancestor of the Muscomorpha, but it seems probable that such a form did exist in late Jurassic or early Cretaceous times, and that it arose from a stem that also gave rise to the Stratiomyoidea. It differed from that stem and all other orthorrhaphous Brachycera by possessing at least 28 conditions already discussed under "Autapomorphic characters."

SUBORDINATE GROUPS OF THE MUSCOMORPHA: ASCHIZA AND SCHIZOPHORA

Muscomorpha is composed of two easily definable but somewhat unequal subgroups or sections: the Schizophora, (Becher 1882) (= Myodaria Robineau-Desvoidy (1830), Muscoidea of authors), a large taxon whose monophyly is readily established; and the Aschiza (Becher 1882), a smaller possibly paraphyletic taxon (Griffiths 1972), which nevertheless is usually accorded equal rank with the Schizophora. In adopting this traditional classification of the Muscomorpha, a problem is encountered that recurs in all the main categories of the Diptera. The common problem is to establish the monophyly of the older, more generalized subgroup. In the suborder Brachycera, the orthorrhaphous Brachycera, as opposed to the Muscomorpha, presents the same problem; and in the order Diptera, the same situation exists in the Nematocera as opposed to the Brachycera. In each case, the older subgroup is founded mainly on plesiomorphic (diagnostic) rather than apomorphic (constitutive) characters. Until autapomorphies are demonstrated for such groups, we cannot dismiss the possibility that they are paraphylectic taxa. At the same time, however, the possibility remains that they are, in fact, monophyletic groups, for which proof of their monophyly has not yet been advanced.

Griffiths (1972), following de Meijere (1900*b*) and Hennig (1952, 1954), divided the Muscomorpha into two sister groups, the Acroptera (Brauer 1883) (= Anatriata, de Meijere 1900*b*) containing only the Lonchopteridae, and the Atriata (de Meijere 1900*b*) containing the remainder. He rejected the Aschiza as "a residual paraphyletic assemblage" and admonished that "such a group has no place in a phylogenetic classification." I am not fully convinced on these points and feel that the question of the monophyly of the Aschiza should be reexamined. I agree with Holmes' (1980) view that the recognition of certain readily recognizable and easily definable subgroups is sometimes useful and acceptable, even though such groups may not be clearly demonstrated to be monophyletic. Perhaps Aschiza is such a group. The main characteristics of both groups are compared in Table 116.1. Those conditions that are believed to be plesiomorphic with respect to the ground plan of the Muscomorpha are marked P, and those that are believed to be apomorphic are marked A.

The best starting point for establishing the polarities of the ground-plan character states of both Aschiza and Schizophora is Hennig's (1958) study. In that study and in his later works on the same subject (Hennig 1971*a*; 1973; 1976*a,b*) he synthesized and updated data from earlier works by himself and others such as de Meijere (1900*b*, 1904), Hendel (1916, 1922–1923), Frey (1921), Malloch (1933*b*, 1948), and Crampton (1942). Singlehandedly, Hennig advanced our knowledge on the evolution of these groups more than anyone else. Griffiths' (1972) classification of the Muscomorpha, based mainly on characters of the male terminalia, is also a good source for much comparative data.

The comparative list of ground-plan characters of Aschiza and Schizophora (Table 116.1) was compiled from the works mentioned and from our own studies (J. F. McAlpine 1981*a,b*; Teskey 1981*a,b*). It is a selected list and contains those features that seem most useful for discussions on the evolution of the subordinate groups of each main section of the Muscomorpha. Some 31 apomorphies shown in the ground plan of the Schizophora prove beyond reasonable doubt that it is a monophyletic taxon. The problem of ascertaining apomorphies for the Aschiza and establishing its monophyly is more difficult (see "Monophyly of the Aschiza").

Table 116.1 Comparison of character states in ground plans of Aschiza and Schizophora

Character state[1]	Aschiza	Schizophora
Larva with (P), without (A) median labial tooth	P	A
Larval mandibles short and blunt (P), long and falciform (A)	P	A
Hypopharyngeal sclerite free from (P), fused with (A) tentopharyngeal sclerite	A	P
Openings of anterior larval spiracle sessile (P), at ends of projecting papillae (A)	P	A
Puparium thin walled and lightly tanned (P), thick walled and heavily tanned (A)	P	P
Puparium with (P), without (A) a middorsal cleavage line	P	A
Puparium with two (P), with one (A) transverse dorsal cleavage line(s)	P	A
Puparium without (P), with (A) horizontal cleavage line on each side extending to abdominal segment 1	P	A
Prothoracic pupal respiratory horn small (P), large (A)	A	P
Prothoracic pupal respiratory horn penetrating (P), not penetrating (A) puparium in abdominal tergite 1	P	P
Adult body relatively hairy (P), nearly bare (A)	P	P
Bristles of adult head, thorax, and abdomen not or weakly differentiated (P), strong (A)	P	A
Feathered bristles present (P), absent (A)	P	A
Male eyes holoptic (P), dichoptic (A)	P	P
Ptilinum absent (P), present (A)	P	A
Ptilinal fissure absent (P), present (A)	P	A
Lunule absent (P); present little exposed, weakly setulose (Strickland 1953) (A)	P	A
Frons uniformly sclerotized (P), with desclerotized frontal vitta (A)	P	A
Fronto-orbital plates undifferentiated (P), differentiated (A)	P	A
Ocellar plate not or weakly differentiated (P), strongly differentiated (A)	P	A
Median occipital plate with (P), without (A) distinct reinforcement ridges	P	P
Outer vertical bristles absent (P), present (A)	P	A
Inner vertical bristles absent (P), present (A)	P	A
Fronto-orbital bristles undifferentiated (P), with four reclinate fronto-orbital bristles (A)	P	A
Proclinate orbital bristles absent (P), present (A)	P	P
Frontal bristles undifferentiated (P), differentiated (A)	P	P
Frontal vitta setulose (P), bare (A)	P	P
Interfrontal bristles undifferentiated (P), differentiated (A)	P	P
Ocellar bristles undifferentiated (P), differentiated (A)	P	A
Divergent postocellar bristles undifferentiated (P), differentiated (A)	P	A
Convergent postocellar bristles undifferentiated (P), differentiated (A)	P	P
Postoccipital bristles weak (P), strong (A)	P	P
Scape, pedicel, and flagellum setose (P), at least flagellum bare (A)	P	A
Pedicel without (P), with (A) a dorsolateral notch	P	A
Apex of pedicel deeply inserted (P), not deeply inserted (A) into base of first flagellomere	P	A
First flagellomere not decumbent (P), decumbent or nodding (A)	P	A
Arista arising dorso-apically (P), apically or dorsobasally (A)	P	P
Arista shortly pubescent (P), plumose or bare (A)	P	P
Face evenly sclerotized and convex (P), desclerotized in middle or carinate (A)	P	P
Vibrissa undifferentiated (P), differentiated (A)	P	P
Clypeus relatively small (P), large (A)	P	P
Hyoid sclerite weak or absent (P), strongly developed (A)	P	P
Pseudotrachea opening directly into external mouth opening (P), opening into one or two main trunks (A)	P	P
Prestomal teeth absent (P), present (A)	P	P
Scutum with transverse suture incomplete (P), complete (A)	P	P
Scutum with acrostichal, dorsocentral supra-alar and postalar series of bristles not or weakly differentiated (P), strongly differentiated (A)	P	A

(continued)

Table 116.1 Comparison of character states in ground plans of Aschiza and Schizophora *(concluded)*

Character state[1]	Aschiza	Schizophora
Scutellar bristles weak or absent (P), with two pairs of scutellar bristles (A)	P	A
Postpronotal bristles absent (P), present (A)	P	A
Notopleural bristles weak, scattered, (P); strong, arranged 1–1 (A)	P	A
All pleural sclerites haired (P), bare (A)	P	P
Main pleural sclerites without bristles (P), with bristles (A)	P	A
Greater ampulla undeveloped (P), developed (A)	P	P
Coxopleural streak present (P), absent (A)	P	P
Precoxal bridge absent (P), present (A)	P	P
Prosternum haired (P), bare (A)	P	P
Metasternal area haired (P), bare (A)	P	P
Costal breaks absent (P), present (A)	P	P
Sc complete (P), abbreviated (A)	P	P
M_{1+2} forked (P), unforked (A)	P	A
A_1 reaching wing margin (P), abbreviated (A)	P	P
Pterostigma present (P), absent (A)	P	P
Cell dm present, long (P); short or absent (A)	P	P
Cell cu*p* present, long (P); short or absent (A)	P	P
Crossvein sc-r present (P), absent (A)	P	P
Anal angle well developed (P), reduced (A)	P	P
Alula well developed (P), reduced (A)	P	P
Lower calypter moderately developed (P), linear or very broad (A)	P	P
Tibial bristles weak or absent (P), strongly differentiated (A)	P	P
Tarsomere 1 of hind tarsus simple (P), swollen or enlarged (A)	P	P
Abdominal bristles weak or absent (P), strong (A)	P	P
Abdominal tergites 1–8 of both sexes haired (P), at least partly bare (A)	P	P
Abdominal sternites 1–7 of male and 1–8 of female haired (P), at least partly bare (A)	P	P
Abdomen with 7 spiracles in both sexes (P), less than 7 (A)	P	P
Abdominal spiracles in membrane (P), in tergites (A)	P	P
Male with abdominal sternite 5 unmodified (P), modified (A)	P	P
Tergite and sternite of segment 6 relatively large, symmetrical, and free (P); relatively reduced, asymmetrical, and partially fused in left side (A)	P	P
Tergite 7 free from segment 8 (P), partially fused in left side with segment 8 (A)	P	A
Sternite 8 large, setulose (P); small, bare (A)	P	P
Gonostylus evident (P), atrophied or indistinct from gonocoxite (A)	P	A
Paramere small, feeble (P); large, strongly sclerotized (A)	P	P
Aedeagus simple, one-segmented (P); specialized (A)	P	P
Basiphallus without (P), with epiphallus (A)	P	P
Aedeagal apodeme rod-like (cuneiform), mostly free from hypandrium (P); fultelliform, extensively fused with hypandrium (A)	P	P
Sternite 10 plate-like, undivided (P); divided in middle (A)	P	P
Surstyli present, movable (P); absent or fused with epandrium (A)	P	P
Surstyli loosely linked (P), closely linked (A) with cerci	P	P
Cerci membranous and free (P), sclerotized and fused (A)	P	P
Female with terminalia flexible, tubular (P); rigid, piercing-type (A)	P	P
7th tergite and sternite free (P), fused at sides (A)	P	P
7th tergite and sternite undivided (P), divided (A)	P	P
Cerci membranous, free (P); sclerotized, fused (A)	P	P
Three sclerotized spermathecae (P), less than or more than three spermathecae (A)	P	P
Three spermathecal ducts opening independently into genital chamber (P); one or more spermathecal ducts fused, with two or one opening(s) into genital chamber (A)	P	P

[1] P = plesiomorphic; A = apomorphic.

SUBORDINATE GROUPS OF THE ASCHIZA

Aschiza consists of seven distinctive families: Platypezidae (including *Opetia* and *Melanderomyia*), Lonchopteridae, Ironomyiidae, Sciadoceridae, Phoridae (including Termitoxeniinae), Syrphidae, and Pipunculidae. Together these families are easily distinguished from the remainder of the Muscomorpha, i.e. the Schizophora, by the absence of a well-developed ptilinum, a ptilinal suture, and a lunule. But, because these three negative features are plesiomorphic characters, they provide no assurance that the Aschiza is monophyletic. The separate conditions of the male tergite 7 and sternite 7 in the ground plan of the Aschiza (as opposed to fused in the ground plan of the Schizophora) (Hennig 1976*b*, p. 5) is in the same category.

One step toward resolving this problem is first to determine the monophyletic subgroups within the Aschiza and then to see whether they share any apomorphic ground-plan conditions. Hennig (1976*b*), who studied the evolutionary aspects of the question in greater depth than anyone else, recognized six, as follows: Lonchopteridae (Acroptera, Anatria, Anatriata); Ironomyiidae; Sciadoceridae + Phoridae; *Opetia* (= *Melanderomyia*?); Platypezidae (including *Microsania* Zetterstedt); and Syrphidae + Pipunculidae (Syrphoidea). In my view there are only two *main* monophyletic subgroups, the superfamilies Platypezoidea and Syrphoidea (Fig. 116.1).

Superfamily Platypezoidea

One synapomorphic character of the five families (Platypezidae, Lonchopteridae, Ironomyiidae, Sciadoceridae, and Phoridae) comprising the Platypezoidea is the strongly bristled condition of the frons, including erect ocellar bristles. Another is the number and arrangement of the notopleural bristles. In the platypezoid ground plan about five such bristles are arranged in several irregular rows. (There are probably two notopleurals, an anterior one, and a posterior one, in the ground plan of the Muscomorpha including the Aschiza.) An additional synapomorphic character of the members of the Platypezoidea is the absence of crossvein sc-r.

Platypezidae. An autapomorphic character of the Platypezidae (Kessel, Ch. 50) (including *Opetia* and *Melanderomyia*) is the reduced, prothoracic, pupal spiracular horns (de Meijere 1900*a*, Kessel *et al.* 1973), which are strongly developed in the remainder of the Aschiza. The absence of a filter apparatus in the pharynx of the larva (Hennig 1976*b*), if true, is an outstanding autapomorphic feature. Possibly the very weakly developed, rather pad-like empodia on all tarsi of Platypezidae is an apomorphic condition, but it can also be argued that this condition reflects the primitive pulvilliform state.

Another possibly autapomorphic feature of the Platypezidae is the way by which the scape and pedicel fit together. The apex of pedicel is never deeply inserted into base of first flagellomere in this family (or in the Lonchopteridae). Because a deeply inserted condition occurs in some Stratiomyidae, I have assumed that this condition is a ground-plan feature of the Muscomorpha, including the Aschiza, and that the condition in the Platypezidae, in which the apex of the pedicel is not deeply inserted into the flagellum, is an apomorphic condition. More study is required to resolve this matter. Males of many Platypezidae do not have an ejaculatory apodeme, which is certainly an apomorphic condition, but it is present in at least two genera, *Opetia* and *Platypezina* (Hennig 1976*b*). Previously (J. F. McAlpine 1981*a*,*b*), I believed that the mid coxal prong was absent (lost secondarily) in all Platypezidae, but subsequent investigations showed that it is present in some genera, e.g. *Opetia*, *Melanderomyia*, and *Grossoseta*. Similarly, the expansion of the hind tarsi in the males, which Hennig (1976*b*) employed as an apomorphic character of the Platypezidae, applies to a portion, only, of the family (absent in *Opetia* and *Melanderomyia*). I agree with Hennig (1976*b*) that Platypezidae is in some ways the most generalized family of the Platypezoidea (Fig. 116.1).

A synapomorphic character of the Lonchopteridae, Ironomyiidae, Sciadoceridae, and Phoridae is the loss of the holoptic condition in males. Although the frons is narrower in the male than in the female in Ironomyiidae and some Phoridae, the males are distinctly dichoptic as opposed to being truly holoptic as are males in the ground plan of the Muscomorpha, including the Aschiza.

Synapomorphic features of the Ironomyiidae, Sciadoceridae, and Phoridae are: fusion of Sc with R_1, and fusion and consolidation of syntergosternite 7 + 8 in the males (tergite 7 free in ground plan of the Muscomorpha, including the Aschiza).

Synapomorphic characters of the Sciadoceridae and Phoridae include many peculiarities of the wing venation (Hennig 1954, p. 361; J. F. McAlpine and Martin 1966), the reduced size and bare condition of the antennal scape and pedicel, and the formation of a precoxal bridge.

Lonchopteridae. Autapomorphic characters of the Lonchopteridae (Peterson, Ch. 49) include: absence of a tracheal air sac in the abdomen (Faucheux 1971); absence of empodia on all tarsi; peculiar, pointed wings; setose wing veins; C secondarily thickened on hind margin of wing; base of M atrophied; crossvein dm-cu elongate and longitudinally aligned; A_1 sexually dimorphic; scape and first flagellomere nonsetose; first flagellomere very short and globose; male tergites 6 and 7 fused; and surstyli reduced.

Ironomyiidae. The peculiar, tongue-like extensions of the antennal scape into both sides of the base of the first flagellomere of Ironomyiidae (J. F. McAlpine 1967, 1973) is a distinctive autapomorphic character of this family.

Sciadoceridae. One autapomorphic character of the Sciadoceridae is the tiny size of cell dm. Another apomorphy is the absence of empodia on all tarsi (as in Lonchop-

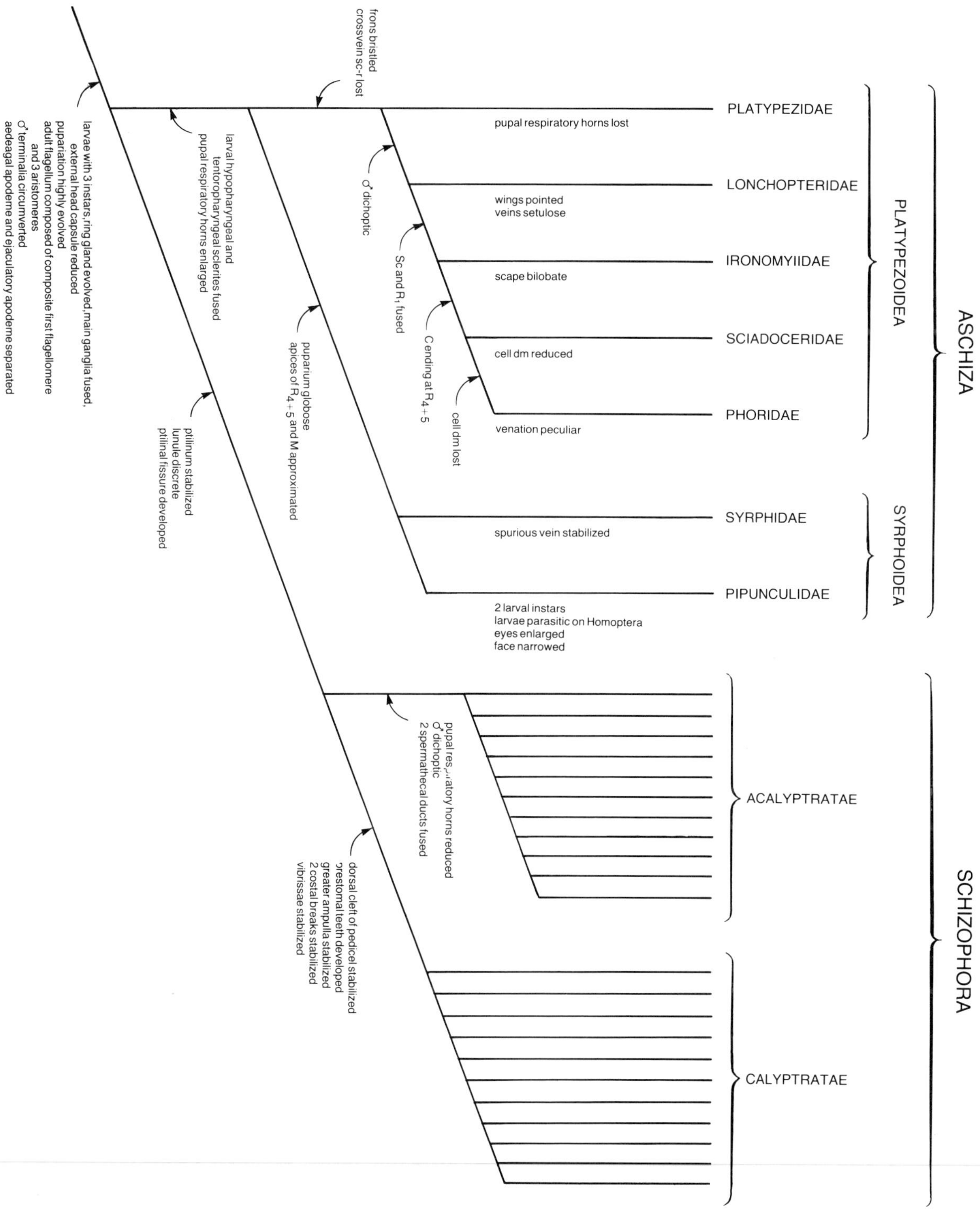

Fig. 116.1. Cladogram of the Muscomorpha summarizing the relationships and apomorphies of the main sections and subsections of the infraorder and the superfamilies and families of the Aschiza. (See text for additional details.)

teridae). Also, the form of the antennal pedicel is unique (Hennig 1976*b*).

Phoridae. In Phoridae (Peterson, Ch. 51) the relatively greater reduction of the wing venation, in particular the loss of cell dm and the base of M, and the peculiar alignment of the veins are autapomorphic characters. The apparent absence of a sperm pump and ejaculatory apodeme in males (Hennig 1976*b*), and the absence of sclerotized spermathecae in females, may be additional autapomorphies.

Superfamily Syrphoidea

Syrphoidea contains two easily recognized families, Syrphidae and Pipunculidae. Autapomorphic characters of the Syrphoidea include the following: puparium more or less globose and with peculiar operculum and cleavage lines; and wing with apices of veins R_{4+5} and M_1 joined or nearly so.

Syrphidae. The most evident and best known autapomorphic character of the Syrphidae (Vockeroth and Thompson, Ch. 52) is the spurious vein. An additional autapomorphic feature of this family is that the modification of the terminal abdominal segments of the male begins on segment 5 instead of on segment 6. The Syrphidae are more generalized in their ground plan than the Pipunculidae on the basis of characters of wing venation (crossvein sc-r present), adult morphology, and biology.

Pipunculidae. Autapomorphic features of the Pipunculidae (Hardy, Ch. 53) include the enlarged compound eyes and narrowed fronto-facial area in both sexes, the reduction of the larval instars to two (Jervis 1980) (three in ground plan of Muscomorpha, including the Aschiza), and the parasitic mode of life.

Monophyly of the Aschiza

On balance, the Platypezoidea is perhaps more generalized than the Syrphoidea. Ground-plan characters of the Platypezoidea that are more plesiotypic than those of the Syrphoidea include the following: puparium more elongate, softer, less convex, and with H-shaped cleavage lines, and so on; antennal scape, pedicel, and first flagellomere setose; and male with tergite and sternite of abdominal segment 7 more complete and more freely separated from each other and from segment 8. Characters of the Syrphoidea that are more plesiotypic than those in the Platypezoidea are the absence of conspicuous bristles on the frons and the presence of crossvein sc-r.

The question of proving the monophyly of the Aschiza by showing a synapomorphy between the Platypezoidea and the Syrphoidea remains. Perhaps the most convincing synapomorphy between the Platypezoidea and the Syrphoidea (autapomorphy of the Aschiza) is the fused condition of the hypopharyngeal and tentoropharyngeal sclerites in the larvae of both clades (Teskey, Ch. 3). The H-shaped hypopharyngeal sclerite of the Muscomorpha is probably mainly derived from the hypopharynx and the labium, and the tentoropharyngeal sclerite is probably derived mainly from the tentorial phragmata (Teskey, Ch. 3). It also seems probable that the anterior line of weakening (articulation) between the hypopharyngeal sclerite and the mandible and the posterior line of weakening between the hypopharyngeal sclerite and the tentopharyngeal sclerite, both of which are evident in all three larval instars of most Schizophora, reflect the ground-plan condition of the infraorder as a whole. If so, the solidly fused condition of the hypopharyngeal and tentoropharyngeal sclerite that occurs throughout the Aschiza is an autapomorphic ground-plan condition of the group. Similar but independently evolved conditions also occur sporadically in certain highly adapted members of the Schizophora, e.g. the Chamaemyiidae and some Agromyzidae. Another probably synapomorphic character relates to the prothoracic, pupal respiratory horns. These processes are unusually large in all families of the Aschiza, except the Platypezidae in which they appear reduced. Therefore the relatively strong development of these structures is assumed to be an apomorphic ground-plan character of the Aschiza. If these assumptions are correct they indicate that the Aschiza is, in fact, a monophylectic taxon (Fig. 116.1).

One consequence of this interpretation of the Aschiza is that the Lonchopteridae probably should be ranked as something less than a superfamily. Unfortunately, neither this aspect nor the more generalized condition of the Platypezidae is reflected in the table of classification presented in Volume 1 of this Manual.

SUBORDINATE GROUPS OF THE SCHIZOPHORA: ACALYPTRATAE AND CALYPTRATAE

Traditionally the Schizophora was divided into three groups: the Acalyptratae Macquart (1835, p. 55, 354) [= Holometopa Brauer (1880), = Haplostomata Frey (1921)]; the Calyptratae Robineau-Desvoidy (1830, p. 20, 22) [= Schizometopa Brauer (1880), Thecostomata Frey (1921)]; and the Hippoboscoidea Samouelle (1819) [= Pupipara Becher (1882)]. The name Calyptratae (as Calypteratae) was first proposed by Robineau-Desvoidy (not Girschner 1893) for the first "gens" of his Myodaria (= Schizophora) and refers to the large calypteres that normally occur in such families as Calliphoridae, Muscidae, and Tachinidae. The name Acalyptratae (as Acalypterae) was first proposed by Macquart (not Girschner 1893) for Robineau-Desvoidy's second "gens," the Mesomydae and seven other superfamilies, all of which have the lower calypter more or less reduced. Becher proposed the name Pupipara for the Hippoboscoidea, because the larvae of the component families are fed within the female abdomen by means of modified appendicular "milk glands" until they are ready to pupate (adenotrophic pupipary). It is now generally accepted that this last group is a subgroup of the Calyptratae (Hennig 1941, 1952, 1965*a*, 1971*a*, 1973; Griffiths 1972).

Both the Acalyptratae and Calyptratae have been retained as subsections of the Schizophora by most dipterists, but it has long been recognized that the size of the lower calypter is too variable in both groups to serve as the main criterion for defining either of them. Consequently many workers have attempted to improve the classification of the Schizophora by employing other characters. For example, Schiner (1864) combined characters used by earlier authors and newly employed characters of the mouthparts and wing venation. Brauer (1880) and Becher (1882) used schizometopy (frons desclerotized in vicinity of frontal vitta, leaving frontal plates continuous with the orbital plates and the parafacials) and holometopy (desclerotized area of frons extending laterally to eye margin, largely obliterating the frontal plates). Girschner (1893) recognized the presence and absence of a presutural intra-alar (= posthumeral) bristle and a postsutural intra-alar bristle. Malloch (1919) went further with the presence and absence of a dorsolateral seam on the antennal pedicel and the position of the abdominal spiracles inside or outside the lateral margins of the tergites. Frey (1921) incorporated the presence or absence of the hyoid sclerite and other characters of the mouthparts; Hendel's (1928) synthesis was of many previously used characters of immatures and adults. Townsend (1935) employed many characters in a typic–atypic, sliding-scale system of classification; Crampton (1944*a*,*b*) studied male terminalia of adults, and Hennig (1952) detailed the immature stages. Hennig's (1958) crowning attempt to divide the Schizophora into monophyletic superfamilies on the basis of a large number of characters is certainly the best synthesis of all previous analyses, and it has become the main starting point for subsequent improvements by himself and others.

The Calyptratae (including the Hippoboscoidea) is one of the best established monophyletic subsections of the Schizophora (Roback 1951; Hennig 1958, 1965; Griffiths 1972), but the monophyly of the Acalyptratae has never been satisfactorily established. Hennig (1958) showed that both holometopy, the basis for Brauer's Holometopa (= Acalyptratae), and schizometopy, the basis for Brauer's Schizometopa (= Calyptratae), arose independently several times and in various ways within the Schizophora; therefore, these conditions do not prove the monophyly of either group. Likewise, the absence or presence of a hyoid sclerite, the basis of Frey's Haplostomata (= Acalyptratae) and Thecostomata (= Calyptratae) do not prove the monophyly of either group because a hyoid sclerite is present in primitive members of both subsections (J. F. McAlpine 1962). Enderlein (1936), probably based largely on Brauer's (1890) and Hendel's (1928) work, proposed that Archischiza (Conopidae) is the sister group of the remainder of the Schizophora (Muscaria). But Hennig (1952) showed that this family must be included in the Acalyptratae and indicated that its closest relatives are in the Tephritoidea. In spite of these remarks, Hennig retained the name Archischiza in two later works (Hennig 1958, 1966*b*), but he abandoned it in his most recent treatments (Hennig 1971*a*, 1973). Griffiths (1972) placed it as a separate family in his "prefamily" Tephritoinea. I agree with both Hennig and Griffiths that the Conopidae cannot be the sister group of the remainder of the Schizophora.

Hennig (1971*a*) addressed the question of monophyly of the Acalyptratae and concluded that "a sister-group relationship between the Acalyptratae and the Calyptratae cannot yet be excluded." At the same time, Griffiths (1972) dismissed the Acalyptratae as a residual paraphyletic group and considered it to be unsatisfactory both from the cladistic standpoint and for purposes of identification.

In my opinion the Acalyptratae, including the Conopidae, is a monophyletic taxon and is the sister group of the Calyptratae, including the Hippoboscoidea. A list of important ground-plan characters of both the Acalyptratae and Calyptratae is presented in Table 116.2. Those that are considered clearly apomorphic in relation to the ground plan of Schizophora are marked with the letter A, and those that are plesiomorphic are marked with the letter P. From a practical standpoint, most Acalyptratae are readily distinguished from members of the Calyptratae and *vice versa* (see J. F. McAlpine, Ch. 4). Therefore both taxons are useful for identification. Justifications for considering that both groups are monophyletic follow.

GROUND-PLAN CHARACTERS AND MONOPHYLY OF THE ACALYPTRATAE

Most characters that distinguish the Acalyptratae from the Calyptratae (Table 116.2) are relatively unchanged conditions retained from the ground plan of the Schizophora (Table 116.1), and the absence of specialized conditions that evolved in the Calyptratae. Although most acalyptrate characters are apomorphic in relation to the ground plan of the Muscomorpha, they are plesiomorphic in relation to the ground plan of the Schizophora. The characters that I consider to be the most important ground-plan characters, which probably occurred in the original acalyptrate fly, are listed in Table 116.2. The only way of knowing of the presence and nature of these characters is by deductive reasoning based on our knowledge of existing (extinct as well as extant) forms. Those that are marked with A are considered to be autapomorphic for the Acalyptratae as a whole.

We are fortunate in having many unusually well-preserved fossil representatives of Acalyptratae in Baltic amber (Oligocene). Hennig (1966*b*, 1967*b*, 1969*b*, 1971*c*, 1972*a*) treated all the specimens that he could assemble (118 specimens representing 29 families), which included revision of material described by earlier authors as well as new material available in European and North American institutions (Larsson 1978). These fossils revealed little if anything more about the ancestor of the Acalyptratae than do living members of the same families. They simply establish that, within the Schizophora, the families they

Table 116.2 Comparison of character states in ground plans of Acalyptratae and Calyptratae

Character state[1]	Acalyptratae	Calyptratae
Egg with annular (P), longitudinal dorsal (A) hatching seam (Hennig 1973, p. 97)	P	A
Larva with (P), without (A) filter apparatus for particle feeding	P	P
Larva with mandibles separate (P), fused (A)	P	P
Larva with (P), without (A) parastomal bars	P	P
Dorsal cornua of larva with (P), without (A) a window	P	P
Puparium tapered anteriorly with flattened opercular area (P), hemispherical on both ends (A)	P	P
Prothoracic pupal respiratory horn penetrating (P), not or weakly penetrating (A) puparium	A	P
Adult male head holoptic (P), dichoptic (A)	A	P
Dorsolateral emargination of pedicel notch-like (P), seam-like (A)	P	A
Arista arising dorsoapically (P), dorsobasally (A)	P	A
Aristal pubescence relatively short (P), long (A)	P	P
First flagellomere porrect (P), decumbent (A)	P	A
Orbital plate extending far forward (P); reduced, replaced anteriorly by frontal plate (A)	P	A
Reclinate orbital bristles numbering 4 (P), 2 or less (A)	P	A
Inclinate frontal bristles numbering 3 or more (P), 2 to none (A)	P	P
Frontal bristles arising laterally to (P), medially to (A) orbitals	P	A
Proclinate orbital undeveloped (P), well developed (A)	P	A
Frontal bristles weak, somewhat reclinate, (P); strong, inclinate (A)	P	A
Postocellar bristles divergent (P), convergent (A)	P	P
Inner vertical bristles strong (P), weak or absent (A)	P	P
Outer vertical bristles strong (P), weak or absent (A)	P	P
Ocelli well developed (P), weak or absent (A)	P	P
Ocellar bristles strong, proclinate (P); reclinate, weak or absent (A)	P	P
Frontal vitta uniformly membranous (P), partly to entirely sclerotized (A)	P	P
Frontal vitta bare to weakly setulose (P), strongly setulose (A)	P	P
Lunule small, narrowly exposed (P); large and widely exposed (A)	P	P
Lunule weakly setulose (P), bare or strongly setulose (A) (Strickland 1953, p. 265)	P	P
Ptilinal scales extending onto face (P), onto frons (A) (Strickland 1953)	P	A
Buccal bladder strongly armed (P), weakly armed (A) (Strickland 1953)	P	A
Face uniformly sclerotized (P), membranized along midline (A)	P	P
Parafacial medium wide (P), very wide or very narrow (A)	P	P
Cheek medium wide (P), very wide or very narrow (A)	P	P
Postclypeus present (P), absent (A)	A?	P
Clypeus medium-sized (P), enlarged or reduced (A)	P	P
Vibrissa undeveloped (P), developed (A)	P	P?
Hyoid sclerite weak or absent (P), well developed (A)	P	A
Prestomal teeth absent (P), present (A)	P	A
Transverse suture incomplete (P), complete (A)	P	P
Coxopleural streak present (P), absent (A)	P	P
Greater ampulla weak or absent (P), well developed (A)	P	A
Precoxal bridge undeveloped (P), developed (A)	P	P
Prosternum bare (A), haired (P)	P	P
Metasternal area haired (P), bare (A)	P	P
Postpronotal bristles, one (P); none or several (A)	P	P
Notopleural bristles, two (P); more than or less than two (A)	P	P
Postsutural supra-alar bristle(s) present (P), absent (A)	P	P
Presutural supra-alar bristle(s) present (P), absent (A)	P	P
Postsutural intra-alar bristle(s) present (P), absent (A)	A?	P
Presutural intra-alar bristle(s) present (P), absent (A)	A?	P
Postsutural dorsocentral bristle(s) present (P?), absent (A?)	P	P

(continued)

Table 116.2 Comparison of character states in ground plans of Acalyptratae and Calyptratae *(continued)*

Character state[1]	Acalyptratae	Calyptratae
Presutural dorsocentral bristle(s) present (P), absent (A)	A	P
Postsutural acrostichal bristle(s) present (P), absent (A)	A	P
Presutural acrostichal bristles present (A), absent (P)	P	A
Scutellar bristles, two pairs (P); more or less than two pairs (A)	P	P
Lower surface of scutellum haired (P), bare (A)	A	P
Upper margins of scutellum haired (P), bare (A)	P	P
Upper disc of scutellum haired (P), bare (A)	P	P
Subscutellum weakly developed (P), strongly developed or absent (A)	P	P
Proepisternum haired (P), bare (A)	P	P
Proepisternal bristle present (A), absent (P)	P	A
Proepimeral bristle present (A), absent (P)	P	A
Anepisternum haired (P), bare (A)	P	P
Posterior anepisternal bristle(s) present (A), absent (P)	P	A
Katepisternum haired (P), bare (A)	P	P
Katepisternal bristle(s) present (A), absent (P)	P	A
Anepimeron haired (P), bare (A)	P	P
Anepimeral bristle(s) present (A), absent (P)	P	A
Katepimeron haired (P), bare (A)	A	P
Meron haired (P), bare (A)	A	P
Meral bristles absent (P), present (A)	P	P
Laterotergite haired (P), bare (A)	A	P
Costa without (P), with (A) costagial break	P	A
Costa without (P), with (A) humeral break	P	A
Costa without (P), with (A) subcostal break	P	A
Pterostigma present (P), absent (A)	P	A
Subcosta complete (P), incomplete (A)	P	P
Subcosta straight (P), inflexed (A) opposite humeral break	P	A
Crossvein sc-r present (P), absent (A)	P	A
R_1 setose above (P), bare above (A)	P	P
Apical portion of M relatively straight (P), anteriorly flexed (A)	P	P
Cell *cup* long, closing near wing margin (P); short, closing far from wing margin (A)	P	A
A_1 reaching wing margin (P), not reaching wing margin (A)	P	A
Cell dm relatively long (P), relatively short (A)	P	A
Anal lobe well developed (P), reduced (A)	P	P
Alula well developed (P), reduced (A)	P	P
Lower calypter moderately large (P), reduced (A)	A	P
Wings unpatterned (P), patterned (A)	P	P
Knob of halter whitish (P), blackish (A)	P	P
Tarsi entirely blackish (P), partly pale (A)	P	P
Tarsomere 1 of hind tarsus simple (P), modified (A)	P	P
Preapical dorsal tibial bristle undeveloped (P), developed (A)	P	A
Abdominal bristles weak or absent (P), strong (A)	P	A?
Abdomen with 7 spiracles in both sexes (P), less than 7 (A)	P	P
Abdominal spiracles in membrane (P), in tergites (A)	P	A
Male with hind margin of 5th sternite regular (P), incised (A)	P	A
Tergite 6 relatively unmodified (P), reduced asymmetrical (A)	P	A
Tergite 6 free (P), fused with segment 7 (A)	P	A
Sternite 6 relatively unmodified (P), reduced and shifted to left side (A)	P	A
Sternite 6 free from tergite 6 (P), fused with tergite 6 in left side (A)	P	A
Segment 7 discrete, free from segment 8 (P); reduced, fused with segment 8 (A)	P	A
Tergite 8 discrete, in ventral position (P); indistinguishable (A)	P	A
Syntergosternite 8 large, setose (P); reduced, bare (A)	P	P
Spiracle 6 present (P), absent (A)	P	P
Left spiracle 6 in membrane (P), enclosed in pregenital sclerite (A)	P	A

(continued)

Table 116.2 Comparison of character states in ground plans of Acalyptratae and Calyptratae *(concluded)*

Character state[1]	Acalyptratae	Calyptratae
Gonopod fused to hypandrium (P), with membranous articulation (A)	P	A
Paramere small, feeble (P); large strongly sclerotized (A)	P	A
Aedeagus simple, one-segmented (Downes 1955, p. 534) (P); specialized (A)	P	A
Aedeagus without (P), with (A) distinct distiphallus containing acrophallus	P	A
Basiphallus without (P), with epiphallus (A)	P	A
Aedeagal apodeme rod-like (P), fultelliform (A)	P	P
Aedeagal apodeme and basiphallus widely (P), narrowly separated (A) (Hennig 1976*a*, p. 97)	P	A
Sternite 10 plate-like, undivided (P); divided into bacciliform sclerites (A)	P	P
Surstylus present (P), absent (A)	P	P
Surstylus movable (P), fused to epandrium (A)	P	P
Surstylus unilobate (P), bilobate (A)	P	P
Surstylus without (P), with prensisetae (A)	P	P
Surstyli loosely (P), closely linked with cerci (A)	P	P
Cerci membranous, free (P); sclerotized, contiguous or fused (A)	P	A
Anus between (P), above (A) base of cerci	P	P
Female with 6th tergite and sternite free (P), fused (A)	P	P
6th tergite and sternite undivided (P), divided (A)	P	P
7th tergite and sternite free (P), fused (A)	P	P
7th tergite and sternite undivided (P), divided (A)	P	P
Tergite 8 undivided (P), divided (A)	P	P
Sternite 8 undivided (P), divided (A)	P	P
Hypoproct with lingulae absent (P), present (A)	P	A
Three spermathecal ducts free, opening independently into genital chamber (P); one or more spermathecal ducts fused, with two or one opening(s) into genital chamber (A)	A	P
Ventral receptacle absent (P), present (A)	A	P

[1] P = plesiomorphic; A = apomorphic.

represent were already well differentiated by Oligocene times.

One of the most important of the 14 autapomorphic characters advanced for the Acalyptratae is the dichoptic conditon of the males, as opposed to the primitive holoptic condition retained in the ground plan of the Calyptratae (van Emden 1965, J. F. McAlpine and Munroe 1968, Griffiths 1972). Hennig (1971*a*) questioned whether male dichopticism is really an apomorphic condition in the Schizophora, but his arguments against it are unconvincing. Holopticism of males is a firmly entrenched ground-plan feature of the order; it is intimately linked with swarming and aerial mating, and probably led to the dipterous condition of the Diptera (J. F. McAlpine and Munroe 1968). It is retained as a ground-plan condition not only in the Muscomorpha, Aschiza, and Schizophora, but also in all the other main categories of the Diptera. All evidence points to the holoptic condition having been lost repeatedly throughout the Diptera (as is also true for the repeated loss of mandibles). Examples of male dichoptism occur in specialized lines in all the main categories of the order including the Aschiza, Schizophora, and even Calyptratae. Even Hennig himself (Hennig 1965*a*) admitted that arguments, to the effect that male holopticism was lost in the ground plan of the Schizophora and then was regained secondarily in the Calyptratae, are difficult to sustain. It is true, however, that the male frons is narrower than that of the female in a few representatives of the Acalyptratae, e.g. Tanypezidae, Lonchaeidae, some Otitidae, some Pallopteridae, some Heleomyzidae, some Lauxaniidae, some Chamaemyiidae, some Curtonotidae, and some Milichiidae. But such cases are not really holoptic when compared with truly holoptic conditions in Anthomyiidae, Muscidae, and Calliphoridae. It should be noted also, that, whereas the tendency toward holoptic males is considered primary in some families of Acalyptratae, e.g. Tanypezidae, Lonchaeidae, and Heleomyzidae, it probably is a secondarily derived condition in others, e.g. *Neoleucopis* Malloch spp. (Chamaemyiidae), *Neomaorina* Miller (Pallopteridae), *Axinota* Wulp (Curtonotidae), and *Holopticander papuanus* Hennig, and *Trypaneoides* Tonnoir & Malloch (Lauxaniidae). According to Stuckenberg (1971) the narrower frons in males of the two lauxaniids suggest crepuscular or nocturnal habits.

The reduced lower calypter in the Acalyptratae (except many Platystomatidae) is usually considered to be an autapomorphic condition (these moderately large in Calyptratae with the exception of Scathophagidae and some Anthomyiidae). A number of chaetotactic characters also appear to be autapomorphic in the ground plan of the Acalyptratae. These include absence of the following bristles and hairs: pre- and postsutural intra-alar bristles, presutural dorsocentral bristles, pre- and postsutural acrostichal bristles, ventral scutellar hairs, katepimeral hairs, meral hairs, and laterotergal hairs (all present in the ground plan of Aschiza and Calyptratae). In addition, there may be an autapomorphic character in the structure of the clypeus in the Acalyptratae. According to Gouin (1949) the postclypeus is absent in the Acalyptratae (except Conopidae), but it is present in the Syrphidae (Aschiza) and the Calyptratae. If the presence of a postclypeus belongs to the ground plan of the Schizophora, its reduction in the Acalyptratae could be an autapomorphic character (Hennig 1971*a*). Finally, the spermathecal ducts of the Acalyptratae are partially fused before joining the common oviduct, so that there are only one or two openings into the genital chamber (Sturtevant 1925–1926). In the ground plan of the Calyptratae, as in the ground plan of the Aschiza and the Muscomorpha as a whole, all three spermathecal ducts reach the common oviduct separately (Sturtevant 1925–1926). If the fused condition of these ducts is truly a ground-plan condition of the Acalyptratae, it also constitutes an important autapomorphic character for the group.

As far as is known, a peculiar seminal vesicle called the ventral receptacle is found only in the Acalyptratae (Sturtevant 1925–1926; J. F. McAlpine, Ch. 2). In its simplest state it occurs as an unmusculated, membranous pocket on the ventral surface of the genital chamber, as in certain Clusiidae and Heleomyzidae; in some specialized cases it is heavily sclerotized, e.g. in the Ephydridae. A thorough survey of this organ in all families is required. If it turns out to belong to the ground plan of the Acalyptratae, it would also be a very significant autapomorphic character for the group.

In almost all Acalyptratae the prothoracic pupal respiratory horns are enclosed within the puparium, but in the majority of Calyptratae the outer lobe of each prothoracic spiracle of the pupa develops into a sclerotized spiracular horn, which protrudes through small circular apertures in the first abdominal tergite of the puparium (Roddy 1955). Hennig postulated that the virtual absence of prothoracic, pupal respiratory horns in the Acalyptratae (present only in a few Heleomyzidae) was an autapomorphic character of the group. Even if these horns were present in the ground plan of the Acalyptratae, they were probably more weakly developed than they were in the ground plan of the Calyptratae and were soon entirely lost except in some Heleomyzidae. Another possible autapomorphy concerns the relative development of tracheal air sacs. These sacs are relatively small in the Acalyptratae (possibly through reduction) compared to their large size and development in the Calyptratae (Faucheux 1971).

These autapomorphic conditions, together with the remainder of the 14 characters marked (A) in Table 116.2, indicate that the Acalyptratae is, in fact, a monophyletic subgroup of the Schizophora.

SUBORDINATE GROUPS OF ACALYPTRATAE

The phylogenetic classification of the Acalyptratae is probably the most difficult chapter in the systematics of Diptera. One reason is that the group is so large; it contains about half of the families (65) recognized in the order. Another reason is that its component families, although individually often small and poorly known, are, as a group, enormously varied and complex. Each family has retained a somewhat different combination of characters that must have prevailed in the ancestral acalyptrate fly, and at the same time, each one has acquired certain apomorphic features of its own. Because of parallelisms and convergences, conditions that appear similar occur commonly in different families. As a result of these factors and the great amount of extinction that has probably occurred, the distribution of many characters among the different families forms a puzzling, mosaic-like pattern, which seems to contradict a continuous phylogenetic sequence. Predictably, these difficulties have given rise to differences of opinion about the placement and ranking of certain species, genera, and families. Despite the fact that many problems are still unresolved, a certain consensus prevails in the conclusions of most serious workers about the main superfamilies. Most classifications now in use recognize eight to ten superfamilial groups, clustered around eight to ten key families. These include the micropezids (Nerioidea), the tanypezids (Diopsoidea), the conopids (Conopoidea), the otitids (Tephritoidea), lauxaniids (Lauxanioidea), sciomyzids (Sciomyzoidea), clusiids (Opomyzoidea), chloropids (Carnoidea), heleomyzids (Sphaeroceroidea), and drosophilids (Ephydroidea). Although the contents and limits of these clusters vary somewhat from worker to worker, they almost certainly reflect a basic framework of cladistic units within the Acalyptratae that has been apparent for a long time.

The most important single attempt at defining the monophyletic superfamilial subgroups of the Acalyptratae and at classifying them in accordance with phylogenetic principles is Hennig's (1958) classic treatment. That work is a synthesis of his own previous works and those of others completed before that time. It still serves as the best starting point for anyone wishing to understand the evolution and classification of the group. In it, Hennig analyzed various sequences of characters from all parts of the body and laid a foundation for judging their plesiomorphic and apomorphic conditions in most acalyptrate families. Based on these interpretations, he proposed a provisional phylogenetic classification of the families and superfamilies. Subsequently, he offered many new ideas, refinements, and improvements, which also must be consulted (see especially Hennig 1965*b*, 1967*b*, 1969*b*, 1971*a*, 1973, and 1976*b*). Griffiths (1972) presented a revision of Hennig's work based mainly on an analysis of the male terminalia, which also contains many interesting and provocative leads that should be considered. However, he underestimated the variability that occurs in all parts of the male terminalia within most acalyptrate families, and he overestimated their reliability as indicators of broad relationships. Consequently, some of his assumptions are unsound. Steyskal (1974) summarized the classifications of Hennig and Griffiths in a dendrogram and tried to solve the problem of naming superfamilial categories throughout the Diptera. The following is one more attempt to elucidate further the evolution of the Acalyptratae and to superimpose upon it an improved system of classification.

No single species, genus, or family of flies is known that embodies all of the gound-plan characters of the Acalyptratae (Table 116.2). If such an example were discovered it would probably be given new family status, because it would not fit in any of those now recognized. But, if evolution is a fact, and if the Acalyptratae is a monophyletic taxon, an ancestral species closely agreeing with these attributes must have existed (probably in the mid to late Cretaceous period), and all the families that we now know must have descended from it. Consequently, the first question to be answered is: What existing flies most resemble the ancestral one? My approach to obtaining an answer is to check the ground-plan characters of all the recognized families and superfamilies of Acalyptratae against those deduced for the acalyptrate prototype (Table 116.2). The next step is to work out and to document plausible cladistic arrangements for the families first within the superfamilies (see Figs. 116.2–8), and then for the superfamilies themselves (see Fig. 116.9). To begin, I organized these families according to the superfamilial groupings already mentioned; these groupings usually correspond to those provisionally adopted in the table of classification given in Volume 1 of this Manual (J. F. McAlpine *et al.*, Ch. 1). But, as will be seen in the discussions of the superfamilies and in the cladistic diagrams (see Figs. 116.2–8), a few families are moved to new positions usually in the same, but sometimes in additional or different, superfamilies.

Superfamily Nerioidea

It is generally agreed that the Nerioidea (= Micropezoidea of most authors) consists of the Micropezidae (including Calobatidae (= Trepidariidae) and Tylidae), Neriidae, and Cypselosomatidae (including Pseudopomyzidae) (Hennig 1958, 1971*a*, 1973; Griffiths 1972; D. K. McAlpine 1974*a*) (Fig. 116.2). According to Sabrosky (personal commun.), Neriidae (proposed as Neriades by Westwood 1840, p. 148) is the oldest family-group name involved and, therefore, is the proper one to use for the superfamily.

This superfamily has retained more plesiomorphic conditions in its ground plan than any other acalyptrate superfamily. It is the only schizophorous group in which the male retained both the tergite and sternite of abdominal segment 6 as large, separate, and relatively unmodified sclerites and, at the same time, also retained both the tergites and sternites of segments 7 and 8 as separate sclerites (Fig. 57.6). Moreover, the female retained three spermathecae. Other important plesiomorphic features in the ground plan of Nerioidea include: fronto-orbital plates continuous and reaching far forward, four fronto-orbital bristles present, vibrissae undeveloped, arista arising near apex of first flagellomere, C unbroken, Sc complete and free from R_1, cell cu*p* long, and with vein CuA_2 meeting vein A_1 at an acute angle, and vein A_1 complete.

The following autapomorphies (synapomorphies of the included families) in the ground plan of Nerioidea attest to its monophyly:

- frons as wide in male as in female;
- epandrium very elongate and trough-like ventrally;
- hypandrium very elongate; together with gonopods and aedeagal guide, forming a posteriorly directed cone with the aedeagus borne at its apex;
- gonopods reduced to a pair of elongate sclerites (sclerite 'X' of Griffiths 1972);
- aedeagus anteriorly directed;
- sternite 10 of male elongate, H-shaped; in rest position, lying ventral to hypandrial cone (see especially, Fig. 57.6);
- sternite 7 and tergite 7 of female lengthened and fused to form a bulbous ovipositor base (oviscape); and
- sternite 8 and tergite 8 of female reduced to a membranous, eversible tube.

No single taxon of the Nerioidea embodies all the ground-plan features of the group. The most plesiomorphic conditions in the form and chaetotaxy of the head (more or less hemispherical in shape, four fronto-orbital bristles, divergent postocellars, strong ocellars) is probably best exemplified in the Cypselosomatidae, but here (except for the fossil species *Cypselosomatites succini* Hennig), it is combined with well-developed vibrissae, which is an apomorphic condition. The most plesiomorphic antennae (subapical arista, notched pedicel) are found in the Micropezidae. The thoracic form and chaetotaxy that is nearest the plesiomorphic ground-plan condition is probably to be found in the ground plan of the Neriidae, except that here the metasternal area is bare (haired in Micropezidae). The most plesiomorphic wing venation (C unbroken, Sc complete, R_1 setulose, R_{4+5} and M_1 not strongly convergent at apices, cells bm and dm separate, cell cu*p* rather long and acute posteroapically, and A_1 complete) occurs in the Micropezidae. The most plesiomorphic conditions of the male terminalia occur in the neriid genus *Odontoloxozus* Enderlein. With regard to females, however, the Micropezidae is the only family in which three spermathecae were retained in the ground plan.

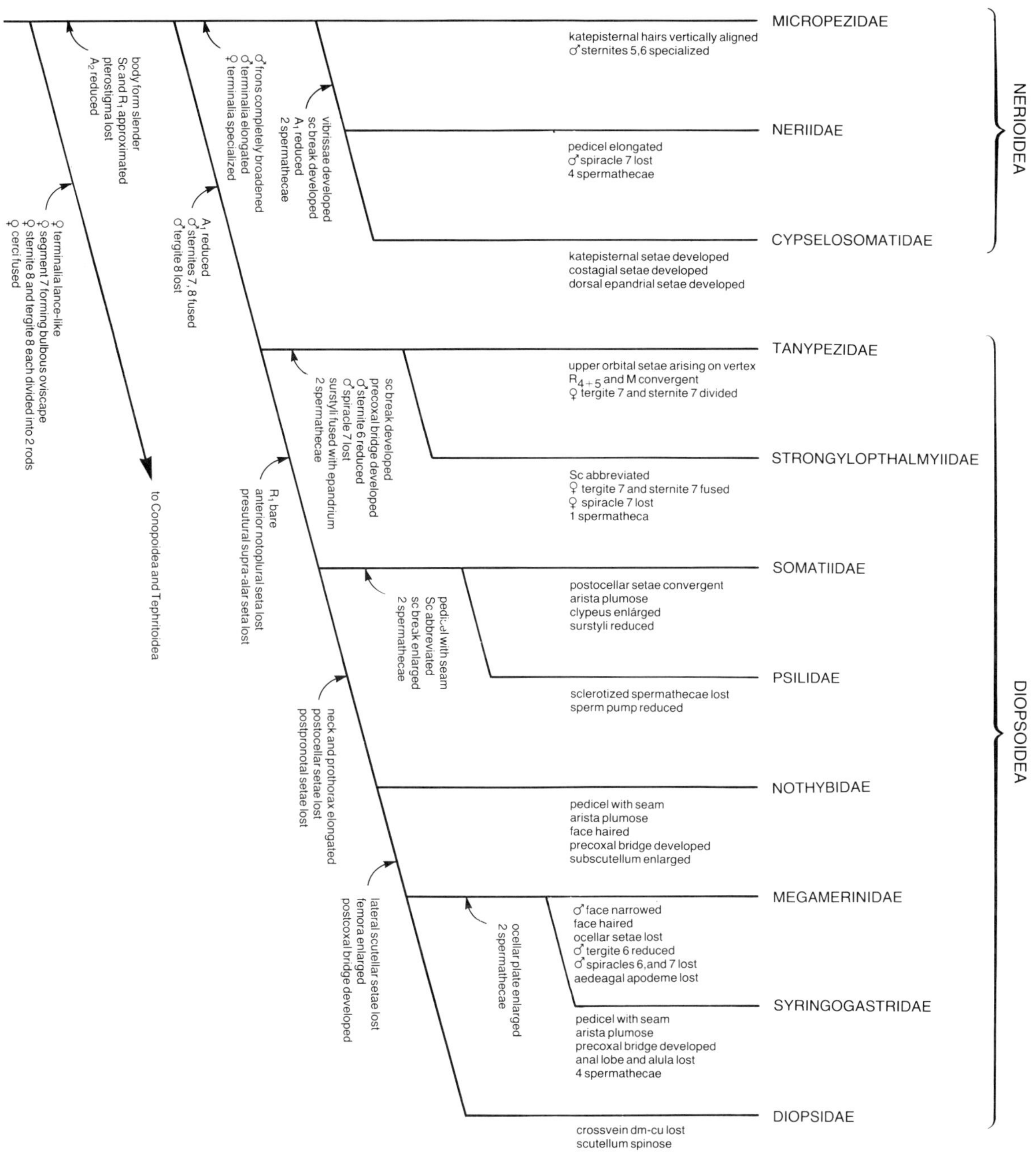

Fig. 116.2. Cladogram of the Nerioidea and Diopsoidea summarizing the relationships and apomorphies of the subgroups and families recognized. (See text for additional details.)

Consequently, I believe the Nerioidea contains two main sister groups, the Neriidae + Cypselosomatidae on the one hand, and the Micropezidae on the other. Neither sister group is strikingly more generalized or specialized than the other. In the line leading to the Neriidae and Cypselosomatidae, sternites 5 and 6 of the male remained in a plesiomorphic condition, and ocellar bristles and four fronto-orbital bristles were retained. Synapomorphies between the Neriidae and the Cypselosomatidae include:

— vibrissae differentiated;
— C weakened and broken at the apex of Sc;
— Sc reduced, failing to reach the wing margin;
— cell cu*p* short and obtusely closed;
— A_1 abbreviated;
— metasternal area bare; and
— spermathecae reduced to two (duplicated to four in Neriidae).

Micropezidae. In the Micropezidae (Steyskal, Ch. 56) the vibrissae remained undifferentiated, C remained unbroken, cell cu*p* remained longer and acutely closed, A_1 remained complete, the metasternal area remained haired, and the female retained three spermathecae. The following autapomorphic conditions (synapomorphic conditions of the component taxa) developed:

— sternite 5 of male bilobed;
— sternite 6 of male highly modified;
— body form long and slender;
— scutellum reduced;
— front legs short and weak; and
— katepisternal bristles vertically aligned.

The family name, Micropezidae, was proposed by Loew (1862). An excellent treatment of the cladistics and classification of the group was provided by D. K. McAlpine (1974*a*). He recognized five subfamilies; the Caleopteryginae, Calobatinae, Micropezinae, Eurybatinae, and Taeniapterinae. European authors, including Hennig (1973 and elsewhere), have often treated the subfamilies Calobatinae, Micropezinae, and Taeniapterinae as separate families. According to Steyskal (Ch. 56), about 500 species, divided among about 40 genera, are known. The family occurs in all zoogeographical regions, but distribution is primarily tropical. Two fossil species from Baltic amber, both in the genus *Electrobata* Hennig (Hennig 1965*b*) are known.

Neriidae. Autapomorphic characters of the Neriidae (Steyskal, Ch. 57) include the following:

— pedicel elongate and porrect, with finger-like extension on inner apex;
— fronto-orbital bristles reduced to two or three;
— postocellar bristles convergent;
— ocellar bristles absent;
— transverse scutal suture complete or nearly so;
— only one pair of scutellar bristles present;
— spiracle 7 of male lost; and
— female with four spermathecae.

The family-group name was proposed (as Neriades) by Westwood (1840). The family is mainly tropical in distribution (Hennig 1937*a*) and consists of about 110 species (Steyskal, Ch. 57) in about 17 genera (Steyskal, 1968*a*, 1977, 1980). Two subfamilies, the Neriinae (circumtropical) and Telostylinae (Old World tropics), are recognized. No fossils are known.

Cypselosomatidae. The Cypselosomatidae (Hendel 1931) and the Pseudopomyzidae (Frey 1941*a*) have often been treated as separate families (see especially, D. K. McAlpine 1966; Hennig 1969*a*, 1971*a*, 1971*b*, 1973; Andersson 1976*b*; Krivosheina 1979*a*), but I agree with Griffiths (1972), Harrison (1976), and Rohdendorf (1977) that they are monophyletic and treat them (J. F. McAlpine, Ch. 55) as sister groups (subfamilies) of a single family.

Autapomorphic characters of the Cypselosomatidae (synapomorphies of the Cypselosomatinae and Pseudopomyzinae) include:

— vibrissae developed;
— arista arising dorsobasally, with subcostal break;
— costagial bristle very strong;
— male with strong paired bristles on sternite 8 and epandrium; and
— female with two spermathecae.

In the ground plan of the Cypselosomatinae, the postocellar bristles are divergent (plesiomorphic condition). Autapomorphies include:

— cells bm and dm confluent;
— A_1 abbreviated;
— R_{4+5} and M_1 convergent;
— lowest fronto-orbital bristles medially located and medioclinate (sometimes called interfrontals); and
— scutellum with only one pair of strong bristles.

Two extant genera, *Cypselosoma* Hendel (D. K. McAlpine 1966, 1978) and *Formicosepsis* de Meijere (Andersson 1976*b*), are known. The fossil genus *Cypselosomatites* Hennig (Hennig 1965*b*, 1971*c*) from Baltic amber probably also belongs here.

In the ground plan of the Pseudopomyzinae, the postocellar bristles are convergent (autapomorphic condition). Plesiomorphically, cells bm and dm remained separate, A_1 is long (practically complete in *Heloclusia* Malloch), R_{4+5} and M_1 are not very convergent, the lowest fronto-orbital bristles are not strongly shifted medially, and the scutellum bears at least two pairs of strong bristles. This subfamily contains seven extant genera (see J. F. McAlpine, Ch. 55) and one fossil species in Baltic amber (Hennig 1971*b*); the most recent treatments of the group are those of Hennig (1969*a*) and Krivosheina (1979*a*).

Superfamily Diopsoidea

My investigations confirm that Hennig (1958, 1965*b*, 1973) was probably correct in regarding the families Tanypezidae, Strongylophthalmyiidae, Psilidae, Nothybidae, Megamerinidae, Syringogastridae, and Diopsidae as a monophyletic superfamily (his Nothyboidea) (Fig. 116.2). This viewpoint is also reflected in the classifications adopted by Stone *et al.* (1965), Papavero (1967*a*), Colless and D. K. McAlpine (1970, 1974), Delphinado and Hardy (1977), Crosskey *et al.* (1980), and Soós and Papp (1984*a*). Griffiths' (1972) reasons for dispersing these same families among several "prefamilies" in two superfamilies are unconvincing.

The oldest familial name in the group is Diopsidae, which dates from Billberg (1820). Therefore, Diopsoidea supersedes previously used superfamily names such as Nothyboidea and Tanypezoidea.

In relation to the ground plan of the Nerioidea, the following autapomorphies (synapomorphies of the included families) can be advanced for the Diopsoidea:

— orbital bristles reduced to two or three;
— arista situated dorsally;
— cell *cup* closed obtusely, i.e. CuA_2 recurved;
— A_1 fading out apically, rarely if ever reaching wing margin;
— sternite 7 of male fused with sternite 8;
— tergite 7 of male largely fused with pregenital segment or absent; and
— tergite 8 of male atrophied.

The sister group of the Diopsoidea is the Nerioidea. Synapomorphies of these two superfamilies with respect to the ground plan of the Acalyptratae are as follows:

— body, legs, and wings slender;
— Sc and R_1 approximated;
— pterostigma lost; and
— A_2 atrophied.

One ground-plan feature of the Diopsoidea that is more plesiomorphic than in the Nerioidea is sexual dimorphism in the head (tendency toward male holopticism). A narrower male frons was retained in one family, the Tanypezidae. A narrower male face appears in the Megamerinidae and may be a related phenomenon. Another more plesiomorphic character of the Diopsoidea is the absence of a differentiated katepisternal bristle in all the component families.

Within the Diopsoidea there are two main sister groups. Group 1 consists of the Tanypezidae and Strongylophthalmyiidae, and group 2 consists of the Somatiidae, Psilidae, Nothybidae, Megamerinidae, Syringogastridae, and Diopsidae (Fig. 116.2).

Synapomorphic characters in the ground plan of group 1 include:

— precoxal bridge present;
— calypteres with long hairs at fold;
— C with a subcostal weakening or break;
— greater ampulla weakly present;
— male with spiracle 7 absent;
— surstyli fused with epandrium;
— epiphallus present;
— sternite 10 divided;
— female with sternites 6, 7, and 8 lengthened to form an elongate, retractible ovipositor; those of segments 7 and 8 longitudinally divided;
— cerci partially fused; and
— spermathecae reduced (two and one, respectively).

Group 1 is more generalized in its ground plan than group 2 in the following characters: male frons narrower than that of female, postpronotal bristles retained, notopleuron with two bristles, chaetotaxy of scutum more complete, and R_1 setose above.

Tanypezidae. The Tanypezidae (Steyskal, Ch. 58) are more generalized than the Strongylophthalmyiidae in having the frons narrower in the male than in the female, three instead of two fronto-orbital bristles, costal break very weak or absent, a complete subcosta, and two (instead of one) spermathecae.

Autapomorphic characters of the family are as follows:

— upper orbital bristle arising on or near vertex;
— first flagellomere relatively elongate;
— R_{4+5} and M_1 convergent at apices;
— CuA_2 strongly recurved;
— male with paramere reduced; and
— female with tergite 7 and sternite 7 each divided into two elongate strips.

The Tanypezidae is primarily a New World group, consisting of *Tanypeza* Fallén (two species, one holarctic) and *Scipopeza* Enderlein (16 Neotropical species) (Steyskal, Ch. 58). The family name was proposed (as Tanypezina) by Rondani (1856). No fossils are known.

Strongylophthalmyiidae. This family is more generalized than the Tanypezidae with respect to the courses of veins R_{4+5} and M_1 and several other characters (Steyskal, Ch. 59). Autapomorphic characters of the family are:

— Sc incomplete;
— female without spiracle 7;
— tergite 7 and sternite 7 of female fused; and
— one spermatheca present.

The family consists of only one genus, *Strongylophthalmyia* Heller, containing 27 species mostly from southern Asia. Hennig (1958) seems to have been the first to rank it as a separate family; previously it was considered a subfamily of the Psilidae. No fossils are known.

Synapomorphies of the six families of group 2 of the Diopsoidea include the following:

— anterior notopleural bristle lost;
— presutural supra-alar bristle lost; and
— R_1 bare.

Group 2 seems to consist of two sister groups, i.e. the Psilidae + Somatiidae, and the Nothybidae + Megamerinidae + Syringogastridae + Diopsidae.

Synapomorphic features in the ground plan of the Somatiidae and Psilidae include:

— frons equally broad in both sexes;
— pedicel with a complete antennal seam;
— arista situated subbasally;
— orbital bristles reduced to two;
— fronto-orbital plate abbreviated;
— disc of propleuron bare;
— C with a strong subcostal break;
— Sc abbreviated;
— pregenital segment of male secondarily symmetrical; and
— hypandrial bridge incomplete.

The ground plan of the Somatiidae + Psilidae remained more generalized than that of the Tanypezidae + Strongylophthalmyiidae in several respects: the prosternum remained free from the propleuron (precoxal bridge absent); in the male, sternite 6 remained relatively large and unmodified, spiracle 7 was retained, the surstyli remained free and movable (lost in Somatiidae), and the aedeagus remained without an epiphallus; in the females, the cerci remained separate (although apomorphically they are fused in some representatives).

Psilidae. The Psilidae (Steyskal, Ch. 60) remained more generalized in body form, wing shape and venation, and chaetotaxy than did the Somatiidae. Within the family Psilidae, great differences are evident in both the male and female terminalia. In males, spiracle 7 may be present or absent, the aedeagus may be very short and simple or very long and complex, the aedeagal apodeme may be rod-like and free (cuneiform) or short and fused with the hypandrium (fultelliform), and sternites 7 and 8 may be well developed or virtually absent. In females, the cerci may be separate or fused.

Two subfamilies, the Psilinae, containing *Psila* Meigen, *Loxocera* Meigen, and *Psilosoma* Zetterstedt, and the Chylisinae, containing *Chyliza* Fallén and *Electrochyliza* Hennig, are recognized (Steyskal, Ch. 60). The most generalized conditions occur in the Psilinae. Autapomorphies (synapomorphies of the subfamilies) that seem to apply to all members of both subfamilies include:

— male with reduced sperm pump and ejaculatory apodeme;
— female without sclerotized spermathecae; and
— sclerites of abdominal segment 8 of female consolidated into a tubiform ovipositor.

The family name, based on *Psila* (Meigen 1803) (unjustifiably changed to *Psilomyia* by Latreille 1829) was proposed (as Psilomydae) by Macquart (1835, p. 416); in the same paper (p. 372), Macquart also proposed Loxoceridae based on *Loxocera* (Meigen 1803). Because of the long and consistent use of Psilidae, however, that name is retained in preference to Loxoceridae. Distribution of the family is basically Holarctic with a few species in the Oriental, Ethiopian, and Neotropical regions. About 200 species are known, including one fossil species in Baltic amber, *Electrochyliza succini* Hennig (Hennig 1965*b*, 1969*a*).

Somatiidae. The Somatiidae are more generalized than the Psilidae in having a large ejaculatory apodeme in the male, two sclerotized spermathecae and a simple, nonretractile ovipositor in the female. Autapomorphies include the following:

— arista plumose;
— interfrontal setulae (lowermost one strong) arranged in a vertical row;
— outer vertical bristle reduced or absent;
— postocellar bristles convergent;
— clypeus enlarged;
— thorax stout;
— transverse suture complete;
— greater ampulla secondarily reduced;
— abdomen very broad;
— male with surstyli absent;
— parameres reduced;
— basiphallus coniform, twisted;
— distiphallus very long, ribbon-like, twisted; and
— female with tergite 7 and sternite 7 fused, enclosing the seventh spiracles.

The family is comprised of a single genus, *Somatia* Schiner, containing only seven very similar species (Steyskal 1968*d*) and is restricted to the Neotropical region. Hendel (1935) was the first to rank it as a separate family. The immature stages and habits are unknown. No fossils are known.

Griffiths' (1972) assignment of *Somatia* to the Periscelididae cannot be sustained. The many synapomorphies, listed above, for the Somatiidae and Psilidae support Hennig's (1971*a*) view that they are sister groups. My reasons for placing the Periscelididae in the Opomyzoidea are given under Asteioinea.

Synapomorphic characters in the ground plan of group 2 (the Nothybidae, Megamerinidae, Syringogastridae, and Diopsidae) are as follows:

— neck and prothoracic region lengthened;
— postocellar bristles absent; and
— postpronotal bristles absent.

The ground plan of this group of families is more generalized than that of the Psilidae + Somatiidae in the following respects: C without a costal break; male with sternite 10 undivided; female with a short, simple ovipositor, i.e. tergites and sternites of segments 6, 7, and 8 short, simple, and free from each other; and three sclerotized spermathecae present.

The Nothybidae are more generalized than the remainder of the group on at least three counts: the scutellum has retained two pairs of scutellar bristles, there is no postcoxal bridge, and none of the femora is enlarged. In contrast, the Megamerinidae, Syringogastridae, and Diopsidae show the following synapomorphies:

— scutellum with only one pair of strong scutellar bristles (basal pair weak or absent);
— postcoxal bridge present; and
— front or hind femora, or both, enlarged and strongly setose below.

Nothybidae. The Nothybidae was first treated as a separate family by Frey (1927). Only eight species are known (Acxél 1955; D. K. McAlpine 1974*b*), all restricted to the Oriental region. The family displays many autapomorphic features as follows:

— body long, with long, slender legs and wings;
— pedicel with a complete dorsal seam;
— arista strongly plumose;
— face haired;
— anterior part of thorax greatly elongate;
— precoxal bridge present;

— laterotergite haired;
— subscutellum greatly enlarged;
— metasternal area bare;
— base of M atrophied before bm-cu crossvein (cells bm and dm confluent);
— anal lobe and alula lost;
— male with tergites and sternites 7 and 8 reduced and secondarily symmetrical; and
— aedeagal apodeme small and fused with hypandrium.

The two species studied, *Nothybus longithorax* Rondani and *N. biguttatus* van der Wulp, show great differences in both male and female terminalia. In males of the former, tergite 6 is much reduced, seven pairs of spiracles are present, the hypandrial bridge is complete, sternite 10 is divided, and the surstyli are simple; females have three spermathecae and appear to be viviparous (one large, well-developed, interuterine larva, with head directed anteriorly, was found inside the enlarged oviduct of the female examined). In *N. biguttatus*, males have a large, unmodified tergite 6, but the seventh pair of spiracles is absent, the hypandrial bridge is incomplete, sternite 10 is undivided, and the surstyli are strongly bilobate; females have only two spermathecae. No information has been published on the immature stages or biology of the family. According to D. K. McAlpine (1974*b*) adults of *N. decorus* de Meijere are able to hover. No fossils are known.

An important key to the evolution of the Megamerinidae + Syringogastridae + Diopsidae line is the fossil species, *Palaeotanypeza spinosa* Meunier. Hennig (1965*b*) assigned it to the Megamerinidae and indicated that all recent members of the family (he included *Syringogaster* Cresson in Megamerinidae) could be derived from it. Judging from the entire complement of characters shown by Hennig, it also qualifies in some respects as an ancestor to Syringogastridae + Diopsidae. As in all three families, it has only one pair of scutellar bristles and a postcoxal bridge. However, *both* the front and hind femora are enlarged and coarsely setose below (hind femur only thus modified in the Megamerinidae and Syringogastridae; front femur only so modified in the Diopsidae). Unlike extant Megamerinidae, abdominal tergite 6 of the male is well developed as in the ground plan of the Syringogastridae and Diopsidae. In addition, it appears as if the pregenital segment is relatively short as in the Syringogastridae and Diopsidae (unusually elongate in extant Megamerinidae). *Palaeotanypeza spinosa* is more generalized than any other member of the Megamerinidae, Syringogastridae, or Diopsidae in having two pairs of orbital bristles. One autapomorphic character shown by *P. spinosa* in relation to extant Megamerinidae, Syringogastridae, and Diopsidae is the convergence of veins R_{4+5} and M_1, a specialization that recurs many times in the Nerioidea and Diopsoidea.

Certainly *Palaeotanypeza* Hennig cannot be *the* direct ancestor of either the Syringogastridae or Diopsidae. For one thing, ocellar bristles are absent in *Palaeotanypeza* but present in the Syringogastridae, indicating that the latter probably originated from an earlier ancestor possibly common to both *Palaeotanypeza* and *Syringogaster*. The convergent condition of veins R_{4+5} and M_1 in *Palaeotanypeza* is more specialized than in the Diopsidae, indicating that the Diopsidae originated before that condition developed. Consequently, Hennig (1965*b*) was probably correct in advocating a broader definition for the Megamerinidae, (restricted to the Oriental region) and assigning *Palaeotanypeza* to that family as an older clade.

At least two lines evolved from an ancestor with many of the qualifications mentioned for *Palaeotanypeza*. One line probably led to the Megamerinidae and Syringogastridae, and the other led to the Diopsidae (Fig. 116.2).

Synapomorphies of extant Megamerinidae and Syringogastridae include:

— ocellar plate enlarged;
— hind femur, only, enlarged and bearing two rows of stout ventral setae; and
— spermathecae reduced to two (doubled in Syringogastridae).

Symplesiomorphically, with respect to the Diopsidae, both groups retained crossvein bm-cu (absent in the Diopsidae) and the single pair of scutellar bristles remained simple (situated on elongate tubercles in the Diopsidae).

Comparison of Megamerinidae and Syringogastridae. These two families differ markedly from each other, and each one has many autapomorphic features. Some of these differences are tabulated in Table 116.3; the autapomorphies are indicated by asterisks.

Megamerinidae. Hendel (1913) first ranked the Megamerinidae as a separate family, but he and most subsequent authors included *Syringogaster* in it. As the family is now defined (Griffiths 1972, Hennig 1973), it consists of only 13 known species in four genera: *Megamerina* Rondani (2 spp., Palearctic); *Gobrya* Walker (= *Syrittomyia* Hendel) and *Texara* Walker (5 spp. in each, Oriental-Papuan); and *Palaeotanypeza* (1 sp., Baltic amber). There seems to be good reason for ranking *Palaeotanypeza* as a separate subfamily of the Megamerinidae, as is now usually accepted for *Centrioncus* Speiser in the Diopsidae (Hennig 1965*b*, Shillito 1971, Steyskal 1972*a*). Steyskal (1977) catalogued most of the known species.

Syringogastridae. I agree with Prado (1969), Griffiths (1972), and Hennig (1973) that *Syringogaster* (restricted to the Neotropical region) represents a separate family. It seems to be more closely related to Megamerinidae than to any other family, but its precise ancestry is unknown. Only eight species are known, all in the genus *Syringogaster* (Prado 1969).

Diopsidae. The line leading to the Diopsidae (Peterson, Ch. 61) displayed the following autapomorphies:

— antennal pedicel without dorsal notch or seam (as in Megamerinidae);
— ocellar bristles absent (as in Megamerinidae);

Table 116.3 Comparison of character states in ground plans of Megamerinidae and Syringogastridae

Character	Megamerinidae	Syringogastridae
Pedicel	* without notch or seam	* with complete dorsal seam
Arista	shortly pubescent	* plumose
Face	* haired	bare
	* narrower in male	equally wide in both sexes
Orbital bristles	one (two in *Palaeotanypeza*)	* none
Ocellar bristles	* absent	present
Precoxal bridge	absent	* present
Postsutural intra-alar bristle	present	* absent
Postsutural dorso-central bristle	present	* absent
Sc and R_1	widely separate	* closely proximate
Anal lobe	present	* absent
Alula	present	* absent
Abdomen (both sexes)		
Tergites 2–3	free	* fused
Spiracle 1	in membrane	* in tergite
Abdomen (male)		
Tergite 6	* reduced (except *Palaeotanypeza*)	large
Sternite 7	asymmetrical	* secondarily symmetrical
Sternite 8	* very large, elongate	* small, short
Left spiracles 6 and 7	* absent	7 present
Aedeagal apodeme	* absent	present
Abdomen (female)		
Ovipositor	* long and rectractible	short and rather exposed
Spermathecae	* two	* four

* Autapomorphic condition.

— dorsocentral bristles absent (as in Syringogastridae);
— scutellar bristles situated on elongate tubercles;
— front femur enlarged and strongly setose below;
— Sc lying close to R_1 (as in Syringogastridae); and
— crossvein bm-cu absent (cells bm and dm confluent).

Plesiomorphically the diopsid line retained a relatively long tergite 6 in the male, seven pairs of spiracles in both sexes, and a simple, relatively short, exposed ovipositor and three spermathecae in the female. Hennig (1965*b*), Shillito (1971), Steyskal (1972*a*), and Feijen (1983) have each reviewed the cladistics of the genera. Two subfamilies, the Centrioncinae (Ethiopian region) and Diopsinae (Holarctic and Ethiopian regions) are usually recognized (Hennig 1965*b*, 1973; Steyskal 1972*a*), but Feijen (1983) treated these as separate families. Two fossil species are known: *Prosphyrocephala succini* (Loew) (Hennig 1965*b*) from Baltic amber and *P. rubiensis* Lewis (Lewis 1971) from Oligocene shales in Montana. All are agreed that the genus *Centrioncus* Speiser, with unstalked eyes, is the most generalized member of the group.

Superfamily Conopoidea

Conopoidea (Fig. 116.3) consists of only one family, the Conopidae (including *Stylogaster* Macquart) (Smith and Peterson, Ch. 54), but phylogenetically it is one of the most puzzling of all the muscomorphan families. Its membership in the Muscomorpha has usually been recognized, but there has always been uncertainty about its position within that group (for a review of early literature, see especially de Meijere 1904). There has been a persistent misconception that it belongs with the Syrphoidea (Aschiza), even though de Meijere (1904), adding to evidence presented earlier by Brauer (1880, 1890), proved that it belongs to the Schizophora. Enderlein (1936) proposed that the family is the sister group (Archischiza) of all other Schizophora (Muscaria). Unfortunately, Hennig (1958, 1966*b*) provisionally accepted Enderlein's name, although he did not recognize it in earlier and later works (Hennig 1952 and 1973, respectively). I agree with Griffiths (1972) that placement of the family as sister group of all other Schizophora cannot be justified, and

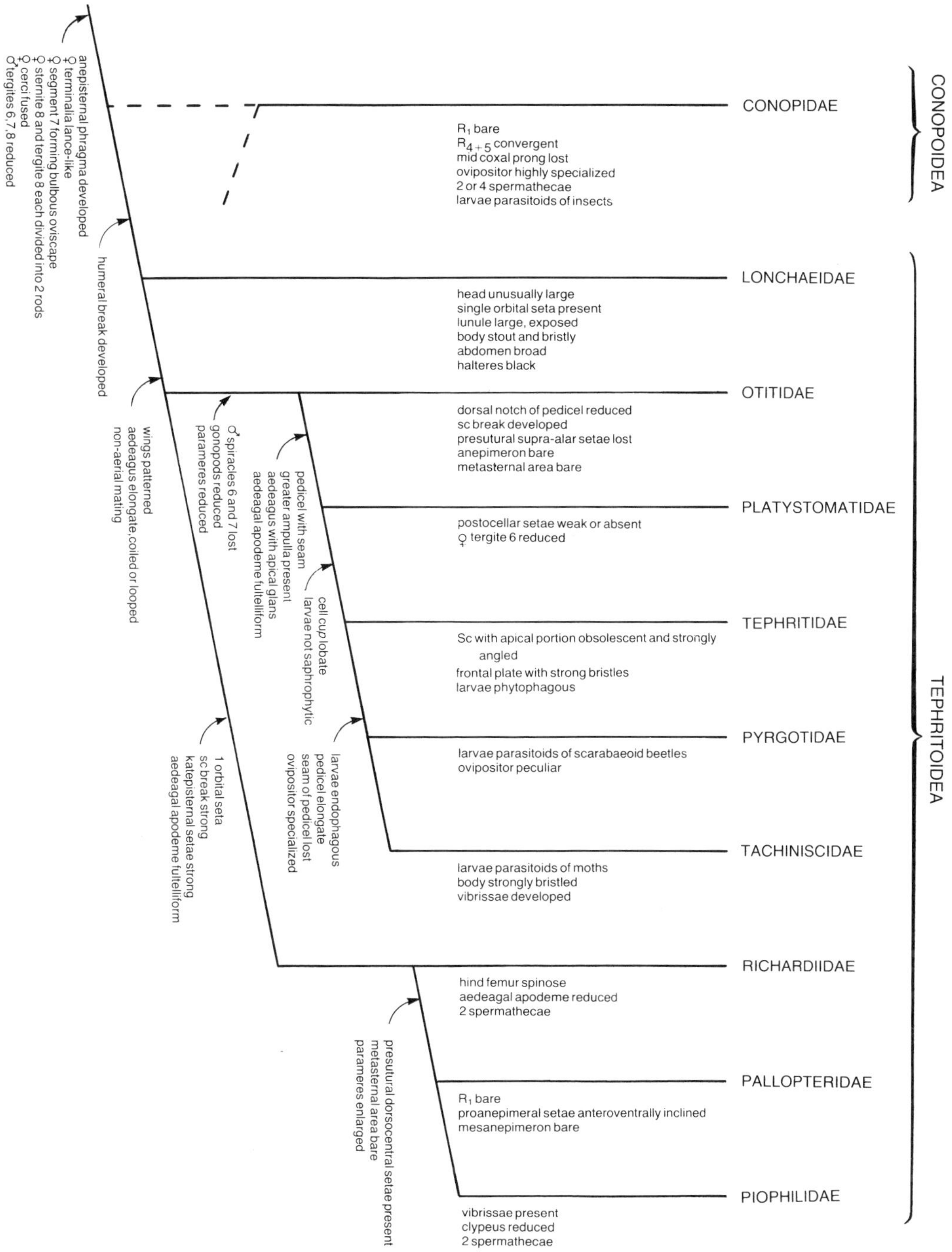

Fig. 116.3. Cladogram of the Conopoidea and Tephritoidea summarizing the relationships and apomorphies of the subgroups and families recognized. (See text for additional details.)

that this idea should be abandoned. Hendel (1928) placed Conopidae in the Schizophora, but did not assign it to either the Calyptratae or the Acalyptratae. Hennig (1952) affirmed that the family belongs in the Acalyptratae and suggested that the Tephritoidea probably contained its closest relatives. Griffiths (1972) referred it to his "prefamily" Tephritoinea as a "family-group" separate from the "Tephritidae family-group" (in which he included all the families of Tephritoidea except Lonchaeidae). This placement more or less conforms with Hennig's (1952) statement that the amazing similarity of Conopidae with Pyrgotidae, although certainly not reflecting a sister group affinity with the Pyrgotidae, perhaps does reflect "a not altogether distant affinity of the

Conopidae with the Tephritoidea." In effect, both Hennig and Griffiths regarded the Conopidae as a sister group of Tephritoidea. This placement is accepted as the most plausible one yet advanced. However, because it cannot be fitted into the Nerioidea or the Diopsoidea, on the one hand, or the Tephritoidea, on the other, it seems preferable to rank it as a separate, coordinate superfamily (Fig. 116.3). Among the most significant plesiomorphic ground-plan characters of the family are the following: fronto-orbital plate continuous and extending far forward, four or more fronto-orbital bristles present, vibrissae absent, C without humeral or subcostal breaks, Sc complete and free from R_1, crossvein sc-r present and discrete, A_1 complete to wing margin, male with aedeagal apodeme rod-like (cuneiform), and surstylus without prensisetae.

Hennig (1966*b*) discussed most of these and other characters in terms of their plesiomorphic and apomorphic states in the Conopidae. He showed that a subapical or apical position of the arista, as opposed to a dorsal position, is an apomorphic condition. Earlier, (Hennig 1958), he provided a very convincing argument that the presence of a drawn out point on cell cu*p* in many Conopidae is an apomorphic condition with respect to the ground plan of the Schizophora. Griffiths (1972) expressed the opinion that the completely unbroken C and the elongate cell cu*p* are also secondary (apomorphic) conditions. Certainly an unbroken C is a plesiomorphic condition in the ground plan of the Schizophora, and I can find no evidence to substantiate the hypothesis that a subcostal break belongs to the ground plan of the Conopidae.

Conditions in the conopid ground plan that are apomorphic with respect to the ground plan of the Acalyptratae are as follows; each of those that is autapomorphic is marked by (AA):

- frons not sexually dimorphic, equally wide in both sexes;
- pedicel elongate, without a dorsal notch or seam;
- posterolateral margins of prosternum produced (AA);
- anepisternum bare;
- metasternal area bare;
- pterostigma absent;
- R_1 bare (AA);
- R_{4+5} and M_1 convergent at apices;
- mid coxal prong absent (AA);
- male with tergite 6 reduced or lost;
- sternite 6, reduced, joined with pregenital segment;
- sternites 7 and 8 fused, forming more or less symmetrical pregenital segment;
- aedeagus with apical glans;
- sternite 10 divided;
- female with segments 5–8 highly modified, anteroventrally directed (AA);
- cerci fused;
- two spermathecae present; and
- larvae developing as internal parasitoids of adult insects (AA).

The precise phylogenetic origin of the Conopidae remains unknown. It appears to have evolved from an early, ecologically specialized acalypterate stock that subsequently also produced the Tephritoidea. The fact that it has retained in its ground plan such primitive features as an unbroken C and a discrete crossvein sc-r, reflect the antiquity of its origins in the Acalyptratae, but the reduction and secondary symmetry in the male pregenital sclerites, preclude membership in either the Nerioidea or Diopsoidea. The presence of an aedeagal glans in the male and the modified abdominal segment 7 and fused cerci in the female seem to align it with the Tephritoidea. More work is needed to pin-point the relationship of this interesting family.

Usually four subfamilies are recognized: the Stylogasterinae, Myopinae, Dalmanniinae, and Conopinae (Smith and Peterson, Ch. 54). Rohdendorf (1964) raised *Stylogaster* to separate family status, but, because it shares the same autapomorphies as the remaining Conopidae, it is best treated as a subfamily within that family. Hennig (1966*b*) discussed the evolution of the subfamilies and later (Hennig 1973) included the Dalmanniinae in the Myopinae. One fossil species, *Palaeomyopa tertiaria* Meunier (= *Palaeosicus loewi* Meunier), is known from Baltic amber (Hennig 1966*b*), and one species, *Poliomyia recta* Scudder, was described from the Eocene stratum of Green River, Wyoming (Scudder 1890).

Superfamily Tephritoidea

Tephritoidea (Fig. 116.3) consists of nine families: the Lonchaeidae, Otitidae, Platystomatidae, Tephritidae, Pyrgotidae, Tachiniscidae, Richardiidae, Pallopteridae, and Piophilidae. It corresponds to the Otitoidea + Pallopteroidea of Hennig (1958, 1971*a*, 1973) and to the Lonchaeidae + Tephritidae family-group of Griffiths (1972). Hendel (1916, 1922) excluded the Piophilidae (including neottiophilids and thyreophorids), but he included the Tanypezidae. Frey (1921) followed Hendel, but he added the Agromyzidae. Crampton (1944*b*) excluded both Tanypezidae and Agromyzidae and was the first to include the Piophilidae; however, he omitted to specify where he placed the Lonchaeidae. My concept of this superfamily was presented and discussed in three previous papers (J. F. McAlpine 1976, 1977, 1981*c*).

Tephritidae as a family-group name predates all other names in the superfamily (Rohdendorf 1977; Sabrosky personal commun.) and therefore Tephritoidea is the correct name to apply to it.

Ground-plan characters of the Tephritoidea were given previously (J. F. McAlpine 1976, 1977, 1981*c*) and need not be repeated here. Autapomorphic ground-plan features (synapomorphic features of the included families) attesting to the monophyly of the superfamily are as follows:

- anepisternum with a more or less perpendicular, linear, internal phragma on posterior half (usually marked externally by a suture-like furrow anterior to

anepisternal bristles; most evident in pale-colored species, but easily seen in all cleared specimens);
— subscutellum moderately well developed;
— C with a humeral weakening or break;
— male with tergites 6, 7, and 8 reduced to fragments or absent;
— sternites 7 and 8 asymmetric, partially fused and shifted into left side of abdomen;
— posterior or inner lobe of surstylus with prensisetae;
— female with segment 7 forming a stout oviscape with an eversible ovipositor sheath;
— tergite 8 and sternite 8 divided longitudinally into paired struts that form the main shaft of an elongate, rigid, piercing-type ovipositor; and
— cerci fused.

A somewhat similar, elongate, retractile ovipositor occurs in the Diopsoidea (in the Tanypezidae, Strongylophthalmyiidae, some Psilidae, and Megamerinidae), but, in the ground plan of these families, tergite 6 of the male is relatively large and the female cerci are not fused. In addition, there is neither a humeral break in C nor an internal phragma on the anepisternum, and the surstyli have no prensisetae. The ground plan of the Tephritoidea is clearly more apomorphic in these respects.

Within the Tephritoidea there are three monophyletic subgroups: one contains the Lonchaeidae alone; another contains the Otitidae, Platystomatidae, Pyrgotidae, Tephritidae, and Tachiniscidae; and a third contains the Richardiidae, Pallopteridae, and Piophilidae (Fig. 116.3).

Lonchaeidae. The Lonchaeidae (J. F. McAlpine, Ch. 62) have retained a more fundamental set of primitive features of the superfamily than any of the other families. All males have a narrower frons than the females, and aerial swarming is a universal mating habit in the family. In males of certain generalized members, e.g. *Dasiops relictus* McAlpine, sternite 6 is large and nearly symmetrical, and tergite 6 is complete and separate (see Fig. 32, J. F. McAlpine 1962). Primitively, the aedeagus, though relatively short, was probably simple (unsegmented and without apical glans), membranous, somewhat convoluted, and posteriorly directed as in some existing species of *Protearomyia* McAlpine (see Fig. 14, J. F. McAlpine 1983). Other plesiomorphic features of the family are the entirely black body color, including the head and legs, a generally hairy condition, including the disc of the mesothoracic anepisternum and anepimeron, relatively unpatterned wings, seven abdominal spiracles in both sexes, and three spermathecae in the female. Although none of these is exclusive to the family, in no other family of the Tephritoidea do they occur together in all representatives. Therefore, the Lonchaeidae appears to be an older sister group of the remainder of the superfamily.

Apomorphic characters of the family, with respect to the ground plan of the Tephritoidea, are as follows; those that are autapomorphic are marked with (AA).

— head unusually large and hemispherical (AA);
— lunule broadly exposed;
— single orbital bristle present;
— thorax and legs bristly;
— proepisternum bare on disc;
— metasternal area bare;
— proepisternal and proepimeral bristles present;
— anepisternal and katepisternal bristles present;
— halteres entirely blackish (AA);
— abdomen short, broad and flattened (AA);
— male with syntergosternite 7 + 8 greatly reduced;
— aedeagus short and eventually rigid (AA); and
— female with sternites 5 and 6 each with a median apodeme on anterior margin.

Lonchaeidae consists of two subfamilies: the Dasiopinae (more ancient lineage) contains one genus *Dasiops*; and the Lonchaeinae (more recent lineage) contains *Protearomyia*, *Chaetolonchaea* Czerny, *Earomyia* Zetterstedt and *Lamprolonchaea* Bezzi (tribe Earomyiini), and *Lonchaea* Fallén, *Setisquamalonchaea* Morge, *Silba* Macquart, and *Neosilba* McAlpine (tribe Lonchaeini) (J. F. McAlpine, Ch. 62; J. F. McAlpine and Steyskal 1982). Cladistics of the family were treated by J. F. McAlpine (1962) and by Morge (1963). The family contains an estimated 700 species, nearly half of which are still undescribed. It occurs on all major land masses except New Zealand, but is probably best represented in the Holarctic region. *Dasiops* and *Lonchaea* are the largest genera and occur in all zoogeographical regions. *Protearomyia* (J. F. McAlpine 1983), *Chaetolonchaea* (J. F. McAlpine 1982) and *Earomyia* are essentially Holarctic in distribution, but *Protearomyia* also occurs in Patagonia. *Lamprolonchaea* occurs mainly in the Ethiopian, Oriental, and Australian regions (where it replaced *Earomyia*), and *Silba* has a similar distribution. *Neosilba* (J. F. McAlpine and Steyskal 1982) is restricted to the Neotropical region, and *Setisquamalonchaea* is endemic to the Palearctic region.

Synapomorphic ground-plan characters of the tephritoid and the piophiloid subgroups are:

— wings patterned and displayed during sexual activity;
— all aspects of mating completed while standing on a substrate; and
— tergite 6 of male further fragmented or lost.

There is little to choose between the relative generalization or specialization of either group. The ground plan of the tephritoid subgroup is more generalized than that of the piophiloid subgroup in the following features: two or more strong orbital bristles retained (one in piophiloid subgroup); katepisternal bristle probably weak or absent (clearly present in piophiloid subgroup), C without subcostal break (clearly present in piophiloid subgroup); crossvein sc-r present (virtually absent in piophiloid subgroup); and aedeagal apodeme rod-like (fultelliform in piophiloid subgroup).

On the other hand, the ground plan of the piophiloid subgroup is more generalized in several other features:

Spiracles 6 and 7 retained in male abdomen (lost in tephritoid subgroup), male with sternite 6 relatively large, free, and unmodified (greatly reduced, asymmetric, and fused with sternites 7 and 8 in tephritoid subgroup); paramere well developed (greatly reduced or absent in tephritoid subgroup); and female with tergite and sternite of segment 7 relatively free (more or less fused in tephritoid subgroups).

Autapomorphic characters of the tephritoid subgroup include the following:

- males with spiracles 6 and 7 lost;
- sternite 6 reduced and asymmetrical, fused with sternites 7 and 8;
- gonopods reduced;
- parameres lost or greatly reduced; and
- females with tergite and sternite of segment 7 fully fused.

Autapomorphic characters of the piophiloid subgroup include the following:

- only one strong orbital bristle present;
- katepisternal bristle strongly differentiated;
- C with a subcostal break; and
- aedeagal apodeme fultelliform.

The tephritoid subgroup is composed of two main sister groups: the Otitidae on the one hand, and the remaining four families (the Platystomatidae, Pyrgotidae, Tephritidae, and Tachiniscidae) on the other.

Otitidae. The Otitidae (Steyskal, Ch. 63) remained closer to the ground plan of the tephritoid subgroup than any other family in several ways: the pedicel is always without a dorsal seam; the anepisternal phragma is frequently weak and incomplete; the surstylus consists of simple anterior and posterior lobes, the latter bearing several prensisetae; the aedeagus, although long and coiled, does not have a discrete, complex apical glans; and, in the ground plan of the family, the aedeagal apodeme remained rod-like and relatively free from the hypandrium. The food habits of the larva continued to be primarily saprophagous.

The following autapomorphies (with respect to the ground plan of the remaining four families) can be advanced for the Otididae:

- pedicel with dorsal notch reduced;
- proepisternal, anepisternal, and katepisternal bristles present;
- anepimeron bare;
- metasternal area bare;
- postsutural acrostichal and intra-alar bristles present;
- presutural supra-alar bristle absent;
- C with a subcostal break;
- crossvein sc-r and pterostigma absent; and
- aedeagus tightly coiled, and stored in right, ventrolateral side of abdomen.

Steyskal (1961; Ch. 63) considers that the family consists of two subfamilies, the Ulidiinae in which the aedeagus is bare and sometimes has an apical glans, and the Otitinae in which the aedeagus is bristled and always has a simple tip. European workers (Hennig 1939*a*, 1973) treat these two taxa and what Steyskal classifies as the tribe Pterocallini, as separate families. A worldwide study of the evolution and classification of the family is needed. Something in the order of 800 to 1000 species are known from all parts of the world. No fossils have been reported.

Autapomorphic characters in the ground plan of the Platystomatidae + Tephritidae + Pyrgotidae + Tachiniscidae subgroup (synapomorphies of the included families), proving its monophyly are as follows:

- pedicel with an elongate dorsal seam;
- greater ampulla more or less developed;
- lower lobe of calypter frequently broadened;
- anepisternal phragma strong and complete;
- posterior notopleural bristle surmounting a tubercle (notopleural callus);
- surstylus consisting of an outer and an inner lobe (as opposed to an anterior and a posterior lobe, respectively), the latter bearing several prensisetae;
- aedeagus elongate and looped, and bearing a complex apical glans that is stored more or less dorsally under tergite 5; and
- aedeagal apodeme fultelliform, i.e. extensively fused with hypandrium.

As in the otitid sister group, some ground-plan characters of these four families remained closer to the ground plan of the tephritoid subgroup as a whole, than they did in the Otitidae: the pedicel has a distinct dorsal notch; C does not have a subcostal break; crossvein sc-r remained present, though weak and indistinct; the pterostigma was retained; strong bristles are not present on the proepisternum, anepisternum, or katepisternum; hairs were retained on the anepimeron and metasternal area; and there are no distinct postsutural, acrostichal, or intra-alar bristles.

Platystomatidae. There is probably a sister-group relationship between the Platystomatidae (Steyskal, Ch. 64) and the Tephritidae + Pyrgotidae + Tachiniscidae. The Platystomatidae remained more generalized in their ground plan than the latter three families in retaining saprophagous larval food habits. In the adults, bristles are frequently relatively weak and sparse; the katepisternum is usually without an outstanding bristle; the anepisternum frequently lacks strong bristles; and cell cu*p* remained obtusely closed (without an acute posteroapical lobe). Apomorphic characters of the family include the following; those that are autapomorphic are marked (AA):

- pedicel with a long dorsal seam;
- postocellar bristles weak or absent; and
- female with abdominal tergite 6 reduced or absent (AA).

This family consists of about 1000 species and occurs in all the main vegetated parts of the world. D. K. McAlpine

(1973) provided the most thorough study of the family yet produced, but it is primarily a treatment of the Australian fauna. A worldwide comparative study of all genera is needed. McAlpine recognized five subfamilies: the Trepherinae (4 genera), the Plastotephritinae (5–10 genera), the Scholastinae (9 genera), the Platystomatinae (26 or more genera), and the Angulitinae (3 genera). The Platystomatidae are most abundant in the Paleotropical regions, with a relatively few species reaching into the New World tropics and the southern parts of the Palearctic and Nearctic regions. One genus, *Rivellia* Robineau-Desvoidy, is virtually worldwide in distribution. No fossils are known.

The line leading to the Tephritidae + Pyrgotidae + Tachiniscidae probably was scarcely distinguishable from the Platystomatidae except that cell cu*p* acquired an acute posteroapical lobe and the larvae were either phytophagous, as in the Tephritidae, or parasitic on other insects, as in the Pyrgotidae and Tachiniscidae. The shift in larval feeding habits from saprophytic to living plants or insects is a significant synapotypic feature of these three families. This line produced two sister groups: the Tephritidae, on the one hand, and the Pyrgotidae + Tachiniscidae, on the other.

Tephritidae. The ground plan of the Tephritidae (Foote and Steyskal, Ch. 66) is more generalized than that of the Pyrgotidae + Tachiniscidae in several ways: the larvae feed in living plant tissues (flowers, stems, leaves, and fruits); the pedicel has a dorsal seam (modified dorsal notch); the hypandrial bridge remained intact in the male; and the typical form of the tephritoid ovipositor was retained in the female. Apomorphic features of the family include the following; those that are autapomorphic are marked (AA):

- pedicel with a complete dorsal seam;
- frontal plate strongly differentiated and bearing a vertical row of strong medioclinate frontal bristles (AA);
- Sc obsolescent at apex and angled sharply forward (AA); and
- greater ampulla relatively strongly developed (AA).

Tephritidae is by far the largest family of the Tephritoidea; an estimated 4000 species are known from all parts of the world. Hering (1947) divided the family into eight subfamilies: the Dacinae, Schistopterinae, Euribiinae, Oedaspinae, Tephritinae, Aciurinae, Terelliinae, and Trypetinae. Hardy (1973, 1974, 1977) reduced this number to four subfamilies: the Dacinae, Schistopterinae, Tephritinae, and Trypetinae. Cogan and Munro (1980), on the other hand, increased it to 11 subfamilies. Clearly a worldwide study of the cladistics and classification of the family is needed. No fossils have been reported.

Synapomorphic characters of the Pyrgotidae and Tachiniscidae include:

- larvae with endophagous (parasitoid) habits;
- pedicel inclined to be elongate strongly decumbent, and with dorsal notch weak or absent;
- postsutural, intra-alar bristle present;
- tendency for loss of mid coxal prong (absent in most Pyrgotidae, absent in *Tachiniscidia* sp.); and
- ovipositor peculiarly adapted for depositing eggs on living insects.

Pyrgotidae. The ground plan of the Pyrgotidae (Steyskal, Ch. 65) is more plesiomorphic than that of the Tachiniscidae. The frontal plate is relatively weakly developed and frontal bristles, if present, are sparse and weak (opposite conditions in Tachiniscidae); the basic wing venation of the Pyrgotidae is probably closer to the ground plan of all the Tephritoidea than in any other family: C without a subcostal break, Sc complete to C, crossvein sc-r clearly present, and R_1 bristled on its entire length. But these characters and others vary greatly in the family, which indicates a potential for the peculiarities that occur in both the Tephritidae and the Tachiniscidae.

Autapomorphic characters in the ground plan of Pyrgotidae include:

- larvae parasitoids in adult scarabaeoid beetles (Coleoptera);
- adaptations in adults for oviposition on beetles;
- crepuscular flight habits; and
- ovipositor with tubular, elongate anteroventrally curved oviscape.

Additional apotypic characters of the family are as follows:

- tendency to lose ocelli;
- frons sometimes narrower in female than male; and
- arista sometimes highly modified.

The family consists of 200 to 300 species, divided among 20 to 30 genera, and occurs mainly in the tropical and temperate regions of the world. Three subfamilies, the Toxurinae, Pyrgotinae, and Lochmostyliinae, are usually recognized. A thorough evolutionary study of the family throughout the world is needed. Key papers include those by Acxél (1956, 1958), Enderlein (1942), Hendel (1934), Hennig (1959), Paramanov (1958), and Steyskal (1972*b*, 1978). No fossils are known.

Tachiniscidae. The Tachiniscidae are usually considered to be a specialized sister group of the Tephritidae (J. F. McAlpine 1977), mainly because the adults have a row of strong medioclinate frontal bristles as in the Tephritidae. However, that character recurs independently several times within the Tephritoidea (some species in each of the families Otitidae, Platystomatidae, Pyrgotidae, Richardiidae, and Piophilidae).

The parasitoid habits of the larvae, surmised by Malloch (1931*a*) and confirmed by Roberts (1969), together with the elongate, nonseamed pedicel, and the tendency for losing the mid coxal prong, show that it is a sister group of the Pyrgotidae. Perhaps when more is known about both families, the Tachiniscidae will prove to be a sister group of a subgroup of the Pyrgotidae rather than of the entire family. If so, it should be ranked as a suprageneric taxon of the Pyrgotidae, but for now it is preferable to treat it as a separate family. Autapomorphic characters of the group include:

- body and legs strongly and relatively densely bristled;
- vibrissa present;
- frontal plates strongly differentiated, and with a row of strong medioclinate bristles (as in Tephritidae);
- postpronotum with several strong bristles;
- anepisternum, katepisternum, and anepimeron with strong bristles;
- anepimeron with a very prominent tubercle from which anepimeral bristles arise.
- C with a subcostal break;
- male with surstyli greatly reduced;
- female with highly modified, dorsally directed ovipositor; and
- sclerotized spermathecae absent (at least in *Tachinisca cyaneiventris* Kertész).

Only three species in three genera, *Tachinisca cyaneiventris* (South America), *Bibundia hermanni* Bischoff (Africa), and *Tachiniscidia africana* Malloch (Africa), are described. Examples of this unusual family are rare. A single specimen of an undescribed species of *Bibundia* Bischof from Nigeria was reared by Roberts (1969) from a caterpillar of a saturniid moth, *Imbrasia nicitans* Fabricius, which is the only recorded data on the biology of the family (Cogan 1980*a*). I examined a female of *T. cyaneiventris* from Colombia (Meta, Villavicencio, 19 June 1976, collected by M. Cooper) and a male of an undescribed species of *Tachiniscidia* Malloch from Angola (Cacola, 23 Dec. 1957–28 Jan. 1958, 1300 m). Malloch (193l*a*) reviewed the family. No fossils are known.

As indicated earlier, the third monophyletic subgroup of the Tephritoidea consists of the Richardiidae, Pallopteridae, and Piophilidae. This subgroup, here referred to as the piophiloid subgroup (= richardioid subgroup of J. F. McAlpine 1977, 1981*c*) together with the tephritoid subgroup forms the sister group of the lonchaeoid subgroup. Autapomorphic characters of both subgroups are listed earlier.

Richardiidae. This family (Steyskal, Ch. 67), which dates from Loew (1873), is considered the most generalized family of the piophiloid subgroup, because it retained the following plesiomorphic characters in its ground plan: middle of proepisternum haired; metasternal area haired; seven pairs of abdominal spiracles present in both sexes; male with a vestige of tergite 8; and aedeagal apodeme free, i.e. not extensively fused to hypandrium.

Autapomorphic characters of the family are as follows:

- hind femur heavily spinose;
- abdominal tergite 2 with strong bristles laterally;
- aedeagal apodeme very reduced;
- gonopods and parameres very reduced; and
- two spermathecae present.

The most generalized section of the family includes the genera *Automola* Loew, *Epiplatea* Loew, and *Omomyia* Coquillett (J. F. McAlpine 1976); Steyskal (Ch. 67) erected a new subfamily, Epiplateinae, for it and considered it to be the older sister group of the remainder of the family (Richardiinae). The family is restricted to the New World and contains about 170 species in 31 genera. Steyskal (1968*c*) provided a catalog of almost all taxa except *Omomyia*. No fossils are known.

Pallopteridae and Piophilidae together comprise the sister group of the Richardiidae (J. F. McAlpine 1976, 1977, 1981*c*). Synapomorphic characters in the ground plan of both families are as follows:

- presutural dorsocentral bristles present;
- proepisternum bare;
- metasternal area bare; and
- parameres relatively large and strongly sclerotized.

Pallopteridae. Separate family status for this group (J. F. McAlpine, Ch. 68) was first proposed by Malloch and McAtee (1924). Its evolutionary relationships and classification were recently reviewed by J. F. McAlpine (1981*c*), and a key to the world genera was provided for the first time.

The Pallopteridae are considered to be more generalized than the Piophilidae in not having vibrissa and in having a proepimeral bristle.

Autapomorphic characters in the ground plan of the Pallopteridae include the following:

- presutural bristles relatively strongly developed (lost secondarily in some genera);
- proanepimeral bristle weak and anteroventrally inclined;
- R_1 bare;
- anepimeron bare; and
- habit of vibrating wings during sexual excitement strongly developed.

The family contains two subfamilies; the Eurygnathomyiinae contains *Eurygnathomyia* Czerny, the only member of the family in which C is spinose, and Pallopterinae contains the 12 remaining genera (see J. F. McAlpine, Ch. 68). Three fossil species are known (see J. F. McAlpine, Ch. 68).

Piophilidae. This family (J. F. McAlpine, Ch. 69) was first proposed by Macquart (1835). In the present concept it includes both the neottiophilids (subfamily Neottiophilinae) and thyreophorids (subtribe Thyreophorina) which some authors treat as separate families. The family was recently revised on a worldwide basis (J. F. McAlpine 1977).

The Piophilidae are considered to be the apotypic sister group of the Pallopteridae. Autapomorphies of the Piophilidae with respect to the ground plan of the piophiloid cluster of families are as follows:

— vibrissae strongly developed;
— clypeus small and withdrawn;
— proepimeral bristle absent;
— surstyli unilobate, without teeth (except in *Amphipogon* Wahlberg);
— aedeagal apodeme extensively fused with hypandrium; and
— two spermathecae present.

The family consists of two subfamilies: the Neottiophilinae, which are the more generalized (R_1 setulose), and the Piophilinae. Neottiophilinae contains only two genera, *Neottiophilum* Frauenfeld and *Actenoptera* Czerny. The larvae of *Neottiophilum*, at least, live in the nests of passerine birds where they are ectoparasitic on the nestlings. The Piophilinae contains two tribes: the Mycetaulini with 6 genera, which are probably all mycetophilous, and the Piophilini with 15 genera, which are probably all necrophilous. The tribe Piophilini may be subdivided into two subtribes: the Piophilina containing *Piophila* Fallén and its allies, and the Thyreophorina containing seven genera, including *Protothyreophora* Ozerov (Ozerov 1984). The supposed phylogeny of all genera, except the last, was discussed in detail by J. F. McAlpine (1977). In all, the family contains 71 species and 23 genera. No fossils are known.

Superfamily Lauxanioidea

This superfamily (Fig. 116.4) was proposed by Hendel (1916, 1922) for the families Lauxaniidae, Celyphidae, and Chamaemyiidae. Hennig (1958) added the Eurychoromyiidae and Periscelididae, but later he (Hennig 1971*a*) excluded the Periscelididae and associated it with the Aulacigastridae (see discussion under Opomyzoidea: Asteioinea). Griffiths (1972) adopted Hennig's (1971*a*) concept of the superfamily, and the same concept is followed here. Latreille's (1804) proposal of Lauxanides is the oldest family-group name applicable to the group; hence Lauxanioidea is the correct name to use for the superfamily.

The monophyly of the Lauxanioidea is based on some 12 autapomorphies (synapomorphies of included taxa), with respect to the ground plan of the Acalyptratae. These characters are listed in Table 116.4, in which they are compared with those of the Sciomyzoidea. Some of these characters are lost in some members of the Lauxanioidea. For example, the convergent postocellar bristles are reduced or absent in the Eurychoromyiidae, in some Chamaemyiidae (*Cremifania* Czerny and *Leucopis* Meigen) and in some Celyphidae; sometimes these are replaced by parallel to slightly divergent ocellar setulae. Preapical dorsal tibial bristles have also been lost in the Eurychoromyiidae and Chamaemyiidae. In an earlier paper (J. F. McAlpine 1963) dealing with the relationships of the families of Lauxanioidea + Sciomyzoidea, the cladistic arrangement that I depicted is wrong (Hennig 1965*b*). A more plausible arrangement is shown in Fig. 116.4.

I agree with Hennig (1965*b*) that the sister group of the Lauxanioidea is probably the Sciomyzoidea. This conclusion is supported by nine synapomorphies with respect to the ground plan of the Acalyptratae (first nine characters listed in Table 116.4). However, some of these conditions may be questionably apomorphic. For example, it is possible that a haired prosternum is a plesiomorphic rather than an apomorphic feature in the Schizophora, and, if so, its occurrence here is of little or no significance.

These two superfamilies also share a suite of plesiomorphic conditions in their ground plans, the components of which are usually listed in characterizations given for the component families. These characters include such things as vibrissae absent; C unbroken; Sc complete; cells bm, dm, and cu*p* discrete; female with three spermathecae (except Eurychoromyiidae) and simple tubular terminalia; and male with freely articulated surstyli. The Lauxanioidea remained more plesiotypic than the Sciomyzoidea in retaining in its ground plan a tergite 6 in the male that is almost as long as tergite 5. Tergite 6 in the ground plan of Sciomyzoidea is seldom more than one-half as long as tergite 5 and is frequently vestigial or absent. In general, however, the Sciomyzoidea are more plesiotypic than the Lauxanioidea, as evidenced, for example, by divergent postocellar bristles, a complete A_1, and a discrete remnant of tergite 8 present in the male (see Table 116.4).

Griffiths' (1972) conclusion that the Lauxanioidea and Sciomyzoidea are remotely related is unconvincing. He divided the Schizophora into two main groups, i.e. the Lonchaeoidea + Lauxanioidea + Drosophiloidea + Nothyboidea, on the one hand, and the Muscoidea (Calyptratae + remaining Acalyptratae including Sciomyzoidea) on the other hand. He based this division both on postulated differences in the relative reduction and coalescence of the tergites and sternites of segments 6 to 8 in the male abdomen, and on the ability or inability of the aedeagus to be swung through a wide arc against the aedeagal apodeme. On the basis of these assumptions, he placed the Lauxanioidea in the more plesiotypic section, in which he considered the ground-plan condition of the aedeagus to be directed rigidly to the posterior, and the Sciomyzoidea in the more apotypic section (Muscoidea), in which he considered the ground-plan condition of the aedeagus to be flexible and capable of being swung in a wide arc against the aedeagal apodeme. Because Griffiths underestimated the great variability of these characters, even within well-established taxa, some of the relationships he postulated are questionable, and his conclusion regarding the relationships of the Lauxanioidea and Sciomyzoidea appears to be a case in point (see further discussion under Cremifamiinae).

The ancestry of the Lauxanioidea + Sciomyzoidea probably traces back to the earliest stem of the Acalyptratae (see Fig. 116.9), but no specific outgroup has yet been established. Together these two superfamilies probably comprise the sister group of the Opomyzoidea + Carnoidea + Sphaeroceroidea + Ephydroidea. More work

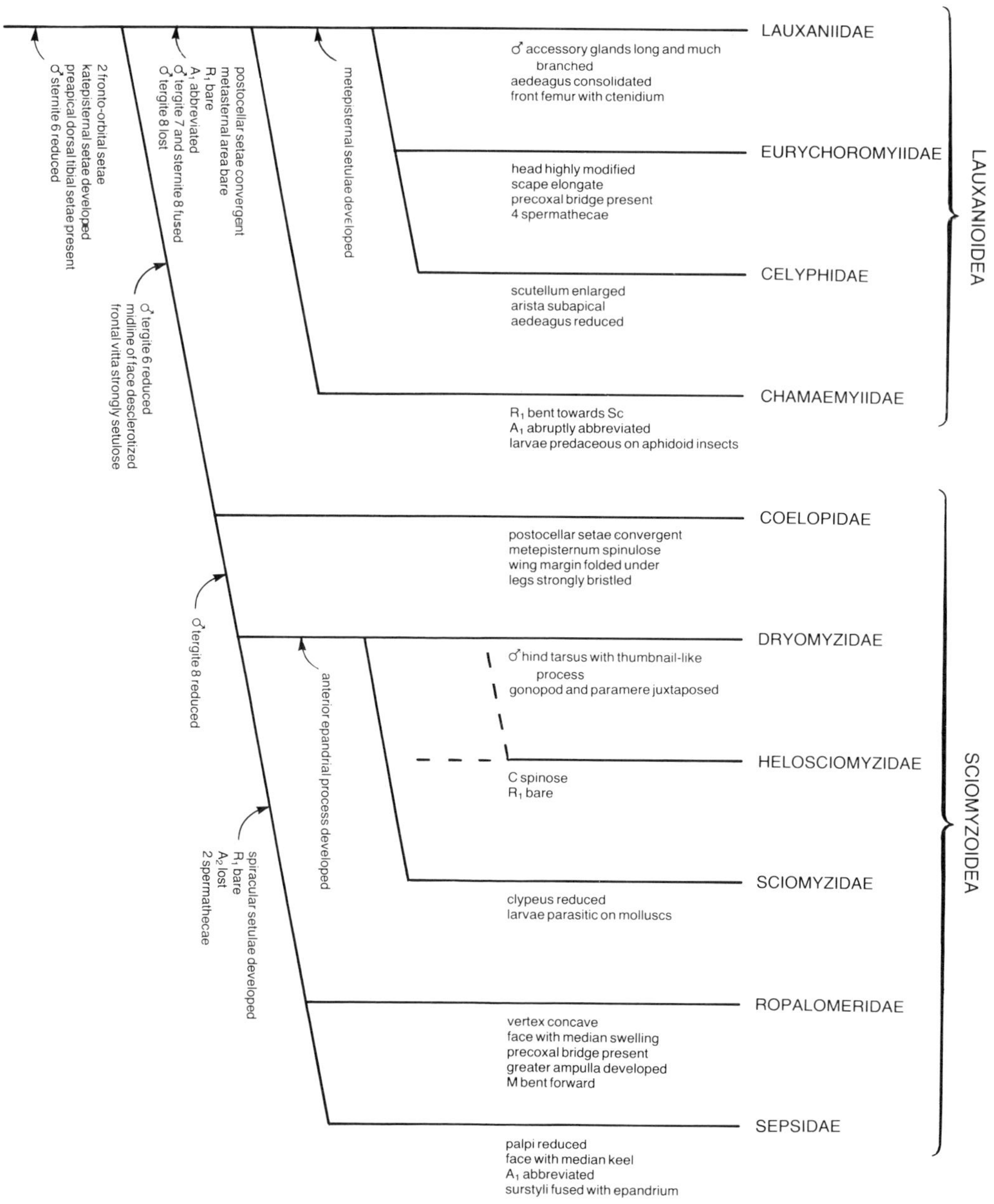

Fig. 116.4. Cladogram of the Lauxanioidea and Sciomyzoidea summarizing the relationships and apomorphies of the subgroups and families recognized. (See text for additional details.)

is required to document these assumptions with clear-cut synapomorphies. The reduction in the fronto-orbital bristles to two and the loss of inclinate frontal bristles seem to be useful characters in this respect.

The Lauxanioidea appears to consist of two main sections; the first consists of the Lauxaniidae, Eurychoromyiidae, and Celyphidae; the second consists of the Chamaemyiidae (including Cremifaniinae). A clearly autapomorphic character of the first section (synapomorphic character of the included families) is exhibited by the form and vestiture of the metepisternum. In the ground plan of these three families, the metepisternum is tiny and depressed, and it bears one to several fine setulae. Attention was drawn to this character in connection with *Eurychoromyia mallea* Hendel (J. F. McAlpine 1968, Figs. 5–6), but in that paper I misinterpreted the tiny metepisternal setula as being homologous with the spiracular bristles present in the Sepsidae and Ropalomeridae. The spiracular bristles in those families are closely associ-

Table 116.4 Comparison of ground-plan characters[1] of Lauxanioidea and Sciomyzoidea

Character	Lauxanioidea	Sciomyzoidea
Fronto-orbital plate reduced anteriorly	+	+
Reclinate fronto-orbital bristles reduced to two	+	+
Inclinate frontal bristles absent	+	+
Lunule bare	+	+
Prosternum haired	+	+
Katepisternal bristle present	+	+
Laterotergite bare	+	+
Preapical dorsal tibial bristle present	+	+
Sternite 6 of male reduced	+	+
Frontal vitta densely and strongly setulose	–	+
Face weakly sclerotized along vertical midline	–	+
Tergite 6 of male reduced	–	+
Postocellar bristles convergent	+	–
Proepisternum bare on disc	+	–
Metasternal area bare	+	–
Meron bare	+	–
R_1 bare	+	–
CuA_2 recurrent	+	–
A_1 abbreviated, not reaching wing margin	+	–
Abdominal bristles strong	+	–
Tergite 7 of male fused with sternite 8	+	–
Syntergosternite 7 + 8 secondarily nearly symmetric	+	–
Tergite 8 of male atrophied	+	–
Gonopod and paramere more or less fused at bases	+	–

[1] Restricted to those characters that are apomorphic with respect to ground plan of Acalyptratae.
+ Condition present.
– Condition not as described.

ated with the peritreme, whereas the metepisternal setulae of the Lauxaniidae and others are well removed from the peritreme. As metepisternal setulae occur in many, though not all, genera of both subfamilies of the Lauxaniidae, their presence appears to be a ground-plan character of the family that has been lost repeatedly within the family. In the Celyphidae they are well developed in *Celyphus* Dalman but appear to have been lost in some other genera. (It is perhaps worth noting here that the metepisternum is enlarged, swollen, and covered with short, stout setulae in all Coelopidae; also, in the Ropalomeridae and in certain Sciomyzidae, the metepisternum is finely setulose over a relatively broad area.)

An autapomorphic character of the chamaemyiid section of the Lauxanioidea is the distinctive curvature towards Sc in the apical bend of R_1.

Lauxaniidae. Adults of the family (Shewell, Ch. 87) frequent low vegetation in moist, semi-shaded habitats. Miller (1977*b*) confirmed that most larvae are saprophagous, living in fallen vegetation. Most adults graze fungal material on leaves (Broadhead 1984). Worldwide, the family contains some 126 genera and about 1500 species (Miller 1977*a*). Representatives are found on all large land masses except Antarctica. In older literature, the family was often called Sapromyzidae, based on Robineau-Desvoidy's (1830) proposal, but Lauxaniidae based on Latreille's (1804) proposal has priority.

In contrast to most acalyptrate families, the Lauxaniidae exhibit an extraordinary morphological plasticity; in this respect they have been referred to as "the acalyptrate equivalent of the orthorrhaphous Stratiomyidae" (Stuckenberg 1971). Nevertheless, there is general agreement that the family is monophyletic (Hennig 1958, Griffiths 1972). Most members are readily distinguishable from other families by a combination of relatively stable characters including vibrissa absent; postocellar bristles convergent; C unbroken; Sc complete; cells bm, dm, and cu*p* discrete; A_1 abbreviated, preapical dorsal tibial bristles present; and female with three spermathecae. Because all these characters are ground-plan conditions of the Lauxanioidea, they are plesiomorphic with respect to the ground plan of the Lauxaniidae. However, the following characters are autapomorphic with respect to the ground plan of Lauxanioidea:

— male accessory glands ("paragonia") branched repeatedly, forming a dense tangle (Sturtevant 1925–1926);
— aedeagus consolidated and reduced into a more or less rigid tubular process;
It seems probable that the aedeagus in the ground plan of the Lauxanioidea is a more or less flexible type, perhaps similar to that of *Cremifania* spp. (J. F. McAlpine 1963). In the Lauxaniidae, the flexible, armed, apical portion appears to have been withdrawn into the enlarged basal portion; in some species of *Homoneura* Wulp, for example, the armed apical portion can be exserted in macerated specimens. Relatively greater consolidation and reduction of the entire aedeagus is apparent in other species of *Homoneura* and in other genera; in some cases it is completely reduced.
— front femur with an anteroventral comb of small spines (ctenidium);
This character is probably a ground-plan condition of the family because it occurs with considerable consistency in the Homoneurinae and also in a few Lauxaniinae.
— lunule unexposed.

It seems probable that in the ground plan of the Lauxanioidea the lunule was moderately exposed, perhaps as in the Chamaemyiini. This condition is also predominant in the Sciomyzoidea and Schizophora as a whole. If this is true, the unexposed condition in the Lauxaniidae is derived (as is the widely exposed condition in the Leucopini).

Possibly the peculiar adaptations for fungal grazing on the labella of Lauxaniidae, i.e. "prongs" attached to the pseudotracheal rings and "scoops" arising from the integument between the pseudotrachea (Broadhead 1984) are sufficiently unique in this family to be considered an autapomorphy. Much more work is needed to establish the occurrence, nature, and homologies of these structures in this and other families of Acalyptratae.

Griffiths' (1972) statement that the parameres (postgonites) are lost in the ground plan of Lauxaniidae is wrong. Both the gonopods (pregonites) and the parameres are clearly distinguishable in many Lauxaniidae, and both are present in the ground plan (Hennig 1948*b*). They are usually more or less fused at their bases, and commonly the gonopods are reduced, relative to the size of the parameres. Similary, Griffiths' (1972) conclusion that the dorsal sclerite between tergite 6 and the epandrium represents tergite 7, only, in Lauxaniidae and other members of the Lauxanioidea is incorrect. Careful study of this sclerite in many representatives shows that it contains elements of segments 7 and 8 as in all Schizophora in which such a sclerite is present (Hennig 1958). Its composite nature in the Lauxaniidae is clearly illustrated by Stuckenberg (1971, Figs. 3, 7). In the Chamaemyiidae it is shown by J. F. McAlpine (1968, Fig. 2).

The suprageneric classification of the Lauxaniidae has not yet been fully resolved. Stuckenberg (1971) concluded that the family divides into two sections, one containing most of the genera in which the black costal setulae extend to R_{4+5} (homoneuriform costa) and the other containing the genera in which the costal setulae end at or near R_{2+3} (sapromyziform costa). He erected the subfamily Homoneurinae for the genera that have a homoneuriform costa and lack a strong, discal, episternal bristle but concluded that the sapromyziform section would probably be divisible into a number of other segregates coordinate with the Homoneurinae. The extension of the black costal setulae to a point well beyond R_{2+3} in the Homoneurinae appears to be a good autapomorphic character of that subfamily, for these setulae end at or near R_{2+3} in all other members of the Lauxanioidea and Sciomyzoidea examined; in fact, these setulae appear to end near R_{2+3} in the ground plan of the Muscomorpha. No autapomorphic character has yet been discovered to prove the monophyly of the remainder of the Lauxaniidae (sapromyziform section), but perhaps such a character will ultimately be found. Shewell (1977) provisionally treated that section as the subfamily Lauxaniinae, and this status was also adopted by Miller (1977*b*, 1980*a*).

Two lauxaniids, *Chamaelauxania succini* Hennig and *Hemilauxania incurviseta* Hennig, were described from Baltic amber (Hennig 1965*b*); another species, *Sapromyza veterana* Melander was described from Miocene shale in Colorado (Melander 1949). Also, *Lonchaea senescens* Scudder (Scudder 1877) from Tertiary shale in British Columbia belongs to Lauxaniidae (J. F. McAlpine 1962).

Eurychoromyiidae. This family contains only the peculiar Bolivian fly, *Eurychoromyia mallea* Hendel (Hendel 1910). It is still known only from the four specimens in the type series. Hendel classified it as "an isolated group of Acalyptratae," and erected the new subfamily Eurychoromyiinae (coordinate with his subfamilies Sciomyzinae and Lauxaniinae) for it. Hennig (1958) provided additional information on the female and ranked it as a full family near the Lauxaniidae and Chamaemyiidae. J. F. McAlpine (1968) published a detailed description of the male lectotype and indicated that it is perhaps related to the Ropalomeridae and Lauxaniidae. Griffiths (1972) placed it as a sister group of the Chamaemyiidae.

On the basis of the wing venation (short A_1), the relatively long tergite 6 in the male, and the nature of the male terminalia, I am now convinced it belongs in the Lauxanioidea. As indicated above, two characters, i.e. the shape of the metepisternum and the presence of a metepisternal setula, appear to align it definitely with the Lauxaniidae and Celyphidae (but not with the Ropalomeridae). It agrees with the Chamaemyiidae in lacking preapical dorsal tibial bristles and in having four spermathecae, but both these conditions were probably arrived at independently from those in the Chamaemyiidae.

The following characters of *Eurychoromyia* Hendel are apomorhic with respect to the ground plan of the Lauxanioidea:

- main bristles of head and thorax reduced or lost;
- fronto-orbital plates broadly enlarged, largely obliterating the frontal vitta;
- fronto-orbital plate demarcated from parafacial area by a lateral extension of the ptilinal suture;
- face strongly convex;
- ocellar triangle small, ocelli very closely crowded together;
- scape elongate, longer than pedicel;
- proboscis stout with broad labella;
- mesonotum broader than long;
- pleuron short, higher than long;
- precoxal bridge present;
- black costal setulae reduced;
- abdomen short and broad, dorsoventrally compressed;
- preapical dorsal tibial bristles lost;
- hind tibia broadened and flattened;
- gonopods and parameres indistinguishably fused and greatly reduced;
- aedeagal apodeme absent; and
- four spermathecae present.

The first 12 of these characters are shared with the Celyphidae, and quite possibly many of them are true synapomorphies. If such is the case, the Eurychoromyiidae could be considered the older sister group of the Celyphidae, and both families together would comprise the sister group of the Lauxaniidae (Fig. 116.4). The male of *Eurychoromyia* is more generalized than any Celyphidae in retaining a relatively large sternite 6, a discrete remnant of tergite 8, and a well-developed aedeagus. It is more derived than any Celyphidae in the last four characters listed.

Celyphidae. In this family the scutellum is enormously enlarged, sometimes completely covering the abdomen, and, in repose, the wings are folded beneath this strongly convex scutellum. These attributes give the adult fly a general appearance remarkably similar to a small, broadly oval beetle. A separate family-group name was proposed by Bigot (1852). The family comprises about 90 species in seven genera, i.e. *Acelyphus* Malloch (Oriental), *Afrocelyphus* Vanschuytbroeck (African), *Celyphus* Dalman, with four subgenera, *Chamaecelyphus* Frey (African), *Idioscelyphus* Malloch (Oriental), *Oocelyphus* Chen (Oriental), and *Spaniocelyphus* Hendel (African and Oriental). Distribution is tropical, predominantly in the Oriental region, but about 15 species and two endemic genera are known in Africa. Adults frequent moist habitats, and the larvae feed on decaying vegetation. No fossils are known.

There is no doubt that the Celyphidae and Lauxaniidae are closely related (Malloch 1929; Frey 1941*b*; Hennig 1958; Tenorio 1969, 1972; Griffiths 1972). Most early authors treated them as a subordinate group of the Lauxaniidae, which course was also followed by Griffiths (1972). However, in the most extensive analysis of the family yet made, Tenorio (1972) concluded that "this group of odd beetle-like flies is both unique and discrete enough to deserve family distinction," and since then separate family ranking has usually been adopted (Tenorio 1979, Miller 1980*b*, Bickel 1982).

As indicated under the Eurychoromyiidae, some 12 apomorphic characters are shared by that family and the Celyphidae, which possibly supports the idea that the Celyphidae is the apotypic sister group of the Eurychoromyiidae. The following autapomorphies of the Celyphidae vouch for its monophyly:

- scutellum greatly enlarged;
- arista arising subapically on first flagellomere;
- aedeagus largely to entirely replaced by fused gonopods and parameres; and
- hypandrium with anteromedian area atrophied.

In addition, the following apomorphic tendencies are characteristic within the family: arista usually widened and flattened at base; clypeus often much enlarged and protruding; crossvein bm-cu commonly absent; longitudinal sutures sometimes formed in the borders of the anterior abdominal tergites (so that each of tergites 1 to 3 are sometimes divided into a dorsal and two lateral plates); and body color commonly brilliant metallic blue, green, or violet.

It may be significant that certain of the peculiarities of the Celyphidae also occur sporadically in the Lauxaniidae, probably as a result of convergent evolution. For example, the scutellum is enlarged just as in the Celyphidae in the Neotropical genus *Celypholauxania* Hendel, but that genus clearly adheres to the lauxaniid pattern in other respects; both the Australian genus *Cerataulina* Hendel and the Papuan genus *Kertéziella* Hendel have the base of the arista widened and flattened as in most Celyphidae, but again both genera clearly belong to the Lauxaniidae rather than to the Celyphidae. The possibility remains that the Celyphidae are in fact a subgroup of the Lauxaniidae, but, until this is clearly demonstrated, it seems prudent to continue to treat them as a separate family.

Chamaemyiidae. Members of this family (J. F. McAlpine, Ch. 88) are often referred to as aphid-killing flies or silver flies because the larvae (and some of the adults) feed on aphidoid insects and the adults are usually a dense silvery gray in color. Distribution is cosmopolitan. The larvae, especially, are useful in the control of aphids and coccids, and, because of deliberate and accidental introductions, many species occur in more than one zoogeographical region. The family comprises about 180 described species in about 20 genera and subgenera in two subfamilies, the Cremifaniinae and Chamaemyiinae. The taxonomy of most genera is poorly known. Except in the Palearctic region, for which about 123 species are listed (Tanasijtschuk 1984), probably less than half of the species are yet described. The vast majority of species belong to the genus *Leucopis* and its various subgenera.

The family name, Chamaemyiidae, is based on Hendel's (1910) proposal of the family-group name Chamaemyiinae. This name is founded on *Chamaemyia* Meigen (Meigen 1803), a senior synonym of *Ochtiphila* Fallén (Fallén 1823). Prior to Hendel's proposal, the family-group name Ochthiphilinae, proposed by Schiner (1862, pp. VI and XIV) and founded on *Ochtiphila* (emended as *Ochthiphila*), and Ochtidae Fallén (Fallén 1823) (emended as Ochthiphilinae) was applied to the same group. The change in name by Hendel was because of the priority of *Chamaemyia* over *Ochthiphila*, and after his proposal both names were often used either preferentially or as equivalents. In response to a petition from J. F. McAlpine and Sabrosky (1965), the International Commission on Zoological Nomenclature (1968, Opinion 847) declared Chamaemyiidae as the official name of the family.

Hendel (1916) was the first to rank the group as a separate family, which ranking is now generally accepted. At first, Malloch (1921) treated it as a subfamily of the Agromyzidae, but later he (Malloch 1930*b*, 1933*a*, 1940) allied it with the Lauxaniidae and Sciomyzidae. The consensus now is that the family belongs to the superfamily Lauxanioidea (Hennig 1958, 1965*b*, 1973; Griffiths 1972). Certainly its ground-plan characters include most of those listed for the Lauxanioidea (Table 116.4). The following characters in the ground plan of the Chamaemyiidae (including the Cremifaniinae) are apomorphic (or apo-oecal) with respect to the ground plan of the Lauxanioidea:

- larvae predaceous on aphidoid insects;
- hypopharyngeal and tentoropharyngeal sclerites fused;
- parastomal bar fused with hypopharyngeal sclerite;
- interspiracular hairs of posterior spiracles unbranched;
- aristomere 3 relatively short (Hennig 1965*b*);
- prosternum bare;
- proanepimeral bristle absent;
- R_1 flexed toward apical bend of Sc;
- A_1 abruptly abbreviated, not attaining wing margin;
- preapical dorsal tibial bristles absent;
- body heavily silvery gray pruinose; and
- tergite 6 of males less than half as long as tergite 5.

The subfamily Cremifaniinae contains at least three extant species, *Cremifania nigrocellulata* Czerny from Europe (introduced in North America), *C. nearctica* McAlpine from western North America, and an undescribed species from Mexico, and one fossil species, *Procremifania electrica* Hennig from Baltic amber. The fossil species is quite similar to *C. nigrocellulata* and to the undescribed Mexican species, but, because of its larger, nearly round eye and its slightly convergent postocellar setulae, Hennig (1965*b*) believed it belongs to the ancestral group of Cremifaniinae. Hennig (1965*b*) strongly agreed with my assumption (J. F. McAlpine 1963) that *Cremifania* is more closely related to the remaining genera of Chamaemyiidae than to any other group of the Chamaemyiinae. This viewpoint is supported by the 12 synapomorphies previously tabulated.

The ground plan of the Cremifaniinae differs from that of the Chamaemyiinae in the following respects:

1. Three pairs of chromosomes (2n = 6) are present (2n = 12 in Chamaemyiinae).
2. Postocellar bristles are weak or absent (present and convergent in Chamaemyiinae).
3. Pterostigmal space is long, subequal in length to cell c (less than one quarter as long as cell c in Chamaemyiinae).
4. Surstylus articulates freely with epandrium (fused or absent in Chamaemyiinae).
5. Epandrium has an anterior process (absent in Chamaemyiinae).
6. Aedeagus is flexible, armed with various processes, and capable of being swung through a wide arc against aedeagal apodeme (rigid and in a fixed posteroventral position in Chamaemyiinae).
7. Epiphallus is present (absent in Chamaemyiinae).
8. Aedeagal apodeme is rod-like and free from hypandrium (fultelliform and fused with hypandrium in Chamaemyiinae).

The first two of these characters are probably autapomorphies of the Cremifaniinae. Characters 3, 4, and 6 are plesiomorphic conditions that occur frequently throughout the Muscomorpha. Character 5 occurs sporadically, but is especially characteristic of Dryomyzidae, most Helosciomyzidae, and Phaeomyiinae. Characters 6 and 7 appear, disappear, and reappear frequently throughout the Muscomorpha and their significance here is uncertain. The absence of a proanepisternal bristle in the Cremifaniinae was mentioned by Hennig (1965*b*) as an autapomorphic character; this character is shared with all other Chamaemyiidae except *Pseudoleucopis* Malloch from Australia. It is uncertain whether the occurrence of a proanepisternal bristle in *Pseudoleucopis* is a plesiomorphic or pseudoplesiomorphic (= special case of apomorphy, Hennig 1971*a*, p. 25) feature; if the latter, then the absence of this bristle in the Cremfaniinae is a synapomorphy between it and the Chamaemyiinae; if not, the presence of it in *Pseudoleucopis* is an autapomorphy of that genus.

At this point it is appropriate to indicate that the separate condition of the three spermathecal ducts in *C. nigrocellulata* illustrated by me (J. F. McAlpine 1963) is probably based on an aberrant specimen. Examination of additional specimens of the same species has shown that the duct leading from the smallest spermatheca joins the larger duct before it enters the oviduct, i.e. only two ducts directly enter the oviduct as in all other Acalyptratae studied.

Griffiths (1972) treated the Cremifaniinae as a separate family with no indication of its sister group. He assigned it to the Sciomyzoidea, rather than to the Lauxanioidea, because the male segments 6 to 8 are asymmetric and the aedeagus has well-differentiated sclerites (paraphalli) in its walls and is able to be swung through a wide arc against the aedeagal apodeme into an anteriorly directed position. He excluded it from the Lauxanioidea because he believed that the conditions of these characters in the ground plan of the Lauxanioidea could not be derived from the basic pattern of the Sciomyzoidea. According to Griffiths, in the basic plan of the Lauxanioidea:

— the pregenital and genital segments are symmetric;
— tergite 6 is similar in length to tergite 5;
— tergite 7 is a well-developed dorsal tergite;
— sternite 8 and tergite 8 are absent; and
— the aedeagus is a uniformly sclerotized structure upcurved along most of its length and always directed more or less posteriorly in relation to the longitudinal axis of the insect, i.e. not able to be swung through a wide arc against the aedeagal apodeme.

My investigations of the Lauxanioidea resulted in three findings. First, the pregenital segments are in fact basically asymmetric. Second, the sclerite that Griffiths called tergite 7 is in fact syntergosternite 7 + 8. Third, the aedeagus in the ground plan of Lauxanioidea might be a complex structure, fundamentally rather similar to that found in *Cremifania* (compare for example, the aedeagi of *Homoneura johnsoni* Miller and *Camptoprosopella borealis* Shewell). I still consider it possible that the plesiomorphic characters of *Cremifania* reflect similar ground-plan conditions for the Lauxanioidea as a whole, and that its many points of agreement with other Chamaemyiidae are true synapomorphies. The points of agreement between it and certain groups assigned to the Sciomyzoidea may, in fact, be the result of parallel or convergent evolution. Pending a convincing demonstration of a sister-group relationship between *Cremifania* and some other group, I consider it best to continue to treat it as an older subfamily of Chamaemyiidae.

The Chamaemyiinae comprises two tribes: the tribe Chamaemyiini containing *Acrometopia* Schiner, *Chaetoleucopis* Malloch, *Chamaemyia* Meigen, *Melanochthiphila* Frey, *Parapamecia* Cogan, *Parochthiphila* Czerny (with two subgenera, *Euestelia* Enderlein and *Parochthiphila sens. str.*), *Plunomia* Curran, *Pseudodinia* Coquillett, *Pseudoleucopis*, and *Toropamecia* Cogan; and the tribe Leucopini containing *Leucopis* (with seven subgenera, *Indioleucopis* Steyskal, *Leucopella* Malloch, *Leucopis sens. str.*, *Leucopomyia* Malloch, *Metaleucopis* Tanasijtschuk, *Neoleucopis* Malloch, and *Xenoleucopis* Malloch), *Lipoleucopis* de Meijere, and *Melaleucopis* Sabrosky. No fossil representative of either tribe is known.

Synapomorphic characters of the Chamaemyiinae with respect to the ground plan of the family include:

— epandrium without discrete surstyli;
— aedeagal apodeme fused to hypandrium (fultelliform);
— aedeagal uniformly sclerotized, rigid, and posteroventrally directed; and
— four spermathecae present.

The ground plan of the tribe Chamaemyiini is more generalized than that of the Leucopini in a number of respects: the lunule is relatively narrow and bare, the frons is less densely setulose, two well-developed fronto-orbital bristles are present, the ocellar and postocellar bristles are well developed, vein A_1 extends well beyond its juncture with CuA_2, and tergite 6 of the male is present and discrete. The only conspicuous autapomorphic character of the Chamaemyiini (synapomorphic character of the component genera) that I have noted is the relatively greater reduction of the gonopods; in the Chamaemyiini they are much shorter and less discrete than the parameres, whereas in the Leucopini they are more or less equal in length and discreteness to the parameres. Autapomorphic characters of the Leucopini (synapomorphies of the component genera) are as follows:

— lunule broadly exposed and setulose;
— ocellar and postocellar bristles reduced or absent;
— fronto-orbital bristles reduced or absent;
— A_1 atrophied beyond juncture of CuA_2; and
— tergite 6 of male lost or fused with syntergosternite 7 + 8.

One ground-plan character of the Leucopini that remained more generalized than in the Chamaemyiini is the relatively greater length and discreteness of the gonopods.

It should be noted here that the genus *Paraleucopis* Malloch, which heretofore has been placed in Chamaemyiidae (Steyskal 1971, 1981; J. F. McAlpine, Ch. 88) actually belongs in the Asteioinea along with two other closely related genera, *Gayomyia* Malloch and *Schizostomyia* Malloch, from Chile (see discussion under Asteiidae).

Superfamily Sciomyzoidea

Besides five long-standing and well-established families: the Coelopidae, Dryomyzidae (including the Helcomyzinae), Sciomyzidae (provisionally including the Phaeomyiinae and Huttonininae), Ropalomeridae, and Sepsidae, the Sciomyzoidea (Fig. 116.4) also contains the Helosciomyzidae. The latter was recently raised from a subfamilial status in the Sciomyzidae to a separate family (Griffiths 1972, Barnes 1981).

The oldest family-group name applicable to this group is Sciomyzides (Fallén 1820*d*); hence, Sciomyzoidea is the correct superfamily name to use.

For the most part, this grouping of families is one of the least contentious of all the superfamilies of the Acalyptratae. Hennig's (1958) concept of it differed little from that of Hendel (1916). I was wrong earlier (J. F. McAlpine 1963) in including the Periscelididae in this complex, for, as shown later, it belongs in the Asteioinea and has nothing to do with the Sciomyzoidea. Colless and D. K. McAlpine (1970) place the Coelopidae in the Heleomyzoidea (= Sphaeroceroidea); later, however, they (Colless and D. K. McAlpine 1975) removed it to the Sciomyzoidea. As discussed under the previous superfamily, the Sciomyzoidea is considered to be the sister group of the Lauxanioidea. Griffiths (1972) treated the Sciomyzoidea (as prefamily Sciomyzoinea of his superfamily Muscoidea) as the sister group of his prefamily Diopsoinea (Diopsidae + Syringogastridae) and included with it the Cremifaniinae (as family Cremifaniidae) and the Megamerinidae. Perhaps the idea that the Cremfaniinae belongs to the Sciomyzoidea has some merit (see discussion under Chamaemyiidae), but I see no justification for removing the Megamerinidae from the Diopsoidea (see discussion under Superfamily Diopsoidea) to the Sciomyzoidea.

The ground plan of the Sciomyzoidea is characterized by the same suite of plesiomorphic characters that occurs in the Lauxanioidea: vibrissa absent; C unbroken; Sc complete; cells bm, dm, and cu*p* discrete; female with three spermathecae and simple tubular terminalia; and male with freely articulated surstyli. The Sciomyzoidea remained even more plesiotypic than the Lauxanioidea in retaining in its ground plan divergent postocellar bristles, a complete A_1, and a discrete remnant of tergite 8 in the male. Characters in the ground plans of both the Sciomyzoidea and the Lauxanioidea that are apomorphic with respect to the ground plan of the Acalyptratae are compared in Table 116.4. Most, if not all, of the first nine characters listed may be considered as synapomorphies of the two superfamilies. Only three rather weak apomorphies, as follows, are exclusive to the Sciomyzoidea:

- frontal vitta relatively densely and strongly setulose;
- face desclerotized along vertical midline; and
- tergite 6 of male relatively reduced.

These three characters provide the best evidence I can offer as proof of the monophyly of the Sciomyzoidea. Certainly, more work is needed in this respect. I agree with Griffiths (1972) that the shortened condition of tergite 6 in the male is particularly significant, but even that character varies considerably within component families, which perhaps indicates that it was nearly as long in the ground plan of the Sciomyzoidea as it was in the ground plan of the Lauxanioidea. Barnes (1981) tabulated the distribution of many important characters in the higher taxa of the Sciomyzoidea and provided the best foundation to date for interpreting the evolutionary relationships of these groups. Although his ideas regarding the cladistics of these taxa are perhaps inferred in the way he arranged them, and in his discussion, he declined to provide a cladogram.

Coelopidae. This well-marked, monophyletic family (Vockeroth, Ch. 82) of the Sciomyzoidea was originally called Phycodromidae (Osten Sacken 1878, Williston 1896, Becker *et al.* 1905, Aldrich 1905), based on *Phycodromia* Stenhammar (Stenhammar 1854) (= *Malacomyia* Haliday 1840). The name Coelopidae has been generally accepted (except Miller 1950) since Hendel's (1910) proposal of Coelopinae. In the interests of stability, usage of Coelopidae should be continued. No divergent opinion has ever been expressed, but the matter has never been referred to the Commission for a formal ruling.

In the ground plan of the Coelopidae, a more complete set of the sclerites of the male terminalia were retained than in any other representative of the Sciomyzoidea. The tergites and sternites of segments 6 to 8 are unusually clearly defined (Fig. 82.5), and the vestiges of tergites 7 and 8 are particularly large and discrete. On this basis the family can be considered the older sister group of the remainder of the Sciomyzoidea.

The following features are apomorphic with respect to the ground plan of the Sciomyzoidea:

- body more or less flattened on dorsum;
- frons and face more or less secondarily sclerotized in middle;
- postocellar bristle convergent;
- laterotergite strongly protruding;
- metepisternum convex and broadly spinulose;
- posterior margin of wing folded under (Fig. 82.2);
- femora swollen;
- legs strongly bristled; and
- tarsomere 1 of fore tarsus of male with a thumbnail-like apicoventral process.

As now defined (Vockeroth, Ch. 82), the Coelopidae contains about 30 species divided among 10 genera: *Coelopa* Meigen, *Coelopina* Malloch, *Malacomyia* Haliday, *Protocoelopa* Malloch, *Coelopella* Malloch, *Dasycoelopa* Malloch, *Chaetocoelopa* Malloch, *Baeopterus* Lamb, *Icaridion* Lamb, and one undescribed genus. Griffiths (1972) included *Heterocheila* Rondani (= *Oedoparea* Loew), but it belongs to the Dryomyzidae (Steyskal, Ch. 83) on the basis of divergent postocellar bristles, presence of a precoxal bridge, the absence of metepisternal spinulae, and the structure of the male terminalia. Hardy (1962) included *Apataenus* Eaton in the Coelopidae, but that genus belongs to Tethinidae (Vockeroth, Ch. 101).

Coelopa occurs in the Holarctic and Ethiopian regions and in Micronesia; *Coelopina* occurs in the southwestern Nearctic region; *Malacomyia* occurs in the western Palearctic region; seven genera occur only in Australia, New Zealand, and Antarctica. The best key to world genera is by Malloch (1933*b*), but it does not include *Baeopterus* and *Icaridion*. Harrison's (1959) key includes both these genera but incorrectly refers some fully winged species of *Icaridion* to *Coelopa*. No fossils are known.

Dryomyzidae. The traditional concept of this family is adopted here (Steyskal, Ch. 83), wherein it is composed of two subfamilies, the Dryomyzinae and Helcomyzinae.

Griffiths (1972) accepted this same definition, except that he incorrectly removed the genus *Heterocheila* from the Helcomyzinae to the Coelopidae. It should be noted, however, that many recent authors (Steyskal 1957*b*, 1958; J. F. McAlpine 1963; Colless and D. K. McAlpine 1970; Barnes 1981) have followed Malloch's (1933*a*) proposal to treat the Helcomyzinae as a separate family. The family name, Dryomyzidae, apparently dates from Schiner's (1862, p. 38) proposal of Dryomyzinae.

The following ground-plan characters of both subfamilies are apomorphic with respect to the ground plan of the Sciomyzoidea:

- tarsomere 1 of the male hind tarsi with a thumbnail-like process, as in Coelopidae (lost in *Heterocheila*);
- tergite 8 of male atrophied, as in all Sciomyzoidea except the Coelopidae and some Phaeomyiinae;
- anterior epandrial process present, as in the Helosciomyzidae (reduced in *Heterocheila* and in the ground plan of the Sciomyzidae); and
- gonopod and paramere juxtaposed (Fig. 83.7).

I do not include the densely haired condition of the aedeagus, as found in the Dryomyzinae (Griffiths 1972, Barnes 1981), as a ground-plan feature of the family. It is more probable that the simpler type of aedeagus found in the Helcomyzinae is closer to the ground-plan condition for the family as a whole. Also, I consider that the large, fan-like, basal lobes on the aedeagus of *Heterocheila* constitute an autapomorphic character of that genus.

In the ground plan of the subfamily Dryomyzinae the prosternum remained narrow and separated from the propleuron, C remained nonspinose, crossvein dm-cu and CuA_2 remained well separated, and three spermathecae were retained.

One apomorphic character that seems to hold for this subfamily is the loss of metasternal setulae; the metasternal area is bare in all species examined. In general, the Dryomyzinae are less hairy throughout than the Helcomyzinae. Abdominal spiracles 2 to 5 are usually enclosed in the margins of the tergites, but this character is not so in at least one species of *Oedoparena* Curran (*O. minor* Suwa from Japan), and in the fossil species *Prodryomyza electrica* Hennig. This subfamily contains about 20 species in two extant genera. *Dryomyza* Fallén is widespread and common, and *Oedoparena* occurs only on Pacific coasts. Two fossil species in Baltic amber, *Palaeotimia l'hoesti* Meunier and *P. electrica* Hennig (Hennig 1965*b*, 1969*b*), also appear to belong to the Dryomyzinae.

The subfamily Heleomyzinae retained a hairier condition throughout, including a setulose metasternal area, but the following apomorphies are present in the ground plan:

- prosternum triangular with well-developed precoxal bridge;
- crossvein dm-cu almost in line with CuA_2; and
- two spermathecae present.

This subfamily contains about 10 species in 4 genera, all of which are restricted to marine coasts. Two genera, *Helcomyza* Curtis and *Heterocheila* are Holarctic and each contains one Nearctic and one Palearctic species. *Paractora* Bigot has four species in extreme southern South America and possibly one on Macquarie Island; *Maorimyia* Tonnoir & Malloch has one species in New Zealand.

Helosciomyzidae. This little family contains 23 species in 9 genera (Barnes 1981), most of which were for many years classified with the Sciomyzidae. The family occurs only in Australia, Tasmania, New Zealand, and South America, and most members have highly restricted relict distributions. Steyskal (1965) proposed the subfamily Helosciomyzinae (in Sciomyzidae) for three genera, *Helosciomyza* Hendel, *Polytocus* Lamb, and *Xenosciomyza* Tonnoir & Malloch, primarily on the basis of their having C spinose, postocellar bristles parallel or nearly so, and three spermathecae. Griffiths (1972) raised the Helosciomyzinae to family rank, but included in it the subfamily Huttonininae from the Sciomyzidae. Barnes (1979) separated the Huttonininae from the Helosciomyzidae to contain the three genera referred to it by Steyskal (1965), plus five new genera: *Cobergius* Barnes, *Dasysciomyza* Barnes, *Napaeosciomyza* Barnes, *Neosciomyza* Barnes, and *Scordalus* Barnes, from Australia and New Zealand. He also included for the first time the peculiar South American genus *Sciogriphoneura* Malloch. Barnes concept of the family appears to be taxonomically sound and is accepted here.

Most members of the Helosciomyzidae possess the following conditions, regarded as apomorphic with respect to the ground plan of the Sciomyzoidea:

- C with at least one row of costal spines;
- R_1 bare above;
- epandrium with an anterior process (lost in some *Helosciomyza* spp., *Scordalus femoratus* (Tonnoir & Malloch), and *Xenosciomyza prima* Tonnoir & Malloch); and
- surstylus with a basal surstylar process (lost in *Helosciomyza, Eurotocus* Steyskal, and *Scordalus* Barnes).

All four of these characters also occur in at least some members of the Dryomyzidae, which would seem to indicate that the Helosciomyzidae is possibly more closely related to the Dryomyzidae than to any other Sciomyzoidea. However, at least the first two characters are universally established in the Helosciomyzidae, and together they provide an autapomorphic combination of characters. Also, the Dryomyzidae has apomorphic conditions that are unknown in the Helosciomyzidae. For example, a thumbnail-like process is present on at least tarsomere 1 of the male hind tarsus, and the gonopods and parameres are peculiarly placed in a side-by-side position, i.e. juxtaposed. Because Helosciomyzidae shares peculiarities with both subfamilies of the Dryomyzidae and does not seem to be more closely related to either one, it seems best to regard it as the sister group of Dryomyzidae as a whole as proposed by Barnes (1981). No fossils are known.

Sciomyzidae. This family (Knutson, Ch. 84) is readily distinguished from its relatives by the following combination of characters: C without costal breaks and not spinose; Sc complete, free from R_1; cell M open not narrowed apically (except in some Salticellinae); A_1 complete (except in Huttonininae); oral vibrissa absent; postverticals divergent to parallel; mid femur with a bristle near middle of anterior surface; and one or more tibiae with a preapical dorsal bristle. As in most families of the Sciomyzoidea, the morphological characters of the Sciomyzidae remained relatively close to the basic plan of the Schizophora, and, although it has a unique combination of characters to differentiate it from related families, little autapomorphic evidence (synapomorphies of the including taxa) has yet been advanced to confirm its monophyly.

Unlike most Sciomyzoidea, the clypeus is relatively small, withdrawn, and separated from the anterior margin of the face by a large membranous area; this character seems to constitute an autapomorphic condition for the whole family (J. F. McAlpine 1963, Hennig 1965*b*, Griffiths 1972, Barnes 1979). The strongest indication that the Sciomyzidae, in the strict sense, are monophyletic is the fact that their larvae feed only on aquatic and terrestrial Mollusca. This one is certainly a derived (autapo-oecal) character with respect to the ground plan of the Sciomyzoidea and the Diptera as a whole.

As indicated under the introduction to the superfamily Sciomyzoidea, Sciomyzides (Fallén 1820*d*) is the oldest family-group name applicable to the family; therefore Sciomyzidae is the correct name to use. Other group names proposed within the family were listed by Steyskal (1965).

Modern papers treating the limits and subdivisions of the Sciomyzidae (Verbeke 1950, 1961; Hennig 1965*b*, 1973; Steyskal 1965; Griffiths 1972) were summarized by Berg and Knutson (1978). Four subfamilies are now usually recognized, basically as proposed by Steyskal (1965), except that his subfamily Helosciomyzinae was redefined and is now ranked as a separate family (as described earlier). The remaining four subfamilies are as follows: Salticellinae contains three species in two genera, *Salticella* Robineau-Desvoidy (Palearctic and South African) and the fossil *Prosalticella succini* Hennig (Hennig 1965*b*); Phaeomyiinae contains three extant species in one genus *Pelidnoptera* Rondani (= *Phaeomyia* Schiner) (Palearctic), and one fossil, *Prophaeomyia loewi* Hennig (Hennig 1965*b*); the Huttonininae contains nine species in two genera (each representing a different tribe), *Prosochaeta* Malloch and *Huttonina* Tonnoir & Malloch (New Zealand); the Sciomyzinae contains about 500 species, including several fossils (Hennig 1965*b*) and represent about 60 genera and two tribes (cosmopolitan).

Barnes (1979, 1981) provided the best and most accurate tabulation of the distribution of important characters in the Sciomyzidae and Sciomyzoidea and clearly established the synapomorphies and autapomorphies of all these "subfamilies." Although he did a better job than anyone else in establishing them as monophyletic units, he declined to present either a cladistic interpretation or a phylogenetic classification.

The monophyly of the Salticellinae and Sciomyzinae (= Sciomyzidae *sensu* Griffiths 1972) was well demonstrated (Knutson *et al.* 1970, Griffiths 1972). Knutson *et al.* (1970) believed that the group consisting of the Salticellinae, Sciomyzinae, and Phaeomyiinae is monophyletic, but Griffiths (1972) treated the Phaeomyiinae as a separate family because its only extant genus, *Pelidnoptera*, seemed to him to be as closely related to the Coelopidae or the Dryomyzidae as to the Sciomyzidae. Barnes (1979, 1981), however, continued to treat the Phaeomyiinae as a subfamily of the Sciomyzidae. He considered that the small size of the clypeus and its separation from the face by a large, membranous area are synapomorphic features, which support the opinion that the Phaeomyiinae, Salticellinae, and Sciomyzinae are monophyletic, but he left the final conclusion open. If these three subfamiles do prove to be monophyletic, the Phaeomyiinae is almost certainly the older sister group of the other two, for unlike them it retained a relatively large remnant of tergite 8 in the male (as in Coelopidae) and three spermathecae in the female. No obvious synapomorphies exist between the Phaeomyiinae and Coelopidae, but the male terminalia of the Phaeomyiinae agree with those of the Dryomyzidae in two derived respects; i.e. a well-developed anterior epandrial process is present, and the gonopods and parameres are peculiarly juxtaposed. However, the first of these characters also occurs in most Helosciomyzidae and in the Cremifaniinae, and the possibility that one or both may occur in some Sciomyzinae should be researched carefully. It is perhaps also worth noting that at least some members of the Phaeomyiinae agree with the Salticellinae in two distinctive ways: first, CuA_2 is sinuate in the three extant species, and in two of them, *Pelidnoptera fumipennis* (Zetterstedt) and *P. fuscipennis* (Meigen), there is a small ventroapical extension in cell cu*p*. Second, the copulatory posture of *P. fumipennis*, at least, is similar to that of *Salticella fasciata* (Meigen), but quite different from that of any other Sciomyzoidea (Knutson *et al.* 1970).

Thus, the affinities of the Phaeomyiinae to other groups within the Sciomyzoidea and to other Sciomyzidae are still uncertain, and more work is required to resolve the matter. A finding that the larvae are malacophagous would support a closer relationship within the Sciomyzidae, but a finding otherwise would almost certainly preclude their membership in the Sciomyzidae. If they do not belong in the Sciomyzidae, they probably should be ranked as a separate family, coordinate with the Helosciomyzidae, as proposed by Griffiths (1972).

The situation regarding the sister group and ranking of the Huttonininae is much the same as for the Phaeomyiinae. Originally, most members of this "subfamily" were included either with the Helosciomyzinae (Steyskal

1965) or with the Helosciomyzidae (Griffiths 1972), but Barnes (1979) removed them from there because he was unable to find evidence to support a monophyletic group that includes both those groups. He redefined the Huttonininae to include the genus *Prosochaeta* (as tribe Prosochaetini) as well as *Huttonina* (as tribe Huttoninini) and concluded that the cladistic relationship of the Huttonininae within the Sciomyzoidea is uncertain. Later, he (Barnes 1981) noncommittally placed them between the Dryomyzinae (as Dryomyzidae) and the Phaeomyiinae as one of ten "well established monophyletic taxa that have been placed in the Sciomyzoidea." Unlike the Phaeomyiinae, Salticellinae, and Sciomyzoidea, but like the Dryomyzinae and others, the clypeus is large and prominent. Also, like the Dryomyzinae especially, the suture between abdominal tergites 1 and 2 is indistinct. The habits of the immature stages are unknown, and, just as for the Phaeomyiinae, such data would be highly significant. Clearly, more work is also required to establish the precise placement and ranking of the Huttonininae within the Sciomyzoidea.

The Ropalomeridae + Sepsidae form a well-founded monophyletic group within Sciomyzoidea (J. F. McAlpine 1963; Hennig 1965*b*, 1971*a*; Griffiths 1972; Barnes 1981). The following synapomorphies with respect to the ground plan of the Sciomyzoidea confirm their sister-group relationship:

- midfacial area heavily sclerotized and prominent;
- one or more subvibrissal setae tending to be enlarged and vibrissa-like;
- prosternum setulose;
- posterior thoracic spiracle with one or more distinct setae on posterior margin;
- R_1 bare;
- A_2 absent;
- aedeagal apodeme fused to hypandrium, i.e. cuneiform; and
- two spermathecae present.

The ancestor to the Ropalomeridae + Sepsidae was probably the same or similar to the one that gave rise to the Helosciomyzidae + Dryomyzidae + Sciomyzidae, as a closer relationship to any particular one of those three families cannot be demonstrated. It is noteworthy, however, that some of the ground-plan characters of the Coelopidae that appear to have been lost in the Helosciomyzidae + Dryomyzidae + Sciomyzidae line were retained in the ground plan of the Ropalomeridae + Sepsidae. These include a flattened scutellum, a strongly bulging laterotergite, and a broadly setulose metepisternum (lost in Sepsidae).

Ropalomeridae. This closely knit little family consists of only 1 subfamily containing 8 genera and about 30 species (Steyskal, Ch. 85). It is restricted to tropical and subtropical America; only one species, *Rhytidops floridensis* (Aldrich), has been found north of Mexico, and only three species occur as far south as northern Argentina. The larvae live in decaying vegetation or sap fluxes. The family name is based on *Ropalomera* Wiedemann (Wiedemann 1824), not *Rhopalomera* (emendation); separate family status was first proposed by Hendel (1916) (incorrectly spelled Rhopalomeridae). The family was revised by Prado (1966) and cataloged by Steyskal (1967).

The Ropalomeridae retained the following plesiomorphic conditions with respect to the ground plan of the Sepsidae: metasternal area setulose (bare in Sepsidae); A_1 reaching wing margin as a fold (incomplete in all Sepsidae except *Protorygma* Hennig); surstyli discrete and freely articulated with epandrium; aedeagal apodeme incompletely fused with hypandrium; abdominal spiracle 6 of male situated in membrane; abdominal sternite 7 of male large; occasionally an abortive third spermatheca present (Prado 1966); and in addition, CuA_2 is straight (recurved in Sepsidae), a condition that Hennig (1965*b*) considered as plesiomorphic with respect to the recurved condition in Sepsidae (see discussion under Sepsidae).

The family is well characterized by the following autapomorphic conditions:

- upper frons and vertex deeply concave, eyes protruding;
- face with a median protuberance;
- precoxal bridge developed;
- greater ampulla well developed;
- apical section of M bent forward in line with crossvein dm-cu, and reaching wing margin anterior to apex of wing;
- hind tibia laterally compressed, strongly expanded, sabre-shaped;
- sternite 6 of male greatly reduced or absent; and
- basiphallus broadly expanded posteriorly forming a pseudophypandrial bridge (this feature was misinterpreted by Griffiths 1972 as a fusion of the hypandrial arms).

No fossils are known.

Sepsidae. Members of this family (Steyskal, Ch. 86) are sometimes called black scavenger flies because the larvae live in excrement, carrion, sewage sludge, and decaying vegetation. The adults occur in large numbers around the same filthy locations and feed on nectar and liquid extracts from dung and decaying organic matter. The family comprises about 240 species in 21 genera (Pont 1979) and is cosmopolitan in distribution. Many species occur on more than one continent and have probably spread in association with people and their livestock, especially cattle.

The family name dates from Macquart's (1835) proposal (as subtribe Sepsidées, Sepsideae). Two subfamilies are recognized here (Steyskal, Ch. 86): the Orygmatinae contains only one Holarctic genus and species, *Orygma luctuosum* Meigen, which lives in wrack along seashores, and the Sepsinae contains three tribes, the Saltellini, Sepsini, and Toxopodini. The Saltellini includes only one Holartic genus and species, *Saltella sphondylii*

(Schrank). The Toxopodini is restricted to the southern parts of the Old World. The Sepsini has a wide distribution and comprises most of the genera and species.

One fossil, *Protorygma electrica* Hennig (Hennig 1965*b*), was assigned by Hennig to the Sepsidae. Its inclusion here very much complicates the definition of the family, because it exhibits certain fundamental conditions, as follows, that do not occur in any other Sepsidae:

- a precoxal bridge is developed;
- R_1 is setulose;
- A_1 is complete;
- four fronto-orbital bristles are present; and
- abdominal spiracle 6 of the females is free in the membrane.

Its membership in the Ropalomeridae + Sepsidae group is firmly substantiated by two synapomorphic conditions, i.e. distinct setae present on the hind margin of the posterior thoracic spiracle and the absence of A_2. As in the Ropalomeridae it has a precoxal bridge, which I now agree is probably a synapomorphy. In addition, it has a complete A_1 and abdominal spiracles 6 and 7 of the female are situated in the membrane. These are clearly plesiomorphic conditions shared with the Ropalomeridae and with several other families of the Sciomyzoidea, but not with the Sepsidae. As in the Sepsidae, CuA_2 is recurved, which may be a synapomorphy, but certainly the same condition prevails throughout the Sciomyzoidea and many other superfamilies. Hennig stressed this feature as evidence for placing *Protorygma* Hennig in the Sepsidae, in which he considered it to be the sister group of the remainder of the family. In my opinion this evidence is not conclusive for a sister-group relationship with the Sepsidae. Plesiomorphically the head of *Protorygma* is more generalized than in either the Ropalomeridae or the Sepsidae (somewhat as in Helcomyzinae). The setose condition of R_1 (as in the ground plan of several sciomyzoid families) is also a more plesiomorphic condition than the bare condition found in both the Ropalomeridae and Sepsidae, and the female abdomen is generalized as in the Coelopidae and Helcomyzinae.

Autapomorphically, *Protorygma* has three additional pairs of fine lateroreclinate fronto-orbital setae and lacks preapical dorsal tibial bristles. On the basis of these considerations its placement in the Sepsidae should probably be reassessed. Perhaps a better case can be made for considering it as the sister group of the Ropalomeridae + Sepsidae. Even Hennig (1965*b*) himself expressed the opinion that "no objection could be raised to derivation of recent Ropalomeridae from *Protorygma*."

Excluding *Protorygma*, the Sepsidae show the following autapomorphies:

- face with a prominent median keel;
- palpi reduced;
- A_1 abbreviated, not reaching wing margin even as a fold;
- abdominal spiracle 6 of male enclosed in tergite 6;
- abdominal spiracles 6 and 7 of female enclosed in their respective tergites;
- sternite 7 of male greatly reduced or absent; and
- surstyli fused with epandrium.

As shown by Hennig (1965*b*), *Orygma* Meigen represents the sister group of all remaining Sepsidae (excluding *Protorygma*). It is the only sepsid genus in which the male retained a large tergite 6 and is also the only one in which spiracles 7 of the male are situated in the membrane.

Superfamily Opomyzoidea

Thirteen families, comprising four subgroups (suprafamiliae *sensu* Kéler 1963, Teilgruppen *sensu* Hennig 1971*a*, prefamilies *sensu* Griffiths 1972), are assembled here as follows:

Suprafamily Clusioinea:	Clusiidae
	Acartophthalmidae
Suprafamily Agromyzoinea:	Odiniidae
	Agromyzidae
	Fergusoninidae
Suprafamily Opomyzoinea:	Opomyzidae
	Anthomyzidae
Suprafamily Asteioinea:	Aulacigastridae
	Periscelididae
	Neurochaetidae
	Teratomyzidae
	Xenasteiidae
	(= Tunisimyiidae)
	Asteiidae

Many of these families have been listed previously either as "unplaced families" (Stone *et al.* 1965), or as "families with unclear affinities" (Hennig 1958, 1965b, 1971*a*). Individually, most of them are relatively distinct and readily definable. At the same time, each relates rather clearly to one or more of the others, and, as a result, the four named suprafamilies are also fairly clearly differentiated. Most of these subgroups have been recognized previously, usually as separate superfamilies (see especially, D. K. McAlpine 1978). Proof for the supposed monophyly of the whole assemblage has never been advanced, which is one of the main tasks of this section. The cladistic arrangement arrived at for all the subgroups (suprafamilies) and families is shown in Fig. 116.5.

Opomyzidae is the oldest family name within this group, having been first used in the form "Opomyzides" by Fallén (1820*c*). Thus, in accordance with articles 35 and 36 of the *International Code of Zoological Nomenclature* (Ride, Sabrosky, *et al.* 1985), Opomyzoidea has priority over all the others as the superfamily name (D. K. McAlpine 1978, Sabrosky personal commun.) For convenience, four suprafamily names, i.e. Clusioinea, Agromyzoinea, Opomyzoinea, and Asteioinea, have been employed in this superfamily. Although the same categorical ranks occur in other superfamilies of the Muscomorpha, it has not been found necessary to refer to them by formal suprafamilial names.

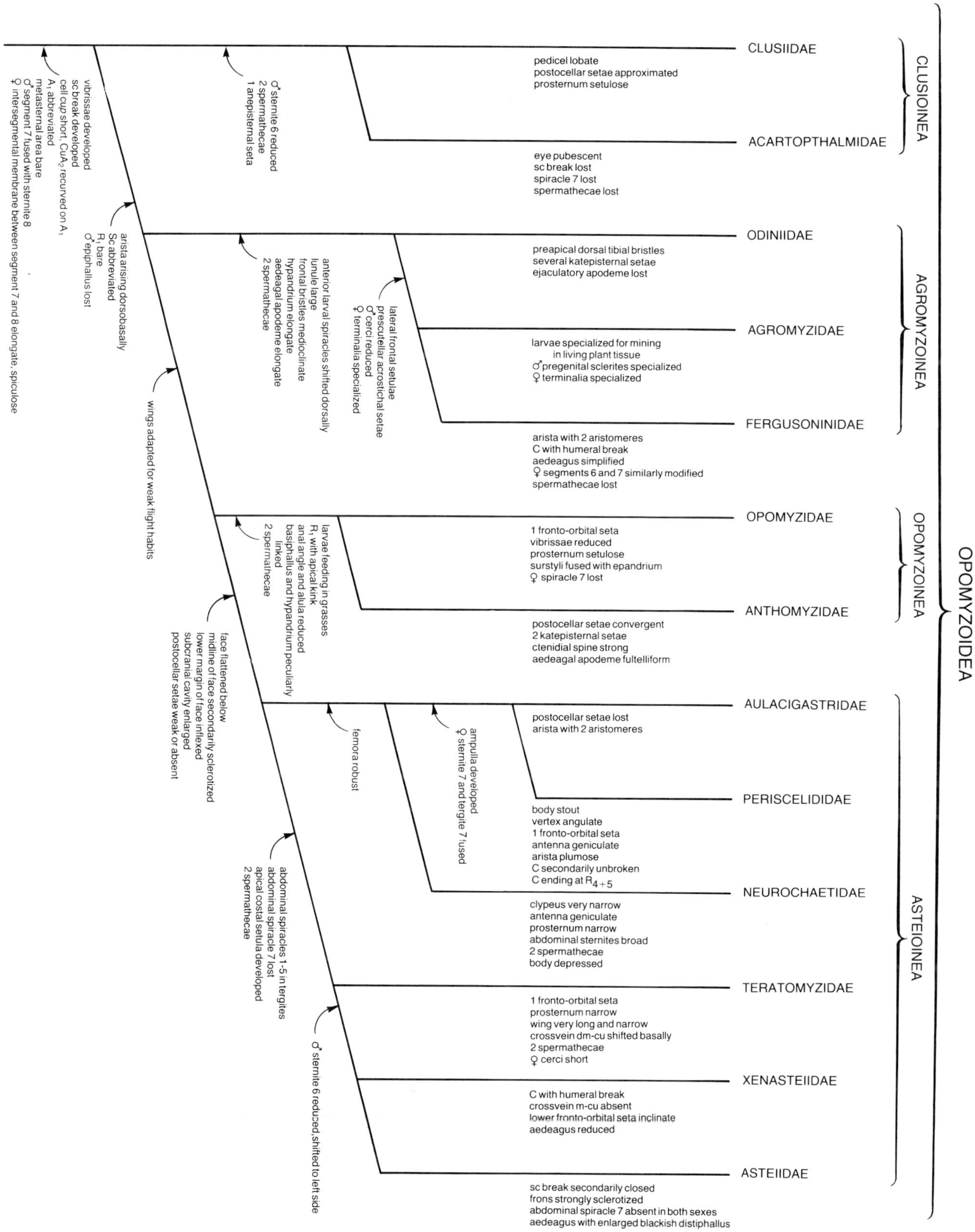

Fig. 116.5. Cladogram of the Opomyzoidea summarizing the relationships and the apomorphies of the subgroups and families recognized. (See text for additional details.)

The most significant plesiomorphic characters in the ground plan of the Opomyzoidea are as follows: larvae saprophytic, probably under bark of trees; scape, pedicel, and flagellum porrect, and main axis of antenna nearly straight; scape setulose; pedicel with a dorsoapical notch; distal membrane of pedicel deeply inserted into base of flagellum; arista arising dorsoapically; postocellar bristles divergent; fronto-orbital plate extending far forward; frontal plate not, or weakly, differentiated from orbital plate; four reclinate fronto-orbital bristles present; lateral margins of frons setulose; two postalar bristles present; postsutural intra-alar bristle absent; katepisternal bristles present; anepisternum setulose; Sc attaining C, free from R_1; R_1 setulose dorsally; cell *cup* complete; alula and anal angle well developed; preapical dorsal tibial bristles absent; precoxal bridge absent; abdomen with 7 spiracles in both sexes; spiracles 1–5 in membrane; male with tergite 6 large and free; sternite 6 large and free; sternite 8 large and haired; aedeagus flexible; epiphallus present; aedeagal apodeme rod-like, mostly free from hypandrium; surstylus free, articulated with epandrium, unilobate; sternite 10 plate-like, undivided medially; female with tergites and sternites 6–8 simple, free from each other; and cerci simple and free.

The task of distinguishing autapomorphic ground-plan characters for the Opomyzoidea is complicated by the great structural diversity shown within and between the 13 families and 4 suprafamilies involved. The difficulty is increased by the rarity of, and consequent lack of knowledge about, many component genera and families. However, I believe that the following conditions, which are apomorphic with regard to the ground plan of the Acalyptratae, belong to the ground plan of the group. Those that appear to be autapomorphic (synapomorphic for the four suprafamilies) are marked (AA):

— first flagellomere short, discoid, more or less correct;
— face membranized along vertical midline (secondarily sclerotized in Agromyzoinea and most Asteioinea);
— vibrissa present (secondarily reduced in some Opomyzoinea and some Asteioinea);
— wing contrastingly patterned (secondarily uniformly clear in many groups) (AA);
— C incised at apex of Sc (secondarily entire, especially in some Agromyzoinea and some Asteioinea);
— cell *cup* short, convexly closed, i.e. CuA_2 recurved on A_1;
— A_1 not attaining wing margin (secondarily lengthened in Teratomyzidae);
— anepisternum with a raised ridge along upper posterior margin (AA);
— katepisternal bristle present;
— metasternal area bare (secondarily setulose in some Periscelididae);
— subscutellum relatively strongly developed;
— male with tergite and sternite of segment 7 reduced, fused with sternite 8 (AA);
— hypandrial bridge weak or absent (secondarily developed in Opomyzoinea);
— female with intersegmental area between segments 7 and 8 elongate and densely spiculose; and
— spermathecae, two (except in some Asteioinea).

These conditions support the contention that the Opomyzoidea is, in fact, a monophyletic taxon. Its sister group probably is the Carnoidea (see discussion under Superfamily Carnoidea).

Suprafamily Clusioinea

This group (Fig. 116.5) contains only the Clusiidae and Acartophthalmidae. Plesiomorphically it differs in its ground plan from the remainder of the Opomyzoidea in having the arista more or less apically placed, C weakly broken at apex of Sc, and Sc widely separated from R_1. Apomorphic characters in its ground plan are:

— pedicel with an angular lobe on outer distal margin (reduced in some Clusiidae and in all Acartophthalmidae);
— anepisternum with one bristle near middle of posterior margin;
— male with sternite 6 reduced; and
— female with two spermathecae (none in *Acartophthalmus* Czerny).

Clusiidae. This family is probably the most generalized (Soós, Ch. 70) of the Opomyzoidea, especially with respect to the form and bristling of the head, the structure of the antenna, the shape and venation of the wing, and in some features of the male and female terminalia. Autapomorphic characters in its ground plan include the following:

— pedicel with a short, angular lobe on outer, distal margin (absent in some members);
— postocellar bristles arising close together (more broadly separated in some representatives, absent in others); and
— prosternum setulose.

The first of these characters is exclusive to the ground plan of the Clusiidae but was lost several times within the family. The relative closeness of the postocellar bristles to each other seems to be accompanied by their weakening and ultimate loss and replacement by fine, convergent hairs. Probably vein C was weakened near the apex of Sc in the ground plan, but this weakening is not always clearly distinct. Hennig (1958) erroneously asserted that C was uninterrupted in some Clusiidae and expressed the belief that this character was a ground-plan condition. However, I agree with D. K. McAlpine (1960) that it is more or less incised near the apex of Sc in all Clusiidae, and, consequently, I would expect this to reflect the ground-plan condition. Perhaps the indistinct subcostal break in some forms, including *Acartophthalmus*, is a secondary condition as appears to be the case in the Periscelididae and Asteiidae. The male terminalia of the Clusiidae are extremely diverse (D. K. McAlpine 1960). Almost always, however, there are two separate pregenital sclerites, tergite 6 and syntergosternite 7 + 8. It seems probable, also, that the aedeagus arose from a basal cone somewhat similar to the type found in the Micropezoidea. The aedeagus is usually long with simple to complex, twisted, convoluted, or coiled distiphallus; the basiphallus usually bears an epiphallus. The hypandrium is frequently desclerotized centrally, and the hypandrial bridge is usually weak or absent. The aedeagal apodeme is of the elongate, cuneiform type. The surstylus is one- or two-lobed. The cerci are usually small and simple.

Two subfamilies, the Clusiodinae, in which all four fronto-orbital bristles are reclinate, and the Clusiinae, in which the anterior (lowermost) fronto-orbital bristle is medioclinate, were proposed by Frey (1960). Although both categories have been accepted by most workers, the monophyly of both groups, especially the latter, is uncertain (Hennig 1965*b*). A thorough evolutionary study of the entire family is needed.

Two fossil species, described from Baltic amber, are known; *Electroclusiodes meunieri* (Hendel) and *E. radiospinosa* Hennig (Hennig 1965*b*, 1969*b*). Another species, also from Baltic amber, *Acartophthalmites tertiariae* Hennig (Hennig 1965*b*, 1969*b*) which was originally assigned to the Acartophthalmidae, probably also belongs here (J. F. McAlpine, Ch. 71). A fourth amber species, *Xenanthomyza larssoni* Hennig (Hennig 1967*b*), originally assigned to the Anthomyzidae, almost certainly belongs to the Clusiidae (see discussion under Anthomyzidae).

Acartophthalmidae. The status of this family (J. F. McAlpine, Ch. 71) is dubious. Until Hennig (1958) raised it to family level, *Acartophthalmus*, containing two Holarctic and one Palearctic species (Máca 1983), was included in Clusiidae (usually as a subfamily). He based this action on his opinion that "it cannot be proven that *Acartophthalmus* is more closely related to the Clusiidae than to other families of Acalyptratae." However, he (Hennig 1965*b*, 1971*a*, 1973) never excluded the possibility that the Clusiidae and Acartophthalmidae are sister groups, and he always treated them as adjacent families. One shared condition that I have noted is the absence of a postsutural intra-alar bristle. This bristle is present in the Agromyzoinea and its absence in Clusiidae and Acartophthalmidae may be especially significant, even though it is probably a plesiomorphic ground-plan character of Acalyptratae and Opomyzoidea.

Similarities of the fossil species, *A. tertiariae* (Hennig 1965*b*, 1969*b*) to both *Acartophthalmus* and to certain Neotropical genera of Clusiidae, e.g. *Trichoclusia* Soós, *Chaetoclusia* Coquillett, and *Chaetoclusiella* Soós, indicate that *Acartophthalmus* is more closely related to the Clusiidae than to any other group. *Acartophthalmites* Hennig corresponds with *Acartophthalmus* in having the extension on the outer distal margin of the pedicel weak, the prosternum very weakly haired (four very fine, short hairs on each side in the specimen in Museum Comparative Zoology, Cambridge, Hennig 1969*b* notwithstanding), C with subcostal break (weakly present in the specimen in Museum of Comparative Zoology, Cambridge, Hennig 1969*b* notwithstanding), and Sc widely separated from R_1. The reduced extension on the pedicel and the absence of or weakly developed subcostal break in both genera could be regarded as secondary developments and, therefore, as synapomorphic conditions. The weakly haired prosternum, the weak subcostal break, and the wide separation between Sc and R_1 are atypical of Clusiidae and support Hennig's association of *Acartophthalmites* with *Acartophthalmus*. However, on the basis of the large size, brown-striped thorax, bare eyes, broad gena, relatively closely placed postocellar bristles, hairy arista, distinct prescutellar acrostichal and proepisternal bristles, presence of a subcostal break, absence of a humeral break, setulose R_1, and setulose prosternum, *Acartophthalmites* corresponds better with the clusiid genera mentioned than with *Acartophthalmus*. Perhaps *Acartophthalmus* will ultimately be shown to be a sister group of a subgroup of the Clusiidae that contains *Acartophthalmites*. Before this problem can be resolved the characters and relationships of all clusiid genera need to be thoroughly analyzed. In the meantime, separate family status is retained for *Acartophthalmus* (J. F. McAlpine, Ch. 71), but it seems reasonably definite that its sister-group relationship lies with the Clusiidae and not with either the Chloropidae or the Carnidae as proposed by Griffiths (1972).

Autapomorphic characters of Acartophthalmidae are:

- compound eye densely pubescent;
- frons with three, instead of four, fronto-orbital bristles;
- postocellar bristles widely separated;
- C with humeral break only and relatively distant from crossvein h;
- spiracle 7 absent in both sexes;
- only one pregenital tergite complex in male, i.e. tergite 6 atrophied;
- ejaculatory apodeme very small; and
- female without sclerotized spermathecae.

The remainder of the Opomyzoidea, i.e. the Agromyzoinea + Opomyzoinea + Asteioinea, comprises the sister group of the Clusioinea. Synapomorphic characters in the ground plan of all three suprafamilies are:

- arista dorsobasally situated;
- Sc with distal portion reduced or absent, merging with R_1;
- R_1 bare; and
- epiphallus reduced or absent.

With respect to the first character, it is difficult to ascertain whether the dorsobasal situation of the arista has resulted from either a shifting of the origin-point of the arista or a lengthening of the lower margin of the first flagellomere. In either case, however, we are dealing with a condition that is apomorphic in relation to the ground plan of the Opomyzoidea.

Suprafamily Agromyzoinea

This suprafamily consists of two sister groups, the Odiniidae and the Agromyzidae + Fergusoninidae (Fig. 116.5). Plesiomorphically, as in the Clusioinea, four fronto-orbital bristles were retained in the ground plan, and tergite 6 in the male, though reduced, remained separate from syntergosternite 7 + 8. The presence of one bristle on the anepisternum and only two spermathecae, two apomorphic characters of Clusioinea, are also shared by the Agromyzoinea; these could be construed as evidence that the Agromyzoinea alone is the sister group of the Clusioinea, but I believe they are convergent developments in both groups. The reduced Sc in the Agromyzoinea argues against a direct connection with the Clusioinea.

Odiniids were originally assigned to the Agromyzidae, and for a long time they were treated as a subfamily of that family. Hendel (1922–1923, 1928) first recognized the Odiniidae as a separate family. A supposed but undocumented sister-group relationship between the two families has usually been accepted (Hennig 1958, 1965*b*, 1971*a*, 1973). Spencer (1969) was more decisive; he reported that the male genitalia of these two families are more similar to each other than to any other family. The following synapomorphies between the Odiniidae and Agromyzidae (autapomorphies of Agromyzoinea) can now be advanced:

- anterior spiracles of larvae shifted towards dorsum;
- lunule large and exposed;
- frontal bristles medioclinate;
- sides of frons with series of tiny setulae lateral to frontal bristles;
- face resclerotized in middle;
- male pregenital tergite complex, especially tergite 6, reduced;
- male with sternite 6 reduced;
- hypandrium long and narrow;
- aedeagal apodeme unusually long;
- segment 8 of female mostly membranized; and
- two sclerotized spermathecae present.

The sclerotized midfacial area in the Agromyzoinea is probably a secondary condition. In the odiniids, *Shewellia agromyzina* Hennig and *Paratraginops pilicornis* Cresson, a more or less membranous area is evident in the anterior midfacial area. Likewise, certain species of *Phytobia* Lioy, which are among the most odiniid-like species of Agromyzidae, the central portion of the face is membranous; in *P. betulivora* Spencer, for instance, the sclerotized portion of the midfacial area is isolated from the parafacials and appears to have developed as secondarily sclerotized areas in what was originally a wholly membranous midfacial area.

Odiniidae. In general this family (J. F. McAlpine, Ch. 72) is less specialized than the Agromyzidae + Fergusoninidae. Structure of the larval mouthparts is simpler and displacement of the anterior spiracles toward the dorsal side of the body is less than in the Agromyzidae (Krivosheina 1979*b*). The same is true for the larval habits. Larvae of almost all odiniids live in galleries of wood-boring insects; those of the Agromyzidae are primary miners in living plant tissue and those of the Fergusoninidae are associated with gall-forming nematodes in *Eucalyptus*. In addition, the structure of the female terminalia in the Odiniidae is much more generalized than is the specialized type found in either the Agromyzidae or the Fergusoninidae. The observation by Máca (1978, p. 150) that females of *Neoalticomerus formosus* (Loew) have three spermathecae is interesting. If three are normally present in this species, it would indicate that three, instead of two, were probably present in the ground plan of the family, and by implication, in the ground plan of the suprafamily Agromyzoinea. However, as in all other Odiniidae examined, only two are present in Nearctic examples of *N. seamansi* Shewell and in specimens identified as *N. formosus* (Loew) in the Canadian National Collection. Perhaps Máca's specimen of *N. formosus* is aberrant in this respect; the small size and different structure of the third spermatheca, as shown by Máca (1978, Figs. 2a,b), seems to indicate that it is not normal.

Autapomorphic characters of Odiniidae include:

- one or more tibiae with preapical dorsal bristle;
- katepisternum with several bristles; and
- ejaculatory apodeme absent.

Two probably monophyletic subfamilies, containing about 50 species divided amongst 10 genera, are recognized (Hennig 1965*b*, 1969*b*, 1973; Cogan 1975; Krivosheina 1979*b*, J. F. McAlpine, Ch. 72). One fossil species, *Protodinia electrica* Hennig (Hennig 1965*b*, 1969*b*), is known from Baltic amber. The family is now sufficiently well known worldwide to warrant a thorough phylogenetic analysis.

Agromyzidae. Almost all the differences between the Agromyzidae (Spencer, Ch. 73) and the Odiniidae can be interpreted as further apomorphic changes that have occurred in the Agromyzidae. One of its few more plesiomorphic conditions is the absence of preapical dorsal tibial bristles.

Autapomorphic characters of the family are:

- larvae feeding in living plant tissue;
- larval mouthparts specially adapted for stem- and leaf-mining, i.e. mandibles toothed and angularly positioned in relation to hypopharyngeal sclerite and hypopharyngeal and tentoropharyngeal sclerites fused;
- anterior larval spiracles lying close together on dorsal surface of thorax;
- cells connected to fat body of larvae containing calcium carbonate crystals (calcospherites) (Frick 1952);
- male with sternite 5 enlarged;
- tergite 6 enlarged;

— sternites 6, 7, and 8 greatly reduced, usually forming a single, small, nearly symmetrical pregenital sclerite;
— aedeagus extremely complex, distiphallus with two gonopores;
— female with segment 7 enlarged and heavily sclerotized, forming a conical nonretractile oviscape, with dorsal apodeme extending far forward into segment 6;
— intersegmental membrane between segments 7 and 8 unusually strongly armed with numerous anteriorly directed denticles on dorsal and ventral surfaces;
— segment 8 modified to form a pair of laterally appressed, serrated blades (egg-guides);
— proctiger unusually elongate, preceded by an elongate membranous tube; and
— cercus with a group (usually four) trichoid sensillae near apex.

Griffiths' (1972) observation to the effect that there is no trace of asymmetry in the pregenital sclerites of male Agromyzidae is an overstatement. Close examination shows not only that the lateral extremities on each side differ from each other but also usually some indication of the characteristic twisting and shifting of the pregenital sclerites and membrane in the left side of the abdomen, e.g. in *Phytobia* spp. Frick's (1952) interpretation that sternites 6, 7, and 8 of the male are greatly reduced and consist of narrow sclerotized strips along the left side (and posterior margin) of tergite 6 is probably essentially true for the ground plan of the family. Griffiths (1964) drew attention to separate remnants of these sternites (which he wrongly called tergites, but see Griffiths 1972) in species of *Phytomyza* Fallén. I believe that tergite 6 is secondarily enlarged because it is commonly as long as, or longer than, tergite 5, (opposite to the general trend in Opomyzoidea and in other Muscomorpha), and also because spiracle 7 is frequently enclosed within it (normally located behind tergite 6, between sternites 6 and 7).

Most workers in the Agromyzidae now recognize two subfamilies, the Agromyzinae and Phytomyzinae, separated mainly on the basis of wing venation. In the Agromyzinae, Sc is traceable throughout its length, although its apical portion is very weak and coalesced with R_1. In the Phytomyzinae, Sc is atrophied apically and does not coalesce with R_1. Except for the more apomorphic condition of Sc, certain members of the Phytomyzinae, e.g. *Selechops* Wahlberg and *Phytobia* whose larvae are cambium miners instead of leaf-miners, are believed to be the most generalized members of the family (Frick 1952, Nowakowski 1962, Spencer 1969).

Agromyzidae is one of the largest families of the Acalyptratae, containing an estimated 3500 species in 27 genera (Spencer, Ch. 73). Although a number of fossil species from Baltic amber were originally assigned to the Agromyzidae, Hennig (1965*b*) concluded that none of these actually belong to the family. One species, *Palaeophytobia platani* Süss & Müller-Stoll, has since been described from feeding channels of the larvae in Upper Miocene wood fragments from *Platanoxylon hungaricum*, a plane tree from Hungary (Süss and Müller-Stoll 1975). It is surprising that agromyzid mines known to exist in fossilized leaves from Tertiary deposits have not been reported in the literature.

Fergusoninidae. This peculiar little family is comprised of about 25 Australian species, all but one of which belong to the genus *Fergusonina* Malloch (Colless and D. K. McAlpine 1974). I know it from examination of adults of two species of *Fergusonina* nr. *microcera* Malloch, and from reports in the literature (Malloch 1924, 1925, 1932*a*; Tonnoir 1937; Frick 1952; Hennig 1958, 1973; Speight 1969; Colless and McAlpine 1970, 1974; Griffiths 1972). *Fergusonina* was originally placed in the Agromyzidae, but Hennig (1958) raised it to family status. Some workers still entertain the possibility that it is a specialized agromyzid (Colless and McAlpine 1970), but usually the question of its relationship to other acalyptrate families is left open (Griffiths 1972, Hennig 1973). Perhaps it is significant that all current specialists on the Agromyzidae have excluded it from that family. No fossils are known.

Some characters of the Fergusoninidae are inaccurately or incompletely reported. For example, Hennig (1958) stated that there are no costal breaks in *F. carteri* Tonnoir, whereas Tonnoir (1937) indicated that a humeral break, but no subcostal break, is present in all species. In the species I studied, the humeral break is strongly developed and there is also a weakening in the usual position of the subcostal break. Griffiths (1972) could find no ejaculatory apodeme in *F. scutellata* Malloch, but a small ejaculatory pump bearing a relatively slender ejaculatory apodeme is present in the species I examined. Griffiths (1972) also reported that cerci are absent in males of *F. scutellata*, but they are clearly present (although reduced and fused medially) in most, if not all, species (see illustrations in Tonnoir 1937). Tonnoir (1937) indicated that both ocellar and postocellar bristles are parallel, but in my material and in published illustrations both sets of bristles are divergent.

The following characters of *Fergusonina* are autapomorphic features of the family:

— larvae living in galls in leafbuds and stems on *Eucalyptus* trees in association with nematodes of the genus *Anguillulina* (= *Fergusobia*) (Colless and McAlpine 1970);
— puparium with a dorsal comb-like plate between abdominal segments 1 and 2 (Tonnoir 1937, Hennig 1958);
— arista with only two aristomeres;
— orbital bristles lateroclinate;
— medioclinate frontal bristles absent;
— parafacials with a series of proclinate setulae (Malloch 1924, Tonnoir 1937, Colless and McAlpine 1970);
— C with humeral break;
— tarsomeres 2, 3, and 4 usually short and broad;
— male with portion of hypandrium (possibly including gonopods) bearing parameres forming an elongate

sheath (aedeagal mantle of Griffiths 1972) around the aedeagus;
— aedeagus in form of simple sclerotized tube, almost entirely withdrawn into aedeagal mantle;
— cerci reduced and fused medially;
— female with abdominal segments 6 and 7 similarly modified and together forming external part of ovipositor, i.e. tergite and sternite of segment 6 fused to form a sclerotized, bulbous tube more or less similar to that formed by tergite and sternite of segment 7, and with remainder of ovipositor telescoped within them;
— spiracle 6 absent;
— tergite 8 and sternite 8 dorsoventrally appressed and forming a long, stylet-like process ending in a dorsal slender point; and
— sclerotized spermathecae absent.

The following characters are shared with the Agromyzoinea and the Agromyzidae; those that are exclusively synapomorphic with similar conditions in the Agromyzidae are marked by asterisks:

— anterior spiracle of larva shifted dorsally on thorax;
— vibrissa present;
— first flagellomere short;
— lunule large and exposed;
— * sides of frons with series of setulae lateral to fronto-orbital bristles;
— postocellar bristles divergent (sometimes nearly parallel);
— * prescutellar acrostichal bristles present;
— anepisternum with one strong bristle (in addition to setulae);
— C with subcostal break or weakening;
— Sc atrophied apically, fused with R_1;
— cell *cup* complete;
— A_1 not attaining wing margin;
— tibiae without preapical dorsal bristle;
— male pregenital complex, especially tergite 6, reduced;
— * cerci of male reduced;
— * segment 7 of female forming sclerotized bulbous oviscape;
— * anterodorsal margin of oviscape with elongate apodeme, projecting far into segment 6; and
— * intersegmental membrane between segments 7 and 8 with numerous stout anteriorly directed spinules.

All these conformations of *Fergusonina* with groundplan characters of the Agromyzoinea, and its clear synapomorphies with the Agromyzidae, indicate that it properly belongs to the Agromyzoinea, probably as a sister group of the Agromyzidae; that is, the Fergusoninidae + Agromyzidae together comprise the sister group of the Odiniidae (Fig. 116.5). I have been unable to decipher the structure of the female proctiger in the Fergusoninidae, and possibly it will provide further clues to its cladistic relationships.

The nature of the aedeagal apodeme, with its broad connection to the hypandrium (through the intermediacy of the strongly developed aedeagal guide) and its transversely flared distal (anterior) end is strangely reminiscent of similar conditions in certain Cypselosomatidae, such as *Heloclusia*, and in some species of *Anthomyza* Fallén. The pair of strong bristles, which frequently arise near the dorsal base of the epandrium in *Fergusonina* (Tonnoir 1937), is another peculiarity that is similarly expressed in the Cypselosomatidae, but the same peculiarity also occurs in many Opomyzoinea. The cladistic significance, if any, of these features is unclear.

Suprafamily Opomyzoinea

This suprafamily consists of only two families, the Opomyzidae and Anthomyzidae. Plesiomorphically its ground plan includes all the conditions listed for the Opomyzoidea (apomorphies with respect to the ground plan of Acalyptratae). I agree with Hennig's (1958, 1971*a*) suggestion that these two families probably are sister groups. The following synapomorphies in the ground plan attest to their monophyly:

— larvae feeding in grass stems;
— wing narrow;
— R_1 with a preapical kink;
— A_1 abruptly abbreviated;
— anal angle and alula reduced;
— ommatidia in anteroventral area of eye enlarged;
— armature of anterior femur peculiar, i.e. with a ctenidium on the anteroventral surface;
— male with sternite 6 reduced;
— epandrium commonly with two strong bristles dorsally;
— hypandrium with hypandrial bridge present (pseudoplesiomorphy?);
— basiphallus linked to hypandrial bridge by complex folding process;
— distiphallus highly complex;
— ejaculatory apodeme reduced; and
— two sclerotized spermathecae present.

Each family is more generalized than the other in some respects, and both have relatively distinct autapomorphic (constitutive) characters, showing that both of them are monophyletic. The occurrence of many of these characters is compared in the Table 116.5; autapomorphic conditions with respect to the ground plan of Opomyzoinea are indicated by asterisks.

Opomyzidae. Griffiths' (1972) interpretation of several parts of the male terminalia of this family (Vockeroth, Ch. 74) is either misleading or incorrect. The aedeagal apodeme is linked to the hypandrium by the aedeagal guide (ventral processes of Griffiths) in all Muscomorpha, so there is nothing distinctive about this condition as it occurs in the Opomyzidae. The cerci are linked to the hypandrium through the intermediacy of the hypoproct and sternite 10 in all Muscomorpha, and again there is nothing distinctive about this character in Opomyzidae. The complex folding structure that links the proximal posterior margin of the basiphallus to the hypandrial bridge (epiphallus of Griffiths 1972) can

Table 116.5 Comparison of character states in ground plans of Opomyzidae and Anthomyzidae

Characters	Opomyzidae	Anthomyzidae
Lateral fronto-orbital setulae	present	* absent
Fronto-orbital bristles	* one	three to four
Oral vibrissa	* undifferentiated	relatively strong
Postocellars (when present)	divergent	* convergent
Prosternum	* setulose	bare
Katepisternal bristle(s)	one	* two (one in *Protanthyomyza* Hennig)
Anepisternal setulae	one	* absent (exception in fossil spp.)
Anepimeral setulae	present	* absent
Postintra-alar bristle	present	* absent
Postalar bristle(s)	two	* one (two in fossil spp.)
Ctenidial spine	absent	* present (absent in some)
Spiracle 7 (♀)	* absent	present
Aedeagal apodeme	rod-like	* fultelliform
Surstylus	* fused	free

* Autapomorphic condition.

scarcely be homologous with the free, spur-like epiphallus that arises from the posteroapical margin of the basiphallus in related families, such as the Clusiidae. Griffiths' statement that parameres (postgonites) are absent is incorrect; in all genera, a narrow, setulose paramere occurs on each side of the basiphallus, closely associated with the complex folding apparatus behind the basiphallus. I do not agree with Hennig (1958, 1971*a*) or Griffiths (1972) that the cerci of the female are fused in the ground plan of the Opomyzidae. Although they are laterally appressed and closely associated with each other, they are quite separate, especially in *Geomyza* Fallén and *Opomyza* Fallén, and, therefore, are probably free in the ground plan of the family.

The limits and contents of the Opomyzidae have varied greatly in the past. Some authors included members of at least three other families in it. For example, Curran (1934) treated the Anthomyzidae, Tethinidae, and Trixoscelididae as members of the Opomyzidae. The family is now usually restricted to four genera, i.e. *Anomalochaeta* Frey, *Opomyza*, *Geomyza*, and *Scelomyza* Séguy, which together contain about 40 species (Vockeroth, Ch. 75; Hennig 1973). I agree with Hennig (1973) that *Anomalochaeta* is probably the sister group of the remainder of the family. It is the most generalized genus in having well-developed, divergent, postocellar bristles, but the cerci of the female are more reduced than in the other genera. No fossil Opomyzidae are known, but it is noteworthy that the two fossil species assigned to the Anthomyzidae share several plesiomorphic conditions with the Opomyzidae (see Table 116.5).

Anthomyzidae. Autapomorphic characters of the family (Vockeroth, Ch. 75) are listed in Table 116.5. Details of the male terminalia are strikingly similar to those of the Opomyzidae, except that the surstyli are free (fused with epandrium in Opomyzidae) and the aedeagal apodeme is shorter and more broadly and closely attached to the hypandrium through the intermediacy of the aedeagal guide. In fact, the aedeagal apodeme in the Anthomyzidae is usually more fultelliform than cuneiform. Similarities to the Opomyzidae are probably most striking in such genera as *Anthomyza* (see Anderssen 1976*a*) and *Amygdalops* Lamb. Both Griffiths (1972) and Anderssen (1976*a*) called the peculiar folding apparatus between the proximal posterior margin of the aedeagus and the hypandrial bridge an epiphallus. This structure is certainly homologous with that in the Opomyzidae, but I doubt whether it is homologous with the separate, spine-like epiphallus that arises from the distal posterior margin of the basiphallus in related groups, such as Clusiidae. Anderssen (1976*a*, Fig. 1) provided an excellent illustration of the male terminalia of *Anthomyza*, including the free, strap-like parameres (postgonites). Griffiths (1972) apparently overlooked these structures in his study of *Anthomyza*; his statement that they are fused with the hypandrium is incorrect.

The family Anthomyzidae, like the Opomyzidae, also lacked definition for many years, and consequently its limits and contents varied considerably according to different workers. Hennig (1971*a*, 1973) transferred to the Aulacigastridae the genera *Stenomicra* Coquillett and *Cyamops* Melander and limited the Anthomyzidae to the following extant genera: *Agnota* Becker, *Amygdalops*, *Anthomyza* (including *Stiphrosoma* Czerny), *Paranthomyza* Czerny (= *Anthomyza?*), *Ischnomyia* Loew, and *Mumetopia* Melander; he also added as possible candidates the genera *Echidnocephalus* Lamb (Lamb 1914) and *Chamaebosca* Speiser (Speiser 1903), the latter on the strength of information given by Wirth (1955). I have not seen any representatives of either of the last two genera. According to D. K. McAlpine (1978), *Apterosepsis basilewskyi* Richards (Richards 1962) belongs to the Anthomyzidae.

It is necessary to mention here several problem genera that Hennig (1958, 1969*b*, 1971*a*) discussed as doubtful relatives of the Anthomyzidae: *Waterhouseia* Malloch (Malloch 1936) was provisionally placed in the Heleomyzidae with an indication that it may represent a new family (Colless and D. K. McAlpine 1970); it is now considered to represent a tribe within the Heleomyzidae (D. K. McAlpine 1985*b*). *Melanthomyza* Malloch (Malloch 1933*b*) is also discussed under the Heleomyzidae. *Nothoasteia* Malloch (Malloch 1936) seems to be correctly placed in the Asteiidae (Sabrosky 1956, Colless and D. K. McAlpine 1970). *Paraleucopis* Malloch (Malloch 1913), *Gayomyia* Malloch (Malloch 1933*b*), and *Schizostomyia* Malloch (Malloch 1933*b*) are treated in detail under Asteiidae.

Representatives of the Anthomyzidae are found in all the main continental areas. Probably less than 50 species are described, but apparently many species still remain undescribed. *Anthomyza* is by far the largest and best known genus (Andersson 1976*a*, 1984). Three fossil genera containing four species, *Anthoclusia gephyrae* Hennig (Hennig 1965*b*, 1967*b*), *Anthoclusia remotinervis* Hennig (Hennig 1969*b*), *Protanthomyza collarti* Hennig (Hennig 1965*b*), and *Xanthomyza larssoni* Hennig (Hennig 1967*b*), were described from Baltic amber and were assigned provisionally to the Anthomyzidae. Hennig (1971*a*) subsequently excluded *Anthoclusia* Hennig from the Anthomyzidae, and D. K. McAlpine (1978) included it in his new family, the Neurochaetidae. *X. larssoni* almost certainly represents a genus of the Clusiidae near *Czerniola* Bezzi, as evidenced by the following characters: characteristic shape of first flagellomere with arista arising apically, frons with four pairs of fronto-orbital bristles, postocellar bristles strongly divergent, back of head strongly concave, katepisternum with one bristle, sc apparently complete, and crossvein dm-cu relatively remote from wing margin. *P. collarti* may actually belong to the Anthomyzidae, but it differs from all recent members of the family in having two postalar bristles, anepisternum with both setulae and bristles, and in not having a distinct subcostal break (although there is a weakening of this point). Perhaps its relationships will become more apparent when more is known about the family as a whole. A world revision of the genera is greatly needed.

Suprafamily Asteioinea

I agree with Hennig (1971*a*) and D. K. McAlpine (1978, 1983) that the Periscelididae, Aulacigastridae, Asteiidae, and Teratomyzidae form a monophyletic unit. D. K. McAlpine (1978) added his new family, the Neurochaetidae, to this group and also questionably included the Anthomyzidae (which I place in the Opomyzoinea). I would now add the Xenasteiidae (Hardy, Mar. 1980) [= Tunisimyiidae (Papp, Aug. 1980)] (new synonymy). Hennig (1971*a*) ranked the group as "1. Teilgruppe, Periscelidea" (as opposed to "2. Teilgruppe, Anthomyzidea") in his superfamily Anthomyzoidea. D. K. McAlpine (1978, 1983) ranked it as a full superfamily, the Asteioidea. I reject Griffiths' (1972) dissociation of the Periscelididae and Teratomyzidae from this group of families; Griffiths' action was based mainly on highly variable characters of the male terminalia, which in this case, misled him into associating them with the Nothybidae and Psilidae.

The name Asteiidae was first used by Loew (1861) and predates all other family group names involved (D. K. McAlpine 1978). Thus, in accordance with articles 35 and 36 of the *International Code of Zoological Nomenclature* (Ride, Sabrosky, *et al.* 1985), Asteioinea is the proper suprafamily name to apply.

The best definition and discussion of the characteristics of the Asteioinea is that given by D. K. McAlpine (1978). Unfortunately, he seems to have confused the inner (posterior) postalar bristle, which is usually absent in the Anthomyzidae, with the postsutural intra-alar bristle, which also is usually absent in members of the Asteioinea (but all three of these bristles may be present in the ground plan of the Opomyzoidea (see, for example, Agromyzoinea). For that reason he wrongly implied that Hennig (1971*a*) incorrectly interpreted the postsutural intra-alar bristle as a postalar bristle. Hennig's interpretation is probably correct (see labeling of these bristles in Hennig 1965*b*, Figs. 142, 161, 176). The more or less sclerotized condition of the face in the Asteioinea is probably a pseudoplesiomorphic condition, resembling to some degree a somewhat similar condition in Agromyzoinea, Hennig's (1971*a*) contrary opinion notwithstanding. I agree with D. K. McAlpine (1978) that the absence of a subcostal break in the Periscelididae and Asteiidae is probably the result of a secondary loss in both families (convergent pseudoplesiomorphies), for a costal weakening in this position is evident in some more generalized members of both families.

Differences in shapes of the pedicel and first flagellomere and the nature of their articulation (Hennig 1971*a*) present one of the most difficult problems to reconcile with the cladistics of the Asteioinea (and the Schizophora for that matter). The Periscelididae and Neurochaetidae have the first flagellomere with its long axis at right angles to that of the pedicel, rather narrowly oval in shape, sharply deflexed (antennae geniculate, *sensu* Sabrosky 1956; hamate, *sensu* Hennig 1971*a*), and with a subbasal dorsal lobe fitting into a notch in the pedicel. The pedicel itself is hood-like (cucullate) and has a dorsal seam as in certain Diopsoidea, Tephritoidea, Anthomyzidae, Aulacigastridae, Ephydroidea, and Calyptratae. Most members of the remaining families of the Asteioinea have the first flagellomere rather broadly discoid and porrect to somewhat drooping, and the pedicel is not strongly hood-like; its distal articular surface faces more or less anteriorly, and the dorsal margin is broadly sinuate and lacks a distinct dorsal seam. Exceptions to this are *Cyamops*, *Planinasus* Cresson, and *Stenomicra*, all of which were placed in the Aulacigastridae (Hennig 1969*a*, 1971*a*). Because of their geniculate (hamate) antennae, D. K. McAlpine (1978, 1983) considered this placement untenable and referred all three genera to the Periscelididae. I agree with him that the loss of the basal segmentation of the arista occurring in various genera that he assigned to both families is probably the result of convergence, but I am not convinced that the geniculate versus porrect antennal structures have not also evolved independently in different genera and families. For example, the difference between these antennal types in certain species of *Cyamops* and *Aulacigaster* Macquart (= *Schizochroa* Hennig) is really not very profound. In fact, within each of the three families, Anthomyzidae, Aulacigastridae, and Asteiidae, one can find almost all degrees of both geniculate and nongeniculate antennal forms.

The ground plan of the Asteioinea differs from that of the Opomyzoidea as follows; characters considered to be autapomorphic are marked (AA):

- size small to minute;
- face secondarily more or less sclerotized in middle (AA);
- face almost flat to convex below (AA);
- anterior (lower) margin of face inflexed so as to be visible from below (AA);
- anteroventral margins of eyes encroaching on face (AA);
- eye with anteroventral ommatidia enlarged;
- subcranial cavity more or less enlarged (AA);
- arista dorsobasally situated;
- inner postalar bristle weak or absent;
- Sc variably weakened and atrophied distally; and
- legs relatively short.

Other significant characters in the ground plan of Asteioinea are as follows:

- fronto-orbital plates with setulae in addition to fronto-orbital bristles (absent in some forms);
- postocellar bristles relatively weak, divergent (commonly parallel, convergent, or absent);
- scape relatively long, setulose (some reduced and bare);
- pedicel not hood-like, with distal articulation deeply inserted into base of first flagellomere (commonly strongly hood-like with a pronounced dorsal slit);
- first flagellomere broadly oval, relatively porrect (commonly narrowly oval, strongly deflexed, i.e. antenna geniculate);
- arista shortly pubescent (commonly with long side branches);
- arista with three aristomeres (some reduced to one or two aristomeres);
- tibiae without preapical dorsal bristles;
- C with subcostal break (Aulacigastridae, Teratomyzidae, Neurochaetidae, and Xenasteiidae, but some secondarily without subcostal break, e.g. Periscelididae and Asteiidae);
- C without humeral break (Teratomyzidae, some Aulacigastridae, some Neurochaetidae, some Periscelididae, and Asteiidae, but some with humeral break, e.g. some Aulacigastridae, some Periscelididae, some Neurochaetidae, and all Xenasteiidae);
- C without an enlarged apical setula (Periscelididae, Aulacigastridae, and Neurochaetidae, but some C with an enlarged apical setula, e.g. Asteiidae, Xenasteiidae, and Teratomyzidae);
- cell cu*p* present (commonly incomplete or absent);
- vein A_1 relatively short, not attaining wing margin (secondarily elongated in some Teratomyzidae);
- male with sternite 6 large and relatively symmetric (some reduced and asymmetric to atrophied);
- aedeagus long, flexible with complex distiphallus (some reduced and secondarily simplified);
- female with three sclerotized spermathecae (usually reduced to two or none); and
- cerci of female long and separate from each other (some short).

As indicated above, many characters show great variation within the Asteioinea, and even within the component families. No doubt some of the seemingly contradictory trends will be resolved as our knowledge of individual families and genera is improved.

Within the Asteioinea (Fig. 116.5) two main subgroups are evident; a more generalized one consisting of the Aulacigastridae, Periscelididae, and Neurochaetidae (subgroup 1), and a more specialized one consisting of the Teratomyzidae, Xenasteiidae, and Asteiidae, (subgroup 2). Some of the main differences are summarized in Table 116.6; clearly apomorphic conditions are marked by asterisks.

It is noteworthy that some convergence of characters occurs in both groups. In the ground plan of subgroup 1 there are three spermathecae, but this number is reduced to two in some Aulacigastridae and in all Neurochaetidae. In the Periscelididae, Teratomyzidae, and some Asteiidae, sternite 6 is relatively large, symmetrical, and unmodified, i.e. seemingly more plesiomorphic than in any other members of the Opomyzoidea. The conclusion reached is that this sternite was reduced independently many times, sometimes even within families, e.g. the Asteiidae.

Subgroup 1

Aulacigastridae. The core genus here is *Aulacigaster* but the question of which other genera should be included, and thus the definition of the family (Tesky, Ch. 76), is unsettled. Hennig (1969*a*, 1971*a*) reexamined this problem and concluded that *Aulacigaster*, *Schizochroa* (= *Aulacigaster*), *Cyamops*, *Stenomicra*, *Pla-*

Table 116.6 Comparison of character states in ground plans of subgroups 1 and 2 of Asteioinea

Characters	Subgroup 1	Subgroup 2
Abdominal spiracles 1–5	in membrane	* in tergites
Abdominal spiracles 7	present	* absent
Apical costal setula	absent	* present
Spermathecae	three (two)	* two
Legs, especially femora	* robust	weak
Antennae	* geniculate (nongeniculate)	nongeniculate (geniculate)

* Apomorphic condition.

ninasus, and the fossil genus and species, *Protoaulacigaster electrica* Hennig (Hennig 1965*b*), probably form a monophyletic group. With the inclusion of *Stenomicra* and *Protoaulacigaster*, however, his originally lengthy list of supposed synapomorphies for the included genera was reduced to one character, i.e. ocellar setae weak or absent. All the other conditions belong to the ground plan of the Asteioinea. D. K. McAlpine (1978) removed *Cyamops* and *Stenomicra* from the Aulacigastridae to the Periscelididae mainly because their geniculate antennal structure resembles that of *Periscelis* Loew more than that of *Aulacigaster*. Later the same author (D. K. McAlpine 1983) reaffirmed this action and added *Planinasus*. He also removed, but left unplaced, the fossil, *P. electrica*. In addition, he synonymized *Schizochroa* under *Aulacigaster*, erected a new subfamily, the Neminινae, for two new genera, *Nemo* D. K. McAlpine from Australia and *Ningulus* D. K. McAlpine from South Africa, and restricted the Aulacigastrinae to the single genus, *Aulacigaster*.

The most important point of contention in D. K. McAlpine's proposals is the dissociation of *Cyamops*, *Stenomicra*, and *Planinasus*, from *Aulacigaster*. These three genera are certainly closely related to each other and to *Aulacigaster*, on the one hand, and to *Periscelis* on the other. In the reduction of the arista, the presence of two fronto-orbital bristles, absence or weak condition of the ocellar bristles, absence of postocellar bristles, and peculiarities of wing venation, they agree better with *Aulacigaster* than with *Periscelis*. But, in the structure of the pedicel and first flagellomere, and to some extent in the bristling of the face, they agree better with *Periscelis* than with *Aulacigaster*. *Cyamops* may provide the key to placement of all three genera. Some of its features indicate a closer relationship with *Aulacigaster* than with *Periscelis*. For example, although the pedicel is geniculate as in *Periscelis*, the first flagellomere is relatively porrect somewhat as in some Neotropical species of *Aulacigaster*; the midfacial hairs and bristles are absent as in *Aulacigaster*; and the palpi are much reduced as in *Aulacigaster* and *Stenomicra*.

As defined by Hennig (1971*a*), the Aulacigastridae are more generalized than the Periscelididae in the following points (corresponding apomorphic conditions in the Periscelididae shown in parentheses): C with subcostal break (secondarily unbroken); C extending to M_{1+2} (ending at R_{4+5}); cell cu*p* complete (incomplete); ocellar bristles, when present, on ocellar plate (at sides of ocellar plate); frons with two or more fronto-orbital bristles (with one fronto-orbital bristle); and greater ampulla relatively weak (relatively strong). Even with the addition of the Nemininae, these factors hold up.

As noted by Hennig (1971*a*), none of the apomorphic conditions of the Periscelididae precludes a close relationship with the Aulacigastridae, but so far no synapomorphic character has been demonstrated for them. The presence of a greater ampulla is possibly such a character. Throughout the Schizophora the occurrence of a well-developed greater ampulla is almost always concurrent with a deeply seamed, geniculate pedicel, and so it occurs in all Periscelididae and in most Aulacigastridae. Perhaps it is an apomorphic condition in the ground plan of the Aulacigastridae + Periscelididae. Another apomorphic point of agreement between the Aulacigastridae and the Periscelididae is the fusion of the abdominal tergite 7 and sternite 7 of the female enclosing spiracle 7. The same conditions also occur in many other families, probably through homoplasy, but they may well be synapomorphic ground-plan features in the Aulacigastridae and Periscelididae. As implied by Hennig (1971*a*), it may be shown eventually that the Aulacigastridae and Periscelididae are subgroups of a single family. If so, the name Periscelididae (Frey 1921, not Hendel 1922) will take precedence over Aulacigastridae (Hendel 1928). If the arrangement proposed by D. K. McAlpine (1983) is upheld, extensive changes will be needed in the definitions of both the Aulacigastridae and Periscelididae. The definition proposed by Hennig (1971*a*) is followed here, as in Teskey (Ch. 76) but certainly D. K. McAlpine's work demonstrates that more work is required to establish a satisfactory arrangement of the genera now included in both the Aulacigastridae and Periscelididae.

Hennig's (1965*b*) remark that the Aulacigastridae have all the distinguishing features of a relict group is noteworthy. Although most of the genera included are very small in terms of species, some have unusually wide distributions, e.g. *Cyamops* and *Stenomicra*. *Stenomicra*, which occurs on all the main continental areas, is by far the largest genus, the family Stenomicridae was recently erected for it (Roháček 1983).

Periscelididae. Following the traditional definition of the family (Frey 1921; Hendel 1922, 1928; Malloch 1926, 1932*b*; Sturtevant 1954; Hennig 1958, 1971*a*, 1973), it contains only *Periscelis*, *Marbenia* Malloch, *Scutops* Coquillett, *Neoscutops* Malloch, and *Diopsosoma* Malloch (J. F. McAlpine, Ch. 77). These five genera certainly comprise a monophyletic group (Hennig 1971*a*, Griffiths 1972). The following are derived conditions with respect to the ground plan of the Asteioinea:

- body unusually short and stout;
- vertex acutely angulate;
- frons with one fronto-orbital bristle;
- ocellar bristles arising beside the ocellar triangle;
- pedicel strongly capiform, with pronounced dorsal seam;
- first flagellomere sharply deflexed, i.e. antenna strongly geniculate;
- arista long, plumose;
- wing unusually short and broad;
- C secondarily unbroken;
- C ending at R_{4+5};
- Sc unusually short;
- delimiting veins of cell cu*p* weak or absent;
- vibrissae displaced high above anterior margin of face;
- all three spermathecae joined to a single excretary duct;
- abdominal tergum 7 and sternum 7 of female fused, enclosing seventh spiracles;
- greater ampulla well developed; and
- eggs blackish.

As indicated in the discussion on the preceding family, the Periscelididae is probably more closely related to the Aulacigastridae, or to a subgroup thereof, than to any other group. If it can be shown that it is the sister group of a subgroup of the Aulacigastridae, e.g. *Cyamops* + *Stenomicra* + *Planinasus* (which D. K. McAlpine (1978, 1983) considers as part of the Periscelididae), then that which is treated here as the Periscelididae is a paraphyletic group and it would be in order to combine both families under one name, i.e. Periscelididae. A thorough evolutionary study of all genera of both families is needed to resolve this matter.

As presently restricted (J. F. McAlpine, Ch. 77) the family contains fewer than 20 species. Only *Periscelis*, which also contains the only known fossil species, *P. annectans* Sturtevant (Sturtevant 1963), from Miocene amber from Mexico, has a wide distribution; all other genera are exclusively Neotropical.

Neurochaetidae. This peculiar little family was erected by D. K. McAlpine (1978) for his new genus of upside-down flies, *Neurochaeta* D. K. McAlpine, in which he included one species from each of Australia, Rhodesia, and Madagascar, and the fossil genus *Anthoclusia* Hennig (Hennig 1965*b*) containing two species, *A. gephyrae* Hennig (Hennig 1965*b*) and *A. remotinervis* Hennig (Hennig 1969*b*), from Baltic amber. Woodley (1982) described two additional species of *Neurochaeta* from the Phillippines and Malaysia.

The ground-plan characters of the family differ from those of the Asteioinea as follows; autapomorphic characters are indicated by (AA):

- clypeus unusually narrow (AA);
- anterior portion of gena with several strong bristles;
- pedicel strongly capiform with pronounced dorsal seam;
- first flagellomere sharply deflexed with a dorsal subbasal tubercle fitting into a cavity in pedicel;
- arista long, plumose;
- katepisternal bristles arising relatively far posteriorly on katepisternum;
- prosternum very narrow (AA);
- abdominal sternites unusually broad (AA);
- aedeagus secondarily simple, uniformly ribbon-like; and
- two sclerotized spermathecae present.

Both D. K. McAlpine and Woodley tabulated the occurrence of other significant characters within the family and indicated general directional changes in them. D. K. McAlpine considered that the family is more closely related to the Periscelididae (in which he included *Cyamops*, *Stenomicra*, and *Planinasus*) than to any other family. It is more generalized than both the Aulacigastridae and Periscelididae in retaining four fronto-orbital bristles, in having sternites 6 and 7 of the male more extensively separated, and in having the sternite 7 and tergite 7 of the female free. On the other hand, it is more specialized than either family in having only two spermathecae. The general form and arrangement of the male terminalia are remarkably similar to those of the Periscelididae. Unfortunately, the presence or absence of a greater ampulla is unknown for most members of the family; there is little or no evidence of it in *N. inversa* D. K. McAlpine, which is not surprising because of the strongly depressed thorax. Probably, Neurochaetidae is the sister group of the Aulacigastridae + Periscelididae (Fig. 116.5).

Subgroup 2

Teratomyzidae. The name Teratomyzidae was proposed by Colless and D. K. McAlpine (1970) for *Teratomyza* Malloch (Malloch 1933*a*) from New Zealand and *Teratoptera* Malloch (Malloch 1933*b*) from Chile and

Australia; they placed it next to the Periscelididae in their superfamily Asteioidea. Previously members of this family were referred either to the Anthomyzidae (Malloch 1933*a*, 1933*b*) or to the Opomyzidae (Séguy 1938; Frey 1958*a*,*b*; Harrison 1959). Hennig (1971*a*) provided the most comprehensive discussion on the characters, relationships, and distribution of the family. He placed it in the Asteioinea (as Periscelidea) and noted that it is similar to the Periscelididae in having only one fronto-orbital bristle and C weak or absent beyond R_{4+5} (both are apomorphic conditions), but he questioned whether these conditions were really indicative of a sister-group relationship of the two families. Hennig also observed that about 12 species (mostly undescribed) are now known from New Zealand, Australia (northward to Queensland), South America (northward to southern Brazil), and Nepal and added to it a third genus, *Neogeomyza* Séguy (Séguy 1938), from Africa and the Philippines (Frey 1958*a*,*b*). Griffiths (1972) treated the Teratomyzidae and Periscelididae as adjacent families (albeit in his superfamily Nothyboidea) and drew attention to the facts that in males of both families spiracles 5 and 6 are in the margins of the respective tergites, that sternite 6 is relatively large and symmetrical, and that the pregenital segment extends symmetrically around the genital segment.

The family Teratomyzidae is seemingly more generalized than the Xenasteiidae and Asteiidae in having a relatively large, symmetric sternite 6 in the male (reduced, asymmetric, and shifted to the left side of abdomen in the Xenasteiidae and Asteiidae). The following apomorphies, with respect to the ground plan of subgroup 2 of the Asteioinea are in evidence; those considered to be autapomorphic are marked with (AA):

- one, only, strong fronto-orbital bristle (as in Periscelididae), but placed relatively far forward (AA);
- prosternum very narrow (as in Neurochaetidae);
- wing long and narrow (AA);
- anal angle and alula reduced;
- costal section between R_1 and R_{2+3} long (as in many Asteiidae);
- C much weakened or absent beyond R_{4+5} (as in Periscelididae);
- crossvein dm-cu strongly shifted toward base of wing (AA);
- spiracles 1–6 situated in margins of tergites;
- sternites 7 + 8 of the male fused forming secondarily symmetric pregenital sclerite (as in Periscelididae);
- tergite 7 and sternite 7 of the female fused (as in Periscelididae);
- two spermathecae present (as in Asteiidae) (AA); and
- cerci of female short (AA).

I agree with Hennig (1971*a*) that a long vein A_1 in some species of the family is a pseudoplesiomorphic, rather than a plesiomorphic, feature. The seemingly unmodified form of sternite 6 in the female indicates that perhaps the Teratomyzidae is the sister group of the Xenasteiidae + Asteiidae (Fig. 116.5). It is one of the few acalyptrate families that shows a typical Gondwanaland distributional pattern. No fossils are known.

Xenasteiidae. This family was erected by Hardy (February 1980) for a new genus, *Xenasteia* Hardy, containing seven new species occurring on islands over a wide range of the Pacific and Indian Oceans. Four months later, Papp (July 1980) described a new family, the Tunisimyiidae, for a new genus and species, *Tunisimyia excellens* Papp, from Tunis. Thanks to the generosity of both authors, I have representatives of both taxa and can find no justification for placing them in separate genera. They agree in habitus, size, color, chaetotaxy, wing venation, and male and female terminalia. At the species level, *T. excellens* seems closest to *X. seychellensis* Hardy, but the elongate surstyli are slender (very broad in *X. seychellensis*) and the halteres are entirely pale (knobs brownish in *X. seychellensis*). Consequently, it is necessary to refer *T. excellens* to *Xenasteia* (new combination) and to sink *Tunisimyia* Papp in favor of *Xenasteia* (new synonymy) and Tunisimyiidae in favor of Xenasteiidae (Papp 1984).

Hardy (1980) referred the Xenasteiidae to the Milichiidae–Carnidae group of families and compared it with the Australimyzidae, Anthomyzidae, and Asteiidae. He showed that its relationship is nearer to the Asteiidae than to any other family, based on the following apomorphies:

- abdominal spiracles 7 lost in both sexes;
- spiracles located in margins of tergites as in certain Asteiidae;
- antennae slightly downwardly flexed as in the Asteiidae; and
- wing lacking crossvein m-cu, veins CuA_2 and A_1, and cell cu*p* as in some Asteiidae.

He also listed nine ways, summarized in Table 116.7, in which the Xenasteiidae differs from the Asteiidae.

Perhaps Hardy attributed greater importance to some of these differences than is warranted. The postocellar bristles are so weak and so slightly convergent that the difference between them seems rather slight. The difference in the width of the parafacials is also one of degree; it varies considerably in the Asteiidae and is scarcely, if any, different in some genera, e.g. *Sigaloessa* Loew, than in *Xenasteia*. The bristle Hardy called the inner postalar is in fact the most posterior intra-alar bristle, which is also present in some Asteiidae, e.g. *Astiosoma* Duda. Nevertheless, there are significant differences between the two families. In the Xenasteiidae, the presence of three fronto-orbital bristles, a subcostal break in the C, and recognizable remains of the tergite 7 of the male are more plesiomorphic conditions than in the Asteiidae. But, the presence of a humeral costal break, the reduced wing venation, the inclinate lower frontal bristle, the reduced parafacials, and the reduced aedeagus are clearly apomorphic features with respect to the ground plan of the Asteiidae. If it cannot be shown that *Xenasteia* is more closely related to a subgroup of the Asteiidae, then it is probably the sister group of all Asteiidae.

Table 116.7 Comparison of character states in ground plans of Xenasteiidae and Asteiidae

Character	Xenasteiidae	Asteiidae
Postocellar bristles	convergent	parallel, divergent (sometimes absent)
Fronto-orbital bristles	3 (lowermost incurved)	1 (upper)
Parafacials	reduced	broad
Posterior intra-alar bristle (= inner postalar bristle of Hardy)	present	absent (sometimes present)
Humeral and subcostal breaks	present	absent
Proepisternal (= propleural) bristle	present	absent
Abdominal sternite 2	divided	undivided
Tergite 7 of the male	present	absent
Aedeagus	reduced	large
Distiphallus	undifferentiated	complex

Asteiidae. As previously noted, the family name Asteiidae (Sabrosky, Ch. 78) was proposed by Loew (1861). Early workers (Frey 1921; Hendel 1922; Malloch 1924, 1927; Duda 1927) regarded the family as a close relative of the Drosophilidae, but Hennig (1958) rejected that idea and later (Hennig 1971*a*) established that it is a member of the Asteioinea (as Periscelidea). Apomorphic characters of the family with respect to the ground plan of subgroup 2 of the Asteioinea are as follows:

- arista with a peculiar zigzag appearance;
- C without a subcostal break (secondarily unbroken);
- cell cu*p* incomplete, i.e. veins CuA_2 and A_1 weak or absent (except in *Phlebosotera* Duda);
- abdominal spiracles 2–6 in margins of tergites (secondarily in membrane in some species);
- abdominal spiracle 7 absent in both sexes; and
- aedeagus with an enlarged, complex, asymmetric, blackish distiphallus.

Hennig's (1971*a*) opinion that the absence of a subcostal break in the Asteiidae is a plesiomorphic character is probably wrong. I agree with D. K. McAlpine (1978) that in small forms with weak wings and a tendency for reduced venation, such as occur in the Asteiidae, there could well be a secondary loss of the subcostal break. The attenuation at this point in *Leiomyza* Macquart seems to indicate the closing up, rather than the formation of a subcostal break.

In the Asteiidae the usual form of the antenna is intermediate between the geniculate type, found in *Periscelis*, *Cyamops*, *Stenomicra*, and *Planinasus*, and the obliquely porrect type, found in *Aulacigaster*. The first flagellomere of most asteiids points obliquely forward and downward; usually it is discoid in lateral profile, but sometimes it is rather pyriform, e.g. some *Asteia* Meigen (Hennig 1971*a*, Fig. 14). The pedicel is more or less notched dorsally, and its distal closing membrane is inserted deeply into the base of the first flagellomere, (but no concealed lobe at the proximal end of the first flagellomere projects deeply into the pedicel as in *Drosophila* Fallén, *Camilla* Haliday, *Curtonotum* Macquart, and so on). Papp (1979, Figs. 11–14) shows that all degrees from the porrect, discoid type to the deflexed, geniculate type, occur in *Asteia*. Hennig (1971*a*) interpreted the intermediate antennal form in the Asteiidae as an apomorphic condition derived from a primitive geniculate type. In my opinion, the reverse is probably true, i.e. the strongly geniculate condition is probably more apomorphic than the porrect discoid type. The geniculate type is usually associated with a plumose arista, which is certainly more apomorphic than a bare or shortly pubescent arista usually associated with a porrect first flagellomere. Also, in the more generalized genera of Asteiidae, e.g. *Leiomyza*, the first flagellomere is more porrect and discoid and the pedicel is less hood-like, than it is in more specialized members, e.g. *Asteia*.

The family consists of about 100 species distributed among at least 11 genera (Sabrosky, Ch. 78), i.e. *Leiomyza*, *Succinasteia* Hennig (fossil from Baltic amber), *Sigaloessa*, *Tucumyia* Sabrosky, *Astiosoma*, *Phlebosotera* Duda, *Anarista* Papp, *Loewimyia* Sabrosky, *Bryania* Aldrich, *Bahamia* Sabrosky, and *Asteia*. The problematic genus, *Nothoasteia* Malloch (1930), containing *N. platycephala* Malloch (from Australia) was originally assigned to this family. This placement was accepted by Sabrosky (1956) and reaffirmed by Colless and D. K. McAlpine (1970), but D. K. McAlpine (1985*c*) suggested that it could perhaps be classified as a genus of the Neurochaetidae. *Asteia* is by far the largest genus.

Sabrosky (1956) divided the family into two subfamilies based on the correlation between two characters of the wings. First, the Asteiinae have crossvein dm-cu absent, alula absent, and wing margin bare in alular region (*Asteia*, *Bryania*, and *Loewimyia*). Second, the

Sigaloessinae have crossvein dm-cu present, alula present (though sometimes narrow), and wing margin haired in alular region (*Sigaloessa*, *Tucumyia*, *Astiosoma*, *Phlebosotera*, and *Leiomyza*). Both *Nothoasteia* and *Anarista* cut across this division.

Hennig (1969*b*), when dealing with his fossil genus, *Succinasteia*, concluded that *Leiomyza* is the most generalized sister group of all other Asteiidae. As stated by Sabrosky (1977), the relationships in this family need to be restudied in the light of more and better material from all parts of the world. On the basis of present information the Asteiidae appear to be the sister group of the Xenasteiidae (Fig. 116.5).

The genera *Paraleucopis* Malloch (Malloch 1913), *Gayomyia* Malloch (Malloch 1933*a*), and *Schizostomyia* Malloch (Malloch 1933*a*) comprise a natural group (Hennig 1971*a*) of the Asteioinea. *Paraleucopis* was originally assigned to the Chamaemyiidae and this placement has always been followed (Steyskal 1971, 1981). For convenience, it was left there in the key to genera of that family (J. F. McAlpine, Ch. 88). *Gayomyia* was also referred to the Chamaemyiidae by Colless and D. K. McAlpine (1970). However, the incomplete Sc, the absence of spiracle 7 in both sexes, the radically different male and female terminalia, and the nonaphidophagous habits of the larvae (some species of *Paraleucopis* live in birds' nests) exclude them from the Chamaemyiidae. In my opinion all three genera are more closely related to primitive Asteiidae than to any other family.

As in all Asteiidae, spiracle 7 is absent in both sexes, the frons is relatively strongly sclerotized, the orbital bristles are reduced, the prosternum is of a size and form characteristic of the Asteiidae, vein C has a subcostal weakening but is without a clear break, the apex of Sc is evanescent and more or less fused with R_1, and the cerci of the female are relatively long and slender. Depending upon the species, two or three sclerotized spermathecae are present in both *Gayomyia* and *Paraleucopis*, and three are present in *Schizostomyia*. Also, depending on the species, spiracles 1 to 5 may or may not be enclosed in the margins of the tergites in *Gayomyia*, but they are always in the membrane in *Paraleucopis* and *Schizostomyia*. An outstanding setula is present at the apex of the wing in *Schizostomyia*, but it is absent in *Gayomyia* and *Paraleucopis*. Some of the costal setulae in the anteroventral row are distinctly enlarged in all species of *Paraleucopis* (giving the impression of a weakly spinose C), and, to a lesser degree, the same character is evident in some species of *Gayomyia*.

Within the Asteiidae, all three of these genera most resemble *Leiomyza*. As in that genus, the arista lacks a zigzag appearance, the vibrissae are extremely weak, the anterior margin of the face is rather deeply notched, the clypeus is enlarged and relatively protuberant, the compound eyes do not impinge on the face, and the anterior ommatidia are scarcely enlarged. In all three genera, cell cu*p* is well developed, a character that is absent in all Asteiidae except *Phlebosotera* Duda (Chile).

Perhaps *Gayomyia sensu lato* (really a complex of genera from southern South America and Australia), *Paraleucopis* (southern Nearctic and Neotropical regions), and *Schizostomyia* (Chile) represent a separate subfamily within the Asteiidae, or alternatively, a separate family annectent to the Asteiidae. Further study is needed to resolve their taxonomy and systematic relationships.

Superfamily Carnoidea

The Carnoidea (= Chloropoidea) consists of nine families: the Australimyzidae, Braulidae, Carnidae, Tethinidae, Canacidae, Milichiidae, Risidae, Cryptochetidae, and Chloropidae (Fig. 116.6). The focal families are the Milichiidae and Chloropidae; their relationships to each other and to other Diptera have always been puzzling and uncertain. Frey (1921) placed them as adjacent families in his "Conopiformes." Hendel (1922, 1936–1937) placed them in annectent superfamilies in his "Drosophilomorphae" and "Gruppe Drosophilides," respectively. Sturtevant (1925–1926) placed them in a special group, "Chloropiformes," stating that "the rudimentary seminal receptacles with long fine ducts and pocket-like ventral receptacle indicate that these two groups are close to each other." Curran (1934) and Harrison (1959) also treated them as adjacent families. Hennig (1958) relegated both to a section containing many families with confused or vague affinities, where he assigned them to separate superfamilies, the Milichioidea and Chloropoidea. At that time he considered some of the remaining families were related to the Chloropidae, whereas others may be more closely related to the Ephydroidea (as Drosophiloidea). As for the Chloropidae, he was unable to find decisive evidence for or against the opinion that it is more closely related to the Drosophilidae or to the Milichiidae. Speight (1969) noted that the shape of the precoxal bridge in the Chloropidae and most Milichiidae is similar to that of the Ephydroidea (as Drosophiloidea). He suggested that the Canacidae, Tethinidae, and Chloropidae be added to the Ephydroidea, that the Milichiidae and Carnidae be added to a superfamily containing the Heleomyzidae, and that the Braulidae be left in isolation. Later, Hennig (1971*a*, 1973) agreed with Sturtevant (1925–1926) and treated the Milichiidae and Chloropidae as sister groups in a single superfamily, Chloropoidea; he rejected the idea that either family belongs to the Ephydroidea, because of their different antennal structure. Colless and D. K. McAlpine (1970) placed both families in Ephydroidea (as Drosophiloidea); later Colless and D. K. McAlpine (1975), mainly on the basis of Hennig's (1971*a*, 1973) action, recommended separation of the Chloropoidea (Chloropidae, possibly Cryptochetidae, Milichiidae, Tethinidae, and Canacidae) and the Brauloidea (Braulidae) from the Ephydroidea (= Drosophiloidea). D. K. McAlpine (1982) reaffirmed this concept of the Chloropoidea, with the possible exclusion of the Cryptochetidae (D. K. McAlpine 1976). Griffiths (1972) placed the Chloropidae

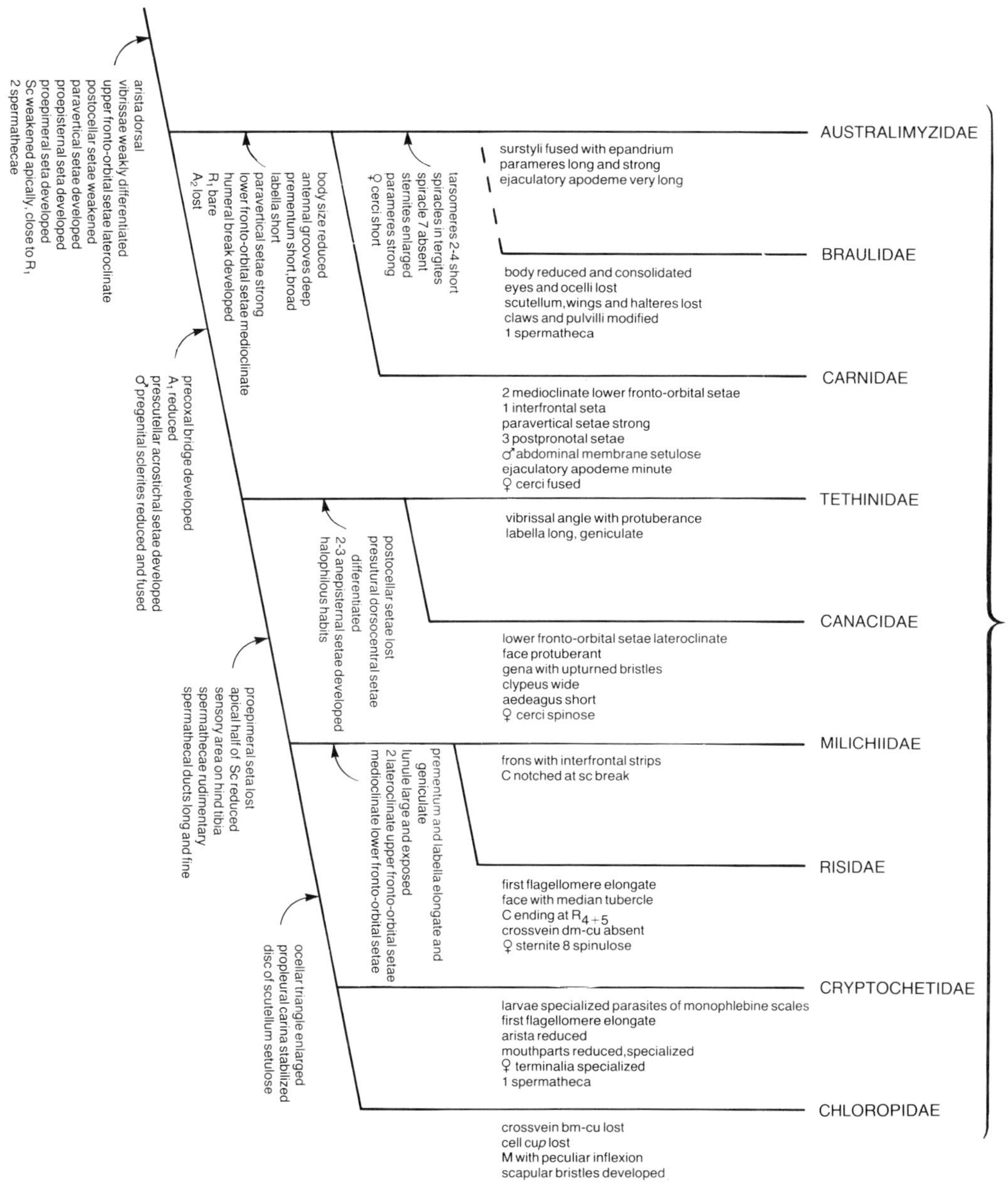

Fig. 116.6. Cladogram of the Carnoidea summarizing the relationships and apomorphies of the subgroups and families recognized. (See text for additional details.)

and Milichiidae, along with the Acartophthalmidae and Carnidae, in his "Chloropidae family-group" in the prefamily Tephritoinea. Andersson (1977) argued convincingly against the inclusion of the Chloropidae in the Tephritoinea, but agreed that the Milichiidae has many features in common with the ground plan of the Chloropidae. Bickel (1982) placed both families in the Chloropoidea along with the Carnidae, Tethinidae, and Canacidae.

From these various interpretations I conclude that, broadly speaking, the Milichiidae and Chloropidae are sister groups in a separate superfamily. This conclusion is borne out by a number of synapomorphies now apparent. Consequently, the provisional arrangement of families proposed in Volume 1 of the Manual (whereby the Milichiidae, Carnidae, and Braulidae were assigned to the Opomyzoidea, and the Chloropidae, Cryptochetidae, Tethinidae, and Canacidae were assigned to the Ephydroidea (= Drosophiloidea) is untenable. The most plausible solution is to recognize the Chloropoidea as a separate superfamily.

According to articles 35 and 36 of the *International Code of Zoological Nomenclature* (Ride, Sabrosky, *et al.* 1985), the proper name for any superfamily is the oldest included family-group name. Therefore, the proper name for this superfamily is Carnoidea based on "Carnites" Newman (1834), rather than Chloropoidea based on "Chloropina" Rondani (1856).

More of the plesiotypic and apotypic ground-plan characters of the Carnoidea were retained in the Australimyzidae and Carnidae than in any other families of the superfamily. The male pregenital sclerites (segments 6, 7, and 8) are less reduced and less fused in the Australimyzidae (see Griffiths 1972) and the Carnidae (see Sabrosky, Ch. 80) than in any other members. Also, in both these families the prosternum remained free from the propleuron (precoxal bridge present in ground plan of all other families). The Australimyzidae are more generalized than the Carnidae in having tergite 6 of the male little reduced (strongly reduced in Carnidae), tergite 7 of the male present and largely free, although reduced (more reduced or atrophied in Carnidae), cerci of the female largely free but reduced (completely fused in Carnidae). On the other hand, they are more specialized than the Carnidae in having three fronto-orbital bristles (four in Carnidae), spiracle 7 absent in both sexes (present in ground plan of Carnidae), surstylus more or less fused with epandrium (free in Carnidae), and aedeagus more or less bare and ribbon-like (large, convoluted, and hairy in Carnidae). *Meoneurites* Hennig (fossil) and *Neomeoneurites* Hennig (Chile) are the most generalized members of the Carnidae (see Hennig 1972*a*). They are more generalized than the Australimyzidae in not having a humeral costal break (although a weakening may be present at this location). Interestingly, both genera agree with the Australimyzidae in having a possibly secondarily bare anepisternum (both setulae and bristles present in all other Carnidae); probably this character must be regarded as a convergence. From this I conclude that the Australimyzidae possibly together with the Braulidae (see later discussion) is the plesiotypic sister group of the remainder of the Carnoidea (Fig. 116.6). Because of the plesiotypic relationship of the Australimyzidae to the remainder of the Carnoidea, they also provide the best clues to the identity of the ancestral sister group of the Carnoidea.

Comparison of the ground-plan characters of the Carnoidea with those of the Opomyzoidea shows a remarkable coincidence of conditions in both superfamilies. The following apomorphic conditions with respect to the ground plan of the Acalyptratae are shared:

— first flagellomere short, discoid, more or less porrect;
— face membranized along vertical midline (secondarily sclerotized in most Canacidae and some Milichiidae;
— vibrissa present;
— C with a subcostal break;
— cell cu*p* short, convexly closed, i.e. CuA_2 recurved on A_1;
— A_1 not attaining wing margin (secondarily lengthened in some Tethinidae);
— katepisternal bristle present;
— metasternal area bare;
— subscutellum relatively well developed; and
— hypandrial bridge weak or absent.

On the other hand, the following differences are noted for the Carnoidea (conditions that are more plesiomorphic than in the ground plan of the Opomyzoidea are indicated by (P), and conditions that are more apomorphic than in the Opomyzoidea are indicated by (A):

— wing not (or seldom) contrastingly patterned (P);
— anepisternum without a raised ridge along posterior margin (P);
— tergite 7 of the male present and relatively free, although reduced (P);
— flexible distiphallus relatively simple (P);
— one or more of uppermost fronto-orbital bristles outwardly curved (A);
— one or more of lowermost fronto-orbital bristles inwardly curved (A);
— postocellar bristles relatively weak (A);
— paravertical bristles present (A);
— vibrissa weak, poorly differentiated from subvibrissal bristles;
— proepisternal bristle present (A);
— proepimeral bristle present (A);
— Sc weak or absent apically, and contiguous or fused with R_1 (A);
— R_1 bare (setulose in *Apataenus*) (A);
— epiphallus absent (A); and
— two spermathecae present (A).

The many apomorphic conditions (A) listed for the Carnoidea attest to its monophyly. The four conditions that are more plesiomorphic than in the Opomyzoidea preclude its inclusion within the Opomyzoidea, but the extensive list of apomorphies shared by the Carnoidea and Opomyzoidea indicate they are sister groups. It is perhaps noteworthy that many similar specializations occur (as convergences) in both superfamilies, e.g. reduction of fronto-orbital bristles, reduction of vibrissae, appearance of humeral break, loss of cell cu*p*, reduction and loss of A_1, loss of crossvein bm-cu, and shortening of cell dm.

Within the Carnoidea two main evolutionary lines are evident; a more generalized line leads to the Australimyzidae, Braulidae, and Carnidae and a more specialized line leads to the Tethinidae, Canacidae, Milichiidae, Risidae, Cryptochetidae, and Chloropidae (Fig. 116.6). Plesiomorphically in the Australimyzidae–Carnidae line the prosternum remained free from the propleuron (weak precoxal bridge present in Braulidae, possibly related to consolidation of all thoracic sclerites), vein A_1 remained sclerotized on basal half, and the pregenital sclerites (tergites 6, 7) of the male remained relatively large and discrete. Apomorphically in this line, the body size was reduced, deep antennal grooves developed, vein Sc became very weak and more or less joined with R_1 at the apex, and vein A_2 was lost. The Tethinidae–Chloropidae line retained in its ground plan several more generalized conditions; namely, larger body size, weaker antennal grooves, a stronger, freer vein Sc, and a visible A_2. On the other hand, this line acquired several apomorphic conditions as follows:

- precoxal bridge present (weak or absent in some Milichiidae);
- vein A_1 desclerotized; and
- pregenital sclerites of male reduced to one composite syntergosternite.

The cladistics of both groups are discussed in detail under the individual families.

Australimyzidae. Separate family (and prefamily) status was proposed by Griffiths (1972) for *Australimyza* Harrison (Harrison 1953). The genus contains five known species from New Zealand, Campbell Island, Antipodes Island, Macquarie Island (Harrison 1959), and Australia (Colless and D. K. McAlpine 1970, 1975). It was originally assigned to the Milichiidae, but Colless and D. K. McAlpine (1970, 1975) referred it to the Carnidae. I agree that it is related to both the Milichiidae and Carnidae (long combined with Milichiidae), but I also concur with Griffiths (1972) that because of its peculiar combination of plesiomorphic and apomorphic characters it cannot satisfactorily be placed in either group. The highly plesiomorphic structure of the male terminalia (tergite 6 large, tergite 7 largely free) and the free cerci of the female indicate that it arose prior to either the Carnidae or Milichiidae, and it is therefore best treated as a separate family. No fossils are known.

Australimyzidae is characterized by the following apomorphies with respect to the ground plan of the Carnoidea:

- lower medioclinate fronto-orbital bristles reduced, i.e. only one strong pair present, and with a very small medioclinate pair below them;
- prementum short and broad;
- labella short;
- anepisternum bare;
- katepisternum with two bristles;
- C with both humeral and subcostal breaks;
- tarsomeres 2, 3, and 4 shortened;
- spiracles 7 absent in both sexes;
- spiracles 2–6 in margins of tergites;
- sternites very broad, especially sternites 4 and 5;
- epandrium reduced (but not divided into two halves as reported by Griffiths 1972);
- surstyli more or less fused with epandrium;
- parameres long and strongly developed;
- distiphallus very long and slender, i.e. ribbon-like;
- ejaculatory apodeme unusually elongate; and
- cerci of the female very short.

Braulidae. A separate family (Peterson, Ch. 81) for *Braula* Nitzsch (Nitzsch 1818), the only known genus, has long been recognized; according to Rohdendorf (1977), Gerstäcker (1863) was the first to apply a family-group name to it. It contains at least one species, the well-known bee louse, *B. coeca* Nitzsch, which seems to occur wherever people utilize the honey bee. It is so extremely reduced and so highly specialized that for most people it is unrecognizable as a fly. Four other species, including two subspecies, were described by Örözi Pal (1966), but they are scarcely distinguishable. It is now generally agreed that the family belongs to the Acalyptratae, but nearly everyone who has studied it has offered a different idea about its relationships to other families (see Peterson, Ch. 81). Its autapomorphic characters include the following:

- compound eyes lost;
- ocelli lost;
- arista unsegmented;
- thorax reduced and consolidated;
- precoxal bridge present;
- scutellum lost;
- wings and halteres lost;
- claws and pulvillae highly modified;
- abdomen with six pairs of pleural sclerites (probably derived from tergites);
- ejaculatory apodeme extremely reduced or absent; and
- female with one spermatheca.

None of these peculiarities is helpful in resolving the relationships of the family to other Schizophora. However the following plesiomorphic (P) and apomorphic (A) conditions seem to suggest a closer relationship within the Carnoidea, especially with the Australimyzidae:

- theca short (A);
- labella short (A);
- tarsomeres 1–4 uniformly short (2–4 short in *Australimyza* (A);
- first abdominal segment reduced, fused with segment 2 (A);
- abdomen with 5 pairs of spiracles, i.e. spiracles 2–6 inclusive (A), (usually six pairs in Australimyzidae, i.e. spiracles 1–6 inclusive);
- spiracle 7 absent in both sexes (A);
- spiracles enclosed in margins of tergites (A);
- sternites 2–5 in both sexes large (A),
- tergite 6 of male large (P);
- remnant of tergite 7 of male present (P);
- hypandrium elongate, becoming narrow anteriorly (A);
- paramere strongly developed, elongate (A);
- distiphallus long and slender (A);
- aedeagal apodeme long, rod-shaped, free from hypandrium (P); and
- cerci of female reduced (A), but largely separate (P).

Most of these characters agree fairly well with *Australimyza* and may indicate that *Braula* is a highly derived descendant from the same stock that gave rise to *Australimyza*. No fossils are known.

Carnidae. Most authors have regarded this little family (Sabrosky, Ch. 80) as a subfamily of the Milichiidae, probably because some representatives of both families are so similar in general appearance. For example, the remarkable similarities in the size, form, color, chaetotaxy, and wing venation between *Hemeromyia* Coquillett (Carnidae) and *Madiza* Fallén (Milichiidae) can be misleading until one examines the mouthparts, the prosternum, and the male and female terminalia. Frey (1921) is sometimes credited as being the first to treat the Carnidae as a separate family, but, as noted by Rohdendorf (1977), Newman (1834) was the first to propose a special family-group name (Carnites) for it. Hennig (1937*b*, 1958) rejected full family rank for it and was of the opinion either that the Carnidae and Milichiidae were subfamilial sister groups or that the Carnidae were more closely related to a subgroup of the Milichiidae. Later he (Hennig 1965*b*, 1972*a*) showed both these alternatives to be wrong and accepted separate family status for it. This concept is now generally followed (Colless and D. K. McAlpine 1970, 1975; Griffiths 1972; Steyskal 1974; Sabrosky, Ch. 80; and so on).

Hennig's (1972*a*) last analysis of the Carnidae contains the most advanced evolutionary information on the family, but he still left open the question of its closer relationship within the Acalyptratae. Griffiths' (1972) opinion that the Acartophthalmidae are the primitive sister group of the Carnidae + Milichiidae + Chloropidae seems highly improbable. I believe that the following synapomorphies between the Carnidae and Australimyzidae indicate a sister group relationship between them:

— body size reduced;
— antennae in rather deep antennal grooves;
— prementum relatively short and swollen;
— labella short, round, and with few pseudotracheae;
— paravertical bristles present in addition to weak postocellar bristles;
— lower fronto-orbital bristles medioclinate;
— humeral break strongly developed;
— R_1 bare; and
— A_2 absent.

The following ground-plan autapomorphies support the monophyly of the family:

— frons with two pairs of lower, incurved fronto-orbital bristles;
— frons also with one pair of relatively strong interfrontal bristles;
— paravertical bristles relatively strong;
— postpronotum with three bristles, each pointing in a different direction;
— pleural membrane of male abdomen strongly setulose;
— distiphallus densely pubescent;
— ejaculatory apodeme minute; and
— cerci of female fused.

Carnidae contains five genera, *Meoneurites* (fossil in Baltic amber), *Neomeoneurites* (Chile), *Hemeromyia* Coquillett (Holarctic), *Meoneura* Rondani (Holarctic), and *Carnus* Nitzsch (Holarctic). Based on Hennig's (1972*a*) analysis, *Meoneurites* and *Neomeoneurites*, together, are the sister group of *Hemeromyia*, *Meoneura*, and *Carnus*. Both *Meoneurites* and *Neomeoneurites* have no clear humeral break in C (symplesiomorphy), and both have a bare anepisternum (synapomorphy), neither of which condition occurs in the remaining genera.

Tethinidae. According to Rohdendorf (1977), family status for this group (Vockeroth, Ch. 101) dates from "Becker 1895," but I have been unable to verify this. Its monophyly was affirmed by Hennig (1958) and reaffirmed by Griffiths (1972), Sabrosky (1978), and D. K. McAlpine (1982). Ground-plan conditions for it and the annectent Canacidae are compared in Table 116.8; autapomorphic conditions are marked by (A), and synapomorphic conditions are marked by (SA).

Five of the characters listed are considered synapomorphies between Tethinidae and Canacidae; these are clear indications of a sister-group relationship between them, as intimated by Hennig (1958), Griffiths (1972), and D. K. McAlpine (1982), and may even indicate that they are subgroups of a single family (D. K. McAlpine 1982, 1985*a*). Most of the differences between them relate to conditions that are more apomorphic in the Canacidae than in the Tethinidae. However, the dorsal pregenital sclerite of the male is less reduced in the Canacidae than in the Tethinidae. A thorough cladistic analysis of all genera of both families is needed.

I do not agree with Hennig (1958), Griffiths (1972), Sabrosky (1978), and D. K. McAlpine (1982) that the relatively widely spaced, convergent bristles usually present in the Tethinidae (and Canacidae) lateral to, and behind, the posterior ocelli are true postocellar (= postvertical) bristles. These bristles are paravertical bristles *sensu* Steyskal (1976) [= inner occipital bristles *sensu* Hendel (1928)]. Also, I do not agree with Hendel (1928) and Malloch (1948) that a pair of fine, closely placed, widely divergent bristles, present in some species of *Tethina* Haliday and in all species of *Dasyrhicnoessa* Hendel at the posterior margin of the ocellar triangle, are true postocellar bristles. Variation in the size and strength of these latter setulae among different species of *Tethina* (and in different genera of Canacidae and Ephydridae) shows that in cases in which they are relatively strong, they are simply specialized ocellar setulae and are not homologous with true postocellar bristles. Confusion arising from misidentification of these specialized ocellar setulae (hereby named "pseudopostocellar bristles") and the absence of postocellar bristles is dispelled when one observes the presence of true postocellar bristles in the Carnidae (in which they are parallel) and in some Milichiidae (in which they are usually convergent, as in the Heleomyzidae and some Sphaeroceridae). Convergent paravertical bristles occur in these latter four families (which also have postocellar bristles), as they do in the Tethinidae (which do not have postocellar bristles). In some Tethinidae the paravertical bristles are extremely weak, e.g. species of *Pelomyia* Williston; in others they are relatively strong and close together, e.g. species of *Apetaenus* Eaton; and in still others they may be subparallel, e.g. *Neopelomyia rostrata* (Hendel).

As stated by D. K. McAlpine (1982): "The family Tethinidae is a diffuse one, and several of the included genera seem to be rather distantly related to *Tethina* and its allies." This relationship is especially true in the case

Table 116.8 Comparison of character states in ground plans of Tethinidae and Canacidae

Character	Tethinidae	Canacidae
Postocellar bristles	absent (SA)	absent (SA)
Precoxal bridge	present	present
Proepisternal bristle	present	present
Proepimeral bristle	present	present or absent
Presutural dorsocentral bristle	present (SA)	present (SA)
Prescutellar acrostichal bristle	present (SA)	present (SA)
Anepisternal bristles	present (3) (SA)	present (2–3) (SA)
Katepisternal bristles	one	one
Vein A_1	short (except *Apetaenus*)	short
Vein A_2	long	long
Habits	halophilous (SA)	halophilous (SA)
Dorsal pregenital sclerite (♂)	single, short (A)	single, long
Paravertical bristle	present	absent or reduced
Outer occipital bristle	present	absent
Lower fronto-orbital bristle(s)	inclinate	lateroclinate (A)
Face	desclerotized to weakly convex	very protuberant (A)
Vibrissal angle	with protuberance (A)	unmodified
Gena	without upcurved bristle(s)	with upcurved bristle(s) (A)
Clypeus	narrow (wide in *Apetaenus*)	very wide (A)
Prementum	long, not deeply incised distally (except *Apetaenus*)	short, broad, deeply incised distally
Labella	long, geniculate (A)	short, nongeniculate
Spiracle 7 (♀)	present or absent	present
Hypandrial bridge	absent	present, with spine (A)
Aedeagus	long	short (A)
Cerci (♀)	simple, blunt with fine hairs	strongly sclerotized, curved with one or two apical spines (A)

A Autapomorphic condition.
SA Synapomorphic condition.

of *Apetaenus*, which agrees with most ground-plan characters of Tethinidae, but which also disagrees in several important, perhaps plesiomorphic, and several apomorphic conditions, e.g. vibrissal angle without discrete protuberances, Sc strong throughout and almost entirely free from R_1, R_1 setulose, (at least in *A. australis* Hutton), A_1 virtually reaching wing margin (plesiomorphic conditions), clypeus much enlarged (as in Canacidae), mouthparts relatively short, paravertical bristles strong, relatively closely placed, and outer occipital bristles absent (apomorphic conditions). The male and female terminalia, however, agree completely with those of the Tethinidae.

The only autapomorphic character found for the family is the peculiar, usually shiny, protuberance on the vibrissal angle just above the vibrissa, but even this feature may not be part of the ground plan because it is scarcely visible, if present at all, in *Apetaenus*. Perhaps it applies to only a subgroup of the Tethinidae. The proboscis in the Tethinidae (except *Apetaenus*), as in the Milichiidae and Chloropidae, is relatively long and geniculate. Possibly this character is the result of homoplasy in the Tethinidae, and if so, it could be considered an additional autapomorphic condition.

The family, as now accepted (Sabrosky 1978, D. K. McAlpine 1982, Vockeroth, Ch. 101), is comprised of 11 genera, *Apetaenus*, *Dasyrhicnoessa*, *Horaismoptera* Hendel, *Macrocanace* Tonnoir & Malloch, *Masoniella* Vockeroth, *Neopelomyia* Hendel, *Pelomyia*, *Pelomyiella* Hendel, *Pseudorhicnoessa* Malloch, *Tethina*, and *Tethinosoma* Malloch; in all, it contains less than 100 described species. Representatives occur on all the continents and on many oceanic islands. Most species are littoral, and all species are holophytic (Sabrosky 1978). No fossils are known.

Canacidae. Hendel (1916) may have been the first to rank this little group as a separate family (Wirth,

Ch. 102). At that time and for some time after, it was associated with the Ephydroidea (as Drosophiloidea), especially the Ephydridae. Beginning with Sturtevant (1925–1926), this association was recognized as incorrect. It is now more or less agreed that it is most closely related to the Tethinidae (Hennig 1958, Griffiths 1972, D. K. McAlpine 1982). Synapomorphic characters of both families are tabulated in Table 116.8. At least seven mostly autapomorphic characters, also listed in Table 116.8, attest to its monophyly. Most of these conditions are apomorphic in relation to those of the Tethinidae, but the larger size and more obviously compound structure of the dorsal pregenital sclerite of the male is more plesiomorphic than in any known tethinid. D. K. McAlpine (1982) retained separate family status for both groups, because "this sclerite of the Canacidae cannot be derived from that of Tethinidae, and must be either more primitive than that of Tethinidae, or must have evolved along different lines." I believe that the consolidated dorsal pregenital sclerite developed from remnants of segments 6, 7, and 8, and I consider it probable that both the Tethinidae and Canacidae evolved from a common ancestor in which this sclerite was more similar to that of present-day Canacidae. I agree with D. K. McAlpine's suggestion that perhaps the Canacidae would be better classified as a subgroup of the Tethinidae. As indicated under that family, a thorough study of all genera of both families is required to properly resolve this question.

In the Canacidae, as discussed under Tethinidae, true postocellar (= postvertical) bristles are absent. D. K. McAlpine (1982) incorrectly applied the name "postvertical bristles" in both families. The convergent bristles referred to in the Tethinidae are paravertical bristles, and the nonhomologous divergent ones sometimes present in the Canacidae, (as in Tethinidae) are specialized ocellar setulae herein called pseudopostocellar bristles.

Canacidae, or beach flies, are found throughout the world, primarily along marine shorelines where they breed in various species of algae. In all, about 90 species are known belonging to 12 genera: *Canace* Haliday, *Canacea* Cresson, *Canaceoides* Cresson, *Chaetocanace* Hendel, *Dynomiella* Soika, *Isocanace* Mathis, *Nocticanace* Malloch, *Paracanace* Mathis & Wirth, *Procanace* Hendel, *Trichocanace* Wirth, *Xanthocanace* Hendel, and *Zalea* D. K. McAlpine 1985*a* (new name for *Zale* D. K. McAlpine 1982, homonym of *Zale* Heubner 1818, Lepidoptera). Only three of these, *Canace*, *Canaceoides*, and *Nocticanace*, occur in the Nearctic (Wirth, Ch. 102).

D. K. McAlpine (1982) divided the family into two subfamilies, the Zalinae (corrected to Zaleinae by D. K. McAlpine 1985*a*) and the Canacinae. The same year, Mathis (1982) recognized two "subfamilies," the Canacinae and Nocticanacinae, within the Canacinae of D. K. McAlpine. As pointed out by D. K. McAlpine (1985*a*), these taxa are of a lower rank than the Zaleinae. Thus, if the subfamily Zaleinae is included in the Canacidae, or if the family Canacidae is expanded to include the closely related Tethinidae, Mathis' taxa should be placed as subordinate subgroups of the subfamily Canacinae. No fossils are known.

Milichiidae. This family is probably the most diverse (Sabrosky, Ch. 79) of the Carnoidea. The form of the adults varies from carnid-like genera such as *Desmometopa* Loew, through agromyzid-like genera such as *Aldrichiomyza* Hendel, and chloropid-like forms such as *Madiza*, to stout-bodied, tachinid-like genera such as *Eusiphona* Coquillett and *Pholeomyia* Bilimek. Many characters that are relatively stable diagnostic features in related families vary within the Milichiidae. For example, postocellar bristles may be convergent, parallel, or divergent; the ocellar plate may be enlarged or reduced; the number and inclination of fronto-orbital bristles varies widely among the different genera; the precoxal bridge is sometimes absent (possibly through secondary reduction), the propleural carina may be present or absent; the anepisternum may be setulose or bare, cell cu*p* may be present or absent; crossvein dm-cu may be present or absent; vein A_2 may be strong or weak; a sensory area may be present or absent on the hind tibia, spiracles 1–5 may be in the membrane or enclosed in the margins of the tergites. Despite its diversity, however, it has long been recognized as a separate family (Hendel 1902, 1903), and it is generally argued that it is a monophyletic unit (Hennig 1958, Griffiths 1972).

The most significant ground-plan characters are listed in Table 116.9, in which they are compared with homologous conditions in the Risidae, Cryptochetidae, and Chloropidae. Most of these characters are apomorphic with respect to the ground plan of the Acalyptratae, and the fact that so many of them are shared by all four families affirms their membership in, and the monophyly of, this subgroup of the Carnoidea. The family Milichiidae is certainly the most generalized member of the subgroup.

Autapomorphic characters in the ground plan of the family include the following:

- — lunule relatively extensively exposed;
- — frons with sclerotized interfrontal strips; and
- — C more or less notched at subcostal break.

Griffiths' (1972) statements, that the male tergite 6 is completely lost, and that parameres (postgonites) are absent in the ground plan of the Milichiidae, are incorrect (see Sabrosky, Ch. 79, Figs. 79.10–11).

Two subfamilies, the Madizinae and Milichiinae, have been recognized almost from the outset (Hendel 1913). According to Sabrosky (Ch. 79), they are composed of 19 genera, many of which have worldwide distributions, containing about 190 described and numerous undescribed species. The literature on the family is scattered, and no up-to-date key to, or list of, the world genera is available. From an evolutionary standpoint Hennig's (1937*b*, 1939*b*, 1958) pioneering works are probably the most informative. A thorough study of the systematics of the

Table 116.9 Comparison of character states in Milichiidae, Risidae, Cryptochetidae, and Chloropidae

Characters	Milichiidae	Risidae	Cryptochetidae	Chloropidae
Postocellar bristles	convgt, divgt, or absent	divergent? absent	convergent? absent?	convergent
Upper front-orb. bristles	2 lateroclinate	2 lateroclinate	several? lateroclinate	2 lateroclinate
Lower front-orb. bristles	2 inclinate	2 inclinate	unrecognizable	1 or 2 inclinate
Vibrissa	reduced	reduced	reduced	reduced
Interfrontal plates	present or absent	absent	absent	absent
Ocellar plate	small to large	enlarged	enlarged	enlarged
Prementum	elongate	elongate	reduced	elongate
Labella	elongate, geniculate	elongate, geniculate	short, nongeniculate	elongate, geniculate
Maxillary lacinia	rudimentary	?	?	rudimentary
Propleural carina	absent or present	?	present	present
Proepisternal bristle	present	?	absent	present
Proepimeral bristle	absent	absent	absent	absent
Prescutellar acrostichal bristle	present	present	present	present
Presutural dorsocentral bristle	present	absent	absent	present
Disc of scutellum	bare	setulose or bare	setulose	setulose
Humeral break	present	present	present	absent (present)
Subcostal notch	present (absent)	absent	absent	absent
Subcostal break	present	present	present	absent
Sc	apical half reduced	apical half reduced	apical half reduced	apical half reduced
A_1	reduced	reduced	reduced	reduced
A_2	long, strong	reduced	long, strong	reduced
Cell *cup*	present (absent)	absent	present	absent
Sensory area on hind tibia	present or absent	?	present	present
Pregenital sclerite (♂)	one	none	one	one
Spiracle 7 (♂)	present, absent	absent	absent	present
Distiphallus	reduced, membranous,	bare, reduced	rigid, bare	reduced, membranous, bare
Spermathecae	2 reduced	absent	1 or 2	2 reduced

family on a world basis is badly needed. Two fossil species, both in the genus *Phyllomyza* Fallén, are known from Baltic and Mexican amber (see Sabrosky, Ch. 79).

Risidae. Separate family status for this group seems redundant. It includes only *Risa* Becker (Becker 1907), which contains only three species, *R. longirostris* Becker (Becker 1907) (Tunis), *R. mongolica* Papp (Papp 1977*a*) (Mongolia), and *R. longicornuta* Papp (Papp 1980) (Faroe Islands), and *Achaetorisa brevicornis* Papp (Papp 1980) (Morocco). *Risa* was usually placed in Milichiidae (Hennig 1965*b*, Steyskal 1968*b*) until Papp (1977*b*) erected a separate family for it because "some of the cephalic features, thoracic chaetotaxy and more especially the structure of the abdomen (are) so strikingly different from Milichiidae." He considered it to be closely related to the Ephydridae.

I have not seen any representatives of either *Risa* or *Achaetorisa* Papp, but judging from descriptions (Hennig 1937*b*, Papp 1977*b*) it seems to me the differences noted do not exclude them from the Milichiidae. They agree with the Milichiidae in many significant characters (Table 116.9), including elongate mouthparts with geniculate labella, rather exposed lunule, two medioclinate lower fronto-orbital bristles and two outwardly directed upper fronto-orbital bristles, two costal breaks, absence of sclerotized spermathecae, and presence of large sclerotized ventral receptacle (as in *Pholeomyia*).

Apomorphies of the Risidae with respect to the ground plan of the Milichiidae, include the following:

— face with a median tubercle;
— frons without discrete sclerotized strips;

- postocellar bristles reduced, divergent?;
- first flagellomere elongate;
- clypeus very narrow;
- anepisternum setulose;
- scutum with reduced bristling;
- C ending at R_{4+5};
- crossvein dm-cu absent;
- cell cu*p* absent;
- abdominal spiracles in margins of tergites;
- spiracle 7 absent;
- dorsal pregenital sclerite of male absent; and
- sternite 8 of female spinulose.

All these conditions either occur or might be expected to occur in the Milichiidae. According to Sabrosky's key (Ch. 79), both *Risa* and *Achaetorisa* would fall in the subfamily Madizinae, with other distinctive milichiid genera such as *Paramyia* Williston (with very similar wing venation and relatively long first flagellomere) and *Aldrichiomyza*, (also with rather elongate first flagellomere). The possibility that the Risidae is a sister group of a subgroup of the Milichiidae cannot be ruled out without much more detailed analysis of both groups. Certainly it is excluded from the Ephydroidea on the basis of the different structure of its antennae, its mouthparts, and its frontal bristling. The resemblances with the Ephydridae noted by Papp must result from homoplasy. According to Papp (1980), the larvae of *A. brevicornis* are parasitic on larvae of the moth *Ancyclosis* (*Cobatia*) *lacteicostella* (Ragonot) in stems of *Halogeton sativus*. Probably it is associated with the moth larvae in *Halogeton*, as reported, but it seems doubtful whether it is a true parasite. No fossils are known.

Cryptochetidae. The main genus, *Cryptochetum* Rondani, was referred to at least six different families by various authors before Brues and Melander (1932) ranked it as a separate family (J. F. McAlpine, Ch. 100). The Chamaemyiidae, Lonchaeidae, Milichiidae, and Drosophilidae have been regarded as the nearest relatives of the groups. Hennig (1958) placed it as a family of uncertain relationship, but showed that a close affinity with the Chamaemyiidae can certainly be excluded; he also ruled out the Lonchaeidae and Milichiidae and considered that a closer relationship with the Ephydroidea (as Drosophiloidea) was the most valid, albeit not actually compelling, assumption. Later, he (Hennig 1969*b*) doubtfully referred the family to the Milichioidea and then (Hennig 1971*a*) affirmed that on the basis of antennal structure it could not belong to the Ephydroidea. Finally, he (Hennig 1973) again placed it among families with uncertain relationships. Griffiths (1972) revived the mistaken notion that the Cryptochetidae is a sister group of the Lonchaeidae. D. K. McAlpine (1976) showed that Griffiths' conclusion is untenable on various grounds and assigned it to an isolated position in his Drosophiloidea. His conclusion rested to a large extent on the premise that his new genus, *Librella* D. K. McAlpine from Australia, is a primitive cryptochetid, i.e. the sister group of the fossil genus *Phanerochaetum* Hennig (Hennig 1965*b*, 1969*b*) + *Cryptochetum*. I agree that *Librella* probably belongs to the Ephydroidea, but I am unconvinced that either it or *Phanerochaetum* are closely related to the Cryptochetidae (as discussed later). My own conclusion (J. F. McAlpine, Ch. 100) is that the Cryptochetidae, *sensu stricto* is a member of the Carnoidea, and that it, together with the Chloropidae, is a sister group of the Milichiidae + Risidae (Fig. 116.6). Hennig (1958) listed 13 apomorphic characters for the Cryptochetidae, based on the genus *Cryptochetum sensu lato*. The addition of *Phanerochaetum tuxeni* Hennig (1965*b*, 1969*b*) from Baltic amber and *Librella demetrius* D. K. McAlpine (D. K. McAlpine 1976) from Australia has complicated matters with respect to the definition, characterization, and relationships of the family. According to Hennig's descriptions, *Phanerochaetum* agrees with *Cryptochetum sensu lato* as follows: first flagellomere elongate, vibrissa reduced, fronto-orbital bristles reduced, C with both humeral and subcostal breaks, crossvein bm-cu weak or absent, cell cu*p* small, vein CuA_2 recurved.

The venation in the anal area of the wing is of paramount importance in establishing close relationships in the Carnoidea or the Ephydroidea (= Drosophiloidea). In both groups CuA_2, when present, is recurved on vein A_1. Both A_1 and A_2 are in some cases present and evident and in some absent or indistinguishable, especially in the Carnoidea. Often these two veins are confused by different workers, which has led to wrong assumptions about the relationships of their bearers.

In the ground plan of the Carnoidea, both A_1 and A_2 are present; they are easily distinguishable in the Australimyzidae, Carnidae, Tethinidae, Canacidae, many of the Milichiidae, and the Cryptochetidae. In the Ephydroidea, A_2 is rarely distinguishable (see species of *Apsinota* Wulp), but A_1 sometimes much resembles A_2. Hennig (1958) misidentified A_2 in *Cryptochetum* as A_1, but he drew special attention to its position closer to the margin of the anal lobe. Unfortunately, he apparently overlooked the remnant of A_1, which is clearly present along the hind margin of cell cu*p*, and which extends as a fold well into the membrane anterior to A_2 in most members of the genus. Later, Hennig (1965*b*, 1969*b*) confused A_1, which is relatively strongly developed in *Phanerochaetum*, (but A_2 is apparently not evident), with A_2 in *Cryptochetum*. Colless and D. K. McAlpine (1970) correctly interpreted A_2 in *Cryptochetum* and drew attention to its strong development in the family. In describing *Librella*, D. K. McAlpine (1976) correctly identified the anal vein in *Librella* as A_1 (as in *Phanerochaetum*), but unfortunately, he followed Hennig's lead and also confused it with vein A_2 in *Cryptochetum*. At the same time, however, he noted that A_2 is present in the Canacidae, in the Tethinidae, and in some less reduced Milichiidae.

A wide interruption or weakening midway along the posterior boundary of cell cu*p*, such as is depicted for *Librella*, is a consistent, though frequently overlooked, apomorphic feature of the Ephydroidea (= Drosophiloidea). As shown by Hennig (1958), a similar but shorter weakening is evident in some Nerioidea (e.g. *Pseudopomyza* Strobl), some Tephritoidea (e.g. Lonchaeidae), some Opomyzoidea (e.g. the Clusiidae, Acartophthalmidae, Odiniidae, and Agromyzidae), and probably others, but this type of interruption is not present in any Carnoidea including *Cryptochetum* (see figures in Thorpe 1941). A reexamination of *P. tuxeni* is necessary to establish, if possible, the exact conditions of cell cu*p* and adjoining veins in that species. From Hennig's (1965*b*, 1969*b*) descriptions, *P. tuxeni* differs from *Cryptochetum sensu lato* as follows: arista elongate, pedicel with a pronounced dorsal notch, compound eye bare, ocelli situated in an equilateral triangle and located right on vertex, ocellar triangle not delimited from frons, frons not densely setulose, vertex without accessory bristles, postocellars convergent, cell cu*p* differently shaped, vein A_1 differentiated. In my opinion, the factual information Hennig provided for *Phanerochaetum* contains no compelling evidence for its membership in the Cryptochetidae. The wing venation, in particular, seems to have more in common with the Ephydroidea than with any Carnoidea.

D. K. McAlpine (1976) found *Phanerochaetum* to be essentially similar to *Librella*, but with reduced cephalic chaetotaxy and with an increased number and size of scutellar setae. According to his description of *Librella*, it has, among others, the following characters that are important for establishing its relationships; those characters that represent possible synapomorphies with the Drosophilidae are preceded by an asterisk:

— arista elongate;
— compound eye bare;
— * pedicel with a pronounced dorsal notch;
— * pedicel with a dorsal seam extending almost to its base;
— first flagellomere relatively long and drooping;
— ocelli in an equilateral triangle high on vertex;
— ocellar triangle not enlarged or strongly delimited from frons;
— vertex with normal complement of bristles;
— * postocellar bristles convergent;
— * face uniformly sclerotized, and with a low, slightly angular carina;
— * prosternum with distinctly sclerotized precoxal bridge;
— main thoracic bristles present and distinct;
— * bristle present between posterior notopleural and supra-alar bristles;
— * scutellum broad and convex, not sharply margined;
— * scutellum bare on disc, with two pairs of bristles, the apical ones strongly convergent;
— C with both a humeral and a subcostal break;
— * C with thickened costal spinules in a single anteroventral series extending from humeral break to near insertion of R_{4+5}, and with an anteroventral series of weaker setulae over same extent;
— crossvein bm-cu absent;
— * vein CuA_2 thickened and strongly recurved;
— * cell cu*p* with a large interruption in posterior margin;
— * vein A_1 short, directed posteriorly from posterior margin of cell cu*p*;
— * vein A_2 indistinguishable;
— * tibiae each with a preapical dorsal bristle;
— spiracles 1–6 present, in abdominal membrane; and
— * two spermathecae present, strongly sclerotized, brown pigmented.

Individually some of the characters are indecisive as indicators of close relationship, because they are found in a number of families and superfamilies. In combination, however, they indicate that *Librella* has much more in common with the Drosophilidae than with any other family (see characters that are preceded by an asterisk). In my opinion, the assignment of both *Phanerochaetum* and *Librella* to the Cryptochetidae should be reconsidered.

The following ground-plan characters of the Cryptochetidae (exclusive of *Phanerochaetum* and *Librella*) are noteworthy. Almost all of them are apomorphic with respect to the Acalyptratae; those that are synapomorphic with the Milichiidae, at least in part, are marked with (AM), those that are synapomorphic with the Chloropidae, at least in part, are marked with (AC); and those that are autapomorphic are marked with (AA):

— larvae with a pair of long tails (gills?) on caudal segment (AA);
— anterior spiracles of larvae retractile;
— posterior spiracles of larvae forming posteroventrally curved hooks (AA);
— larvae endoparasitic in monophlebine scale insects (AA);
— first flagellomere elongate and drooping (AM, AA);
— arista greatly reduced (AA);
— pedicel without a dorsal notch or seam (AM, AC);
— distal margin of pedicel with a long, clavate connection penetrating deeply into base of first flagellomere (AM);
— proximal margin of first flagellomere and distal margin of pedicel rounded (AM, AC);
— vibrissae reduced (AA, AC);
— posterior ocelli widely separated (AA);
— ocellar triangle enlarged (AM, AC);
— frons abundantly setulose (AC);
— compound eyes enlarged (AM);
— compound eyes hairy (AC, AM);
— palpi unusually stout (AA);
— labella peculiarly armed (AA);
— precoxal bridge present (AC, AM);
— propleural carina developed (AM, AC);
— scutum and scutellum densely setulose (AC);
— main bristles of scutum reduced (AA, AC);
— scutellum enlarged, convex strongly margined (AC);
— notopleuron with several bristles posteriorly (AC);
— anepisternum bare (AM, AC);
— anepimeron, posteriorly, with deep indentation (AM);
— pleural area below wing base with peculiar ear-like lobes (AM);
— costal spinulae not aligned in anterodorsal and anteroventral rows;

— C with both humeral and subcostal breaks (AM);
— cell cup present, small (AM);
— posterior boundary of cell cup weak but entire (AM);
— vein CuA_2 recurved (AM);
— vein A_1 reduced (AM);
— vein A_2 strong, traceable to wing margin;
— tibiae with preapical dorsal bristles;
— hind tibia with sensory area (AM, AC);
— sternites 1–5 reduced (AA);
— spiracle 6, only, functional in both sexes (AA);
— surstyli fused to epandrium (AA, AM, AC);
— parameres large, setulose partially enveloping aedeagus (AA, AM);
— aedeagus slender, reduced (AM, AC);
— ejaculatory apodeme reduced (AM);
— sternite 8 of female forming peculiar piercer (AA);
— cerci of female reduced (AA); and
— spermatheca one (AA).

This combination of characters establishes that the Cryptochetidae (exclusive of *Phanerochaetum* and *Librella*) belong to the Carnoidea, that the family is a highly specialized monophyletic unit, that it shares a number of synapomorphies with both the Milichiidae and Chloropidae, and that it must have arisen more recently than Milichiidae but prior to Chloropidae (Fig. 116.6).

About 25 species, all belonging to *Cryptochetum sensu lato*, are known (J. F. McAlpine, Ch. 100). These are mainly endemic to the Ethiopian, Oriental, and Australian regions, but a few occur in the southern Palearctic region. None are endemic to the New World. As already indicated, it seems doubtful whether the fossil species *P. tuxeni* or the Australian species *Librella demetrius* D. K. McAlpine actually belong to this family.

Chloropidae. This family is one of the largest and most ubiquitous families (Sabrosky, Ch. 99) of Acalyptratae, but despite this it is also one of the most closely knit, most easily recognized, and most clearly defined members. Its family-group status originates from "Chloropina" Rondani (1856), based on *Chlorops* Meigen (Meigen 1803), the first separate chloropid genus. The family name, Oscinidae, based on *Oscinis* Latreille (Latreille 1804) (= *Chlorops*) and "Oscinides" Fallén (Fallén 1820*a*), was used extensively in the 19th century but was replaced by Chloropidae after the synonymy between *Chlorops* and *Oscinis* was established (Andersson 1977). Two families, Mindidae and Echiniidae, proposed by Paramanov (1956 and 1961, respectively) are synonyms of the Chloropidae, as established by D. K. McAlpine (1958*b*) and Sabrosky (1962), respectively.

Recently, Andersson (1977) subjected Chloropidae to one of the most exemplary systematic and phylogenetic analyses yet carried out on any large acalyptrate family. Following an exhaustive study of character transformations within the family, he drew up a description of the hypothetical stem species. On the basis of this work the following characters, which are apomorphic with respect to the ground plan of the Acalyptratae, can be listed. Those that are synapomorphic with respect to the ground plan of Milichiidae are marked (AM), those that are autapomorphic are marked (AA):

— larvae phytophagous, attacking monocotyledonous plants (AA);
— maxillary lacinia reduced (AM);
— postocellar bristles convergent (AM);
— frons with a large, distinct ocellar triangle (AM);
— frons with three fronto-orbital bristles, the posterior one posterolaterally directed, the two anterior ones anterolaterally directed (AA);
— frons with distinct interfrontal setulae (AM);
— facial carina more or less distinct (AM);
— scapular bristles (anteriormost presutural dorsocentrals, see Andersson 1977, p. 26) present (AA);
— propleuron with well-developed carina (AA);
— anepisternum with a few short bristles (AA);
— katepisternum with one bristle (AM);
— Sc reduced on apical half (AM);
— crossvein bm-cu absent (AA);
— cell cup absent (AA);
— A_1 reduced or absent (AA);
— M with a distinct inflexion at middle of basal section (AA);
— hind tibia with a sensory area (AA);
— distiphallus membranous, without pubescence (AM);
— female with two rudimentary spermathecae on long ducts (AM); and
— female with a pocket-like ventral receptacle (AM).

Sturtevant (1925–1926) first stated that the rudimentary seminal receptacles, with their long, fine ducts, and the pocket-like ventral receptacle indicate that the Chloropidae and Milichiidae are closely related to each other. Hennig (1971*a*) concluded that these similarities noted by Sturtevant, along with other synapomorphic groundplan characters of the Milichiidae and Chloropidae, "do not contradict the assumption that they are sister groups." Griffiths (1972) and Andersson (1977) accepted the same viewpoint. Andersson (1977) noted that the ground plan of the Milichiidae has many features in common with the Chloropidae (see characters marked AM in the previous list), but profered no opinion about a sister-group relationship between them. In my opinion, the sister group of the Chloropidae probably is the Cryptochetidae, and these two families together comprise the sister group of the Milichiidae + Risidae (Fig. 116.6). The Chloropidae remained more generalized than any of the other three families in not acquiring a humeral break in C, but it developed its own set of autapomorphies in the wing.

Three subfamilies, the Siphonellopsinae, Oscinellinae, and Chloropinae, are usually recognized (Kanmiya 1983). Andersson (1977) discussed the phylogenetic relationships within the family and concluded that the latter two subfamilies together are a sister group to the Siphonellopsinae; Narchuk (1983) raised the Siphonellopsinae to full family status. In all, some 2000 species, from all over the world, have been described in over 160 valid genera (Sabrosky, Ch. 99). One fossil species, *Protoscinella electrica* Hennig (Hennig 1965*b*) from Baltic amber is known; other fossil species, most of which were erroneously referred to the family, were discussed by Hennig in the same paper. The report by J. F. McAlpine and Martin (1969) of a chloropid in Canadian amber (Cretaceous) is wrong. Subsequent, more detailed study of this specimen revealed it to be a member of the Sciadoceridae.

Superfamily Sphaeroceroidea

This superfamily (Fig. 116.7) contains only three or four main groups for which separate family rankings seem warranted, i.e. the Heleomyzidae [including Borboropsidae (Griffiths 1972), Chiropteromyzidae (Frey 1952), Cnemospathidae (Enderlein 1938), Heteromyzidae (Griffiths 1972), Notomyzidae (Griffiths 1972), Rhinotoridae (Williston 1896, Hendel 1916), and Trixos-

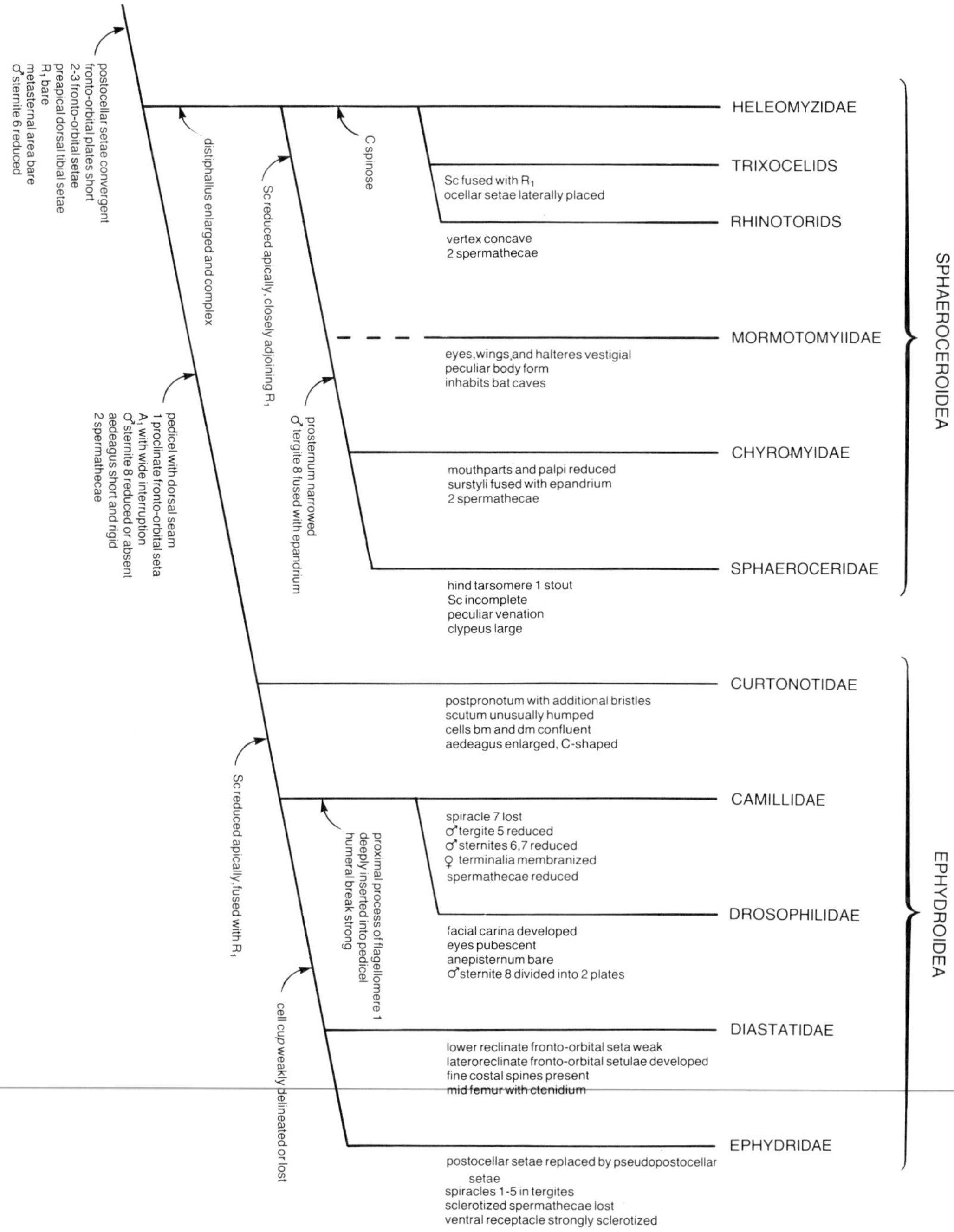

Fig. 116.7. Cladogram of the Sphaeroceroidea and Ephydroidea summarizing the relationships and apomorphies of the subgroups and families recognized. (See text for additional details.)

celididae (Hendel 1916)], the Sphaeroceridae, and perhaps the Mormotomyiidae and Chyromyidae. A similar grouping of families, called Heleomyzoidea, was proposed by Colless and D. K. McAlpine (1970), but they included the Coelopidae and Somatiidae, which they later (Colless and D. K. McAlpine 1975) properly referred to the Sciomyzoidea and Diopsoidea (as Tanypezoidea), respectively.

Under the rules (Ride and Sabrosky 1985), Sphaeroceroidea is the correct superfamily name to apply, because the family-group name "Sphaerocerides" (Macquart 1835) predates both Heleomyzidae (Westwood 1840) and Chyromyidae (Hendel 1916), and all the other family-group names to which I have already referred.

The following ground-plan characters of the Sphaeroceroidea are especially significant. Those that are considered to be plesiomorphic with respect to the ground plan of the Acalyptratae are listed first and are indicated by (P). Apomorphic characters that are shared with ground-plan characters of the Opomyzoidea are marked (OA); those that are synapomorphic for the Sphaeroceroidea and Ephydroidea are marked (AE); those that are autapomorphic for the Sphaeroceroidea (synapomorphic in the ground plans of included families) are marked (AA):

— larvae saprophytic, probably in decaying organic matter (P);
— puparium with slightly protruding respiratory horns (P);
— scape, pedicel, and flagellum porrect, main axis nearly straight (P);
— scape setulose (P);
— pedicel with a dorsoapical notch (P);
— distal membrane of pedicel deeply inserted into first flagellomere (P);
— frons narrower in male than in female (P);
— interfrontal setulae fine, scattered (P);
— proanepisternum haired (P);
— katepisternum haired (P);
— anepisternum haired (P);
— precoxal bridge undeveloped (P);
— Sc complete, free from R_1 (P);
— pterostigma present (P);
— cell *cup* complete, transversely closed (P);
— A_2 present, distinct (P);
— anal angle and alula well developed (P);
— abdomen with 7 spiracles in both sexes (P);
— spiracles 1–5 situated in membrane (P);
— male with tergite 6 large and free (P);
— sternite 8 large and haired (P);
— hypandrial bridge complete (P);
— epiphallus absent (P);
— aedeagal apodeme rod-like, narrowly united with hypandrium (P);
— surstylus freely articulated with epandrium (P);
— sternite 10 plate-like, undivided medially (P);
— female with sternites and tergites 6–8, simple, free from each other (P);
— cerci simple and free (P);
— three spermathecae present, sclerotized (P);
— first flagellomere short, discoid, more or less porrect (OA);
— arista arising dorsally (OA);
— postocellar bristles convergent (AE);
— fronto-orbital plates abbreviated anteriorly (AE);
— fronto-orbital bristles reduced to two or three (AE);
— lower half or more of frontal vitta extending almost to eye margin (AE);
— vibrissa present (OA);
— face membranized along vertical midline (secondarily sclerotized in many members (OA);
— katepisternal bristle present (OA);
— metasternal area bare (OA);
— subscutellum relatively strongly developed (OA);
— preapical dorsal tibiae bristles present (secondarily lost in Chyromyidae) (AE);
— C with subcostal break (OA);
— R_1 bare (AE);
— A_1 not attaining wing margin (secondarily lengthened in a few genera) (OA);
— male with sternite 6 reduced, asymmetric, shifted to left side (AE);
— tergite and sternite of segment 7 reduced, fused with sternite 8 (OA);
— tergite 8 atrophied or combined with sternite 8 (OA); and
— distiphallus enlarged and highly complex (AA).

The seven synapomorphies of the Sphaeroceroidea and Ephydroidea (characters marked AE in this list) attest to a sister-group relationship between these two superfamilies.

The Sphaeroceroidea probably stemmed from the same acalyptrate stock that produced the Opomyzoidea and Carnoidea (see Fig. 116.9). The 11 synapomorphies with respect to the ground plan of the Opomyzoidea (marked by OA in above list) support this contention.

The Sphaeroceroidea retained a greater variety of plesiotypic conditions in their ground plan than did the Ephydroidea, e.g. antennae porrect, scape without a dorsal seam, precoxal bridge undeveloped, pterostigma present, aedeagus long and flexible, and three well-developed spermathecae present. On the other hand, they seem to have developed very few universally distributed autapomorphies (synapomorphies of the included families). The only autapomorphic condition that I have noted is the tendency for the distiphallus to be enlarged and highly complex (aedeagus more or less rigid with short, simple distiphallus in Ephydroidea). The phylogenetic relationships of the component families (outlined in Fig. 116.6) are discussed in detail under each family.

Heleomyzidae. This relatively large, morphologically and biologically diverse family (Gill and Peterson, Ch. 89) is in great need of a thorough systematic analysis similar to the one that Andersson (1977) carried out on the Chloropidae. D. K. McAlpine (1967, 1968, 1982, 1985*b*) is progressing well with such studies on certain segments of the family, but, for most workers, the limits of the entire family and the classification and relationships of its subgroups, remain uncertain.

Various rankings have been applied, especially with respect to the Trixoscelididae, Chyromyidae, and Rhinotoridae (for reviews, see especially Harrison 1959 and Gill 1962). Czerny (1927) treated both the Trixoscelididae and the Chyromyidae as subfamilies of the Heleomyzidae. Curran (1934) included the Trixoscelididae as part of the Chyromyidae, and separated both from the Heleomyzidae. Gill (1962) recognized both groups as families separate from the Heleomyzidae, but later (Gill 1968) in agreement with Harrison (1959) and D. K. McAlpine (1967), he included the Trixoscelididae in the Heleomyzidae. The latter author (D. K. McAlpine 1968) also convincingly showed that the Rhinotoridae is a subgroup of the Heleomyzidae, and reduced it to tribal level. In line with the catalog of North American Diptera (Stone *et al.* 1965), we have treated the Trixoscelididae (Teskey, Ch. 90), the Chyromyidae (J. F. McAlpine, Ch. 91), and the Rhinotoridae (J. F. McAlpine, Ch. 92) as separate families, but we have done so fully realizing that this ranking is very subjective. I now fully agree with D. K. McAlpine (1985*b*) that at least the Trixoscelididae and Rhinotoridae are subgroups (tribes) of the Heleomyzidae.

Certainly, Griffiths' (1972) fragmentation of the Heleomyzidae into eight families, which he dispersed among at least three superfamilies (Heteromyzidae in Diopsoidea; Heleomyzidae, Rhinotoridae, Borboropsidae, and Trixoscelididae in Opomyzoidea; Chiropteromyzidae and Cnemospathidae in Tephritoidea; and Notomyzidae, unplaced), is untenable. The proposal by D. K. McAlpine and Kent (1982) and D. K. McAlpine (1985*b*) to treat these and other monophyletic subgroups as tribes of the Heleomyzidae is a more rational approach, at least as a provisional solution. Further studies of the entire complex on a worldwide basis are required before the problems of subfamilial limits can be resolved. Nevertheless, D. K. McAlpine's (1985*b*) reclassification of the family into tribes is a great step forward in the systematics of this family.

Even the correct name to apply to the Heleomyzidae has its problems. The oldest family-group name referable to the group is either "Micromyzides" Fallén, in which Fallén (1810) included among other genera his new genus *Heleomyza* Fallén, or "Heteromyzides" Fallén, in which the same author (Fallén 1820*b*) included both *Heteromyza* Fallén and *Heleomyza*. Macquart (1835) wrongly assigned *Heteromyza* and *Heleomyza* to his "Scatophagides" and applied the family name Heteromyzidae to another unnatural collection of genera. Westwood (1840) proposed the name "Heleomyzides" for both *Heteromyza* and *Heleomyza*, and this application of Heleomyzidae has received general acceptance. To my knowledge, Micromyzidae has never been used for any family group; Heteromyzidae, on the other hand has had an inconsistent, checkered history. For example, Frey (1921) used it in an old sense for the Dryomyzidae; Griffiths (1972) used it in a new sense for a subgroup of the Heleomyzidae. We have accepted the family-group name Heleomyzidae of Westwood (1840) on the basis of general consistency of usage and for the sake of nomenclatorial stability.

The ground-plan characters of the Heleomyzidae coincide almost entirely with the list given for the superfamily Sphaeroceroidea. Thus, as noted by Hennig (1958), the family as now defined, is based mainly upon symplesiomorphy with respect to the ground plan of the superfamily. This definition renders it suspect as a paraphyletic group. The spinose condition of the C, so widespread and prevalent in the Heleomyzidae, is the best possibility as an autapomorphic character of the family, but that condition is not entirely universal within the family. One indication that it is, indeed, a ground-plan feature, which has been lost a number of times within the family, is the fact that C is spinose in many of the more generalized members of the family, e.g. *Heteromyza* and *Tapeigaster* Macquart.

Members of the Heleomyzidae are distributed mainly in the temperate areas of the world (D. K. McAlpine 1967, 1985*b*), and although the family comprises predominantly Holarctic genera (Cogan 1971), both Northern and Southern hemispheres contain some very primitive elements, e.g. *Heteromyza* (Holarctic) and *Tapeigaster* (Australia). About 56 genera (exclusive of those placed in the Chryomyidae, Rhinotoridae, and Trixoscelididae) are recognized. D. K. McAlpine (1985*b*) classified the family into 22 tribes (including the Rhinotorini and Trixoscelidini) and provided a list of all the living genera and their synonyms.

The genus *Melanthomyza* Malloch (Malloch 1933*b*), known only from a rare Chilean species *M. polita* Malloch, possibly also belongs to the Heleomyzidae. Malloch assigned it hesitatingly to the Anthomyzidae, but it does not seem to fit within the present definition of that family. A single, headless female in the Canadian National Collection (Hda. Illapel, Coquimbo, Chile, 30.X.1954, L. Pẽna) shows the following characteristics not mentioned by Malloch: C minutely, but distinctly, spinose; seven pairs of abdominal spiracles present and situated in the margins of the tergites; tergite 7 and sternite 7 free from each other; tergite 8 and sternite 8 reduced and simple; cerci unusually short, stout, and with broadly rounded apices; and two large heavily sclerotized spermathecae present. It seems to resemble *Heleomicra collessi* D. K. McAlpine, which is a minute, black species of Heleomyzidae from Australia, but in that species the costa is without spines among the hairs.

In all, more than 400 described species of Heleomyzidae are known (Bickel 1982). Czerny's (1927) treatment of the Palearctic fauna is still the best general treatment of the whole family in that region. Gill (1962) revised the Nearctic species. Harrison (1959) reviewed the New Zealand representatives, Cogan (1971) reviewed the family in the Ethiopian region, and D. K. McAlpine (1985*b*) revised the Australian genera and listed the described species. As indicated above, subfamilial limits are not yet clearly established. Five fossil species from Baltic amber are known (Hennig 1965*b*, 1969*b*).

Rhinotoridae. This group consists of seven genera, i.e. *Rhinotora* Schiner, *Rhinotoroides* Lopes, *Neorhinotora* Lopes, *Cairnsimyia* Malloch, *Anastomyza* Malloch, *Apophoneura* Malloch, and *Zentula* D. K. McAlpine (D. K. McAlpine 1968, 1985*b*; D. K. McAlpine and Kent 1982). Williston (1896) treated the type genus as a subfamily, Rhinotorinae, within the Ropalomeridae and thereby established the family-group name for it. Hendel (1916) first ranked it as a separate family, which was adopted by Brues and Melander (1932 and subsequent editions), Wheeler (1954), Steyskal (1957*a*), Hennig (1958, 1971*a*, 1973) D. K. McAlpine (1958*a*, 1958*b*), Foote (1965), Guimarães and Papavero (1966), Papavero (1967*b*), and Griffiths (1972). Malloch (1931*b*) referred the Australian genus, *Cairnsimyia*, and two Neotropical genera, *Anastomyza* and *Apophoneura* (Malloch 1933*b*), to the Heleomyzidae. Hennig (1958) and Guimarães and Papavero (1966) drew attention to similarities between these latter genera (then assigned to Heleomyzidae) and *Rhinotora*. D. K. McAlpine (1968) re-examined the matter, concluded that *Rhinotora*, *Rhinotoroides*, *Neorhinotora*, and *Cairnsimyia* represent an extreme development of that branch of the Heleomyzidae which includes *Anastomyza* and *Apophoneura*, and reduced the group to a tribe, the Rhinotorini, within the Heleomyzidae. Griffiths (1972) included the Australian genus, *Tapeigaster* in the same group (which he treated as Rhinotoridae), but D. K. McAlpine and Kent (1982) rejected this placement on various grounds and assigned *Tapeigaster* to its own tribe, the Tapeigastrini, within the Heleomyzidae.

Elsewhere in this Manual, the Rhinotorini, *sensu* D. K. McAlpine (1985*b*), is treated as a separate family (see especially J. F. McAlpine, Ch. 92) as was done with the Trixoscelididae. It is, however, almost certain that it is a monophyletic subgroup of the Heleomyzidae as shown by D. K. McAlpine (1968, 1985*b*).

The following conditions are apomorphic ground-plan characters of the group with respect to the ground plan of the Sphaeroceroidea. Those that are distinctively autapomorphic are marked (AA):

— puparium without protruding respiratory horns;
— frons equally wide in both sexes;
— vertex weakly to strongly concave (AA);
— postocellar bristles weak or absent (AA);
— clypeus enlarged (AA);
— prosternum reduced, narrow, with deep median groove;
— scutellum weakly to strongly elongate (AA);
— wing conspicuously spotted;
— C weakly and sparsely spinose;
— Sc approximated to R_1;
— pterostigma absent;
— R_{2+3}, R_{4+5}, and M tending to be undulate, and with supernumerary crossveins (AA);
— CuA_1 weakened apically, not or scarcely attaining wing margin;
— A_2 absent;
— femora, especially anterior femur of male, enlarged and spinescent ventrally;
— tergite 6 of male reduced;
— aedeagal apodeme short, relatively broadly fused with hypandrium;
— ejaculatory apodeme minute; and
— two spermathecae present.

The group is mainly limited to the Neotropical region (*Rhinotoroides*, *Rhinotora*, *Neorhinotora*, *Apophoneura*, and *Anastomyza*). One species, *Neorhinotora diversa* (Giglio-Tos) extends into the southern Nearctic region and two genera, *Cairnsimyia* and *Zentula* D. K. McAlpine, are known from the Australian region (including New Guinea) (D. K. McAlpine 1968, 1985*b*). In all, about 35 species are described. No fossils are known.

Trixoscelididae. Hendel (1916), not Frey (1921), was the first to recognize this group as a separate family. As here defined (Cogan 1977; Teskey, Ch. 90), it consists of *Trixoscelis* Rondani, by far the largest genus, and *Spilochroa* Williston, *Stuckenbergiella* Cogan, and *Zagonia* Coquillett. Probably *Paratrixoscelis* Soós (Soós 1977), which seems rather weakly distinct from *Trixoscelis*, belongs here; its type species, *Geomyza oedipus* Becker, has usually been placed in *Trixoscelis* (Hackman 1970), but Soós (1977) created a new genus, *Paratrixoscelis*, for it.

As indicated in the opening discussion of the Sphaeroceroidea, *Trixoscelis* and its relatives have often been placed within the Heleomyzidae (Czerny 1927; Malloch 1930*a*; Collin 1943; Harrison 1959; D. K. McAlpine 1967, 1968, 1982, 1985*b*; Gill 1968; Cogan 1971; Colless and D. K. McAlpine 1970, 1975; Bickel 1982). At other times they were considered as a separate family, sometimes even by the same authors (Frey 1921; Hendel 1928; Séguy 1950; Melander 1952; Hennig 1958, 1973; Gill 1962; Griffiths 1972; Cogan 1977). D. K. McAlpine (1982, 1985*b*) regards the group as one of 22 tribes of the Heleomyzidae. It seems virtually certain that the group, as now defined, is a monophyletic subgroup of the Heleomyzidae, but its precise sister group within that family has not been established.

The following conditions shared by all trixoscelidids are apomorphic with respect to the ground plan of the Sphaeroceroidea:

— puparium probably without protruding respiratory horns;
— frons equally wide in both sexes;
— interfrontal setulae reduced or absent;
— ocellar bristles frequently arising laterally to anterior ocellus;
— proanepisternum bare (except for proanepisternal bristle);
— C deeply incised at end of Sc;
— * Sc becoming fused apically with R_1;

— pterostigma absent;
— A_2 reduced or absent;
— * abdominal spiracles 7 lost in both sexes;
— * male with tergite 6 much shortened; and
— * female with sternite and tergite of segment 7 closely associated or fused.

Only four (indicated by asterisks) of the seven apomorphies listed for the group by Griffiths (1972) and Cogan (1977) have proven to be valid.

Trixoscelidids are found most commonly in arid areas. They are best represented in the Holarctic (54 spp., Soós 1977), Ethiopian (23 spp., Cogan 1977), and Neotropical (7 spp., Gill 1968) regions. They are not recorded in the Oriental and Australian regions. No fossils are known.

Chyromyidae. Like the Trixoscelididae, this family (J. F. McAlpine, Ch. 91), has had a fluctuating history. Older authors placed the type genus, *Chyromya* Robineau-Desvoidy, in the Sapromyzidae (= Lauxaniidae) and the Opomyzidae (Malloch 1914). Hendel (1916) was the first to rank it as a separate family, but Czerny (1927) treated it as a subfamily of the Heleomyzidae. However, for the last half century it has usually been accepted as a discrete family (Hendel 1928, 1933; Curran 1934; Collin 1949; Hennig 1958, 1965*b*, 1971*a*, 1973; Griffiths 1972; Colless and D. K. McAlpine 1970, 1975; Andersson 1971, 1976*c*; Bickel 1982). It consists of only three genera, *Chyromya*, *Gymnochiromyia* Hendel, and *Aphaniosoma* Becker, which contain, in all, about 40 species that occur in the Holarctic, Ethiopian, and Oriental regions (J. F. McAlpine, Ch. 91).

The following ground-plan characters of the Chyromyidae are especially significant; those that are autapomorphic are indicated by (AA), and those that are synapomorphic with the Sphaeroceridae are marked (AS):

— integument and vestiture extensively pale yellowish (AA);
— scape, pedicel, and flagellum porrect; main axis nearly straight;
— scape setulose;
— pedicel with a shallow, dorsal notch;
— flagellum discoid;
— palpi and mouthparts small (AA);
— postocellar bristles convergent;
— interfrontal setulae present, but fine and scattered;
— fronto-orbital bristles three and reclinate (anterior one secondarily weakly incurved in some species);
— vibrissa weakly developed (AA);
— face membranized medially;
— prosternum very narrow, with deep median groove (AS);
— proanepisternum bare except for weak proepisternal bristle;
— postpronotum with an inwardly directed inner bristle (AS);
— anepisternum with both setulae and bristles;
— C with subcostal break;
— costal spines reduced or absent (distinguishable in *Chyromya* spp.) (AA);
— Sc very close to, or joining, R_1 apically (AS);
— A_1 not reaching wing margin (AS);
— A_2 indistinct or absent;
— preapical dorsal tibial bristles absent (AA);
— abdominal spiracles 6 and 7 present in both sexes;
— spiracles 6 and 7 of the male located in tergite 6 (AS);
— sternite 8 of the male united with epandrium (AS);
— surstyli more or less fused with epandrium (AA);
— aedeagus short and stout (AS);
— distiphallus very complex (AS);
— aedeagal apodeme extensively fused wth hypandrium;
— ejaculatory apodeme shortened, broadened, and umbrella-like (AA);
— sternite 8 of the female divided or otherwise modified (AS); and
— two spermathecae present (AA).

Nothing in this list of ground plan characters would exclude the Chyromyidae from the Sphaeroceroidea (or the Heleomyzidae), and many of the conditions are foreshadowed in the Heleomyzidae (including Trixoscelidini). The autapomorphic conditions indicate that the group is a monophyletic unit (Hennig 1958, 1971*a*; Griffiths 1972), but its precise sister group has not yet been established. Hennig (1965*b*) proposed as a working hypothesis that the Chyromyidae and Aulacigastridae are sister groups and that his fossil species, *Gephromyiella electrica* Hennig, may be the common ancestor to both families. In my opinion, *Gephromyiella* Hennig is probably misplaced in the Chyromyidae, and, as discussed earlier, I believe *Aulacigaster* and its allies belong in the Asteioinea (Opomyzoidea). If so, the similarities between the Chyromyidae and Aulacigastridae noted by Hennig (1965*b*) and Griffiths (1972) are the result of homoplasy. The latter author thought it possible that the Sphaeroceridae is the sister group of the Chyromyidae, and I concur with that idea. A number of peculiarities shared by Chyromyidae and Sphaeroceridae (marked AS in the previous list) strengthen the possibility that Sphaeroceridae is the sister group of Chyromyidae, and that together both families form the sister group of the Heleomyzidae (Fig. 116.7).

Sphaeroceridae. This large, easily recognized family (Marshall and Richards, Ch. 93) is one of the best founded monophyletic subgroups of the Schizophora (Hennig 1958, Griffiths 1972). Its most characteristic morphological feature is the shortened, somewhat swollen condition of tarsomere 1 of the hind leg. This character, which also occurs commonly in the Heleomyzidae, especially in males, is a thoroughly established ground-plan character in both sexes of the Sphaeroceridae. The wing venation in the Sphaeroceridae is also very distinctive (see next paragraph). Although several different names, e.g. Cypselidae, Borboridae, and Leptoceridae, have been applied to the family in the past, Sphaeroceridae, first proposed by Macquart (1835) as Sphaerocerides, is the correct name to use.

The following ground-plan characters are apomorphic with respect to the ground plan of the Sphaeroceroidea; those that are autapomorphic are marked with (AA), and those that are synapomorphic with the Chyromyidae are marked with (AC):

— puparium without protruding respiratory horns;
— frons equally wide in both sexes;
— frontal vitta with sclerotized, setulose, interfrontal plates (AA);
— palpi rather small (AC);
— clypeus large and prominent (AA);
— subcranial cavity large (AA);
— inner vertical bristle shifted forward (AA);
— paravertical bristle relatively strong (AA);
— prosternum very narrow, with deep, median groove (AC);
— tarsomere 1 of hind leg short and thick in both sexes (AA);
— Sc reduced and incomplete (AA);
— pterostigma reduced;
— apex of R_{2+3} rather abruptly forwardly curved (AA);
— R_{4+5} usually curved forward and ending before wing tip (AA);
— M curved forward (AA);
— posterior margin of cell dm (preapical section of CuA_1) posteriorly bowed (AA);
— apical section of CuA_1 not attaining wing margin (AA);
— A_2 absent (AA);
— male with spiracle 7 absent on right side (AA);
— spiracle 7 on left side displaced to segment 6 (AA, AC);
— tergites 6 and 7 reduced or absent;
— sternite 8 fused with epandrium (AC);
— female with sternite 8 divided longitudinally or otherwise modified (AC); and
— epiphallus present.

Various opinions have been offered concerning the cladistic relationship of the Sphaeroceridae. Hendel (1916) first placed it in a group with the Tethinidae, Milichiidae, and Agromyzidae, but in later years he (Hendel 1922, 1928) placed it near the Ephydridae, probably because of certain similarities in the adult mouthparts (filter apparatus in esophagus, prementum short and broad, subcranial cavity enlarged) noted by Frey (1921). Crampton (1944b) followed the same idea. Hennig (1958) placed it in his superfamily Milichioidea with the Tethinidae, Milichiidae, Carnidae, Canacidae, and Braulidae and ruled out any close relationship between it and the Ephydroidea (his Drosophiloidea); later he (Hennig 1973) relegated it to a heterogeneous group of families under the title "Systematische Stellung ungeklart." Speight (1969), on the basis of a reduced prosternal basisternum, recommended the removal of the Sphaeroceridae from the Milichioidea–Drosophiloidea–Chloropoidea complex and proposed placing it near either the Heleomyzidae or the Anthomyzidae. Hennig (1971*a*) excluded the possibility of a closer relationship between it and the Anthomyzidae but directed attention to "the great conformity which exists between the Sphaeroceridae and the Milichiidae with respect to the structure of the antennae." Colless and D. K. McAlpine (1970, 1975) placed it between the Heleomyzidae and Chyromyidae in their superfamily Heleomyzoidea. Griffiths (1972) placed it in his prefamily Anthomyzoinea (Heleomyzidae *sensu lato* Anthomyzidae, Asteiidae, Opomyzidae, Sphaeroceridae, Chyromyidae, and Aulacigastridae) and expressed the opinion that the group most closely related to the Sphaeroceridae is probably *Borboropsis* Czerny, the Chyromyidae, or the Aulacigastridae.

As noted under the Chyromyidae (characters marked AS), the Sphaeroceridae share many characters with that family; those that are marked (AC) are especially significant. On these grounds, I believe the Sphaeroceridae and Chyromyidae are sister groups, and that together they form a sister group of the Heleomyzidae (Fig. 116.7).

Sphaeroceridae have a worldwide distribution, and more than 700 species have been described (Hackman 1969). These are divided among about 75 genera and subgenera (for lists, see Hackman 1969 and Hennig 1973) in three subfamilies, the Copromyzinae, Sphaerocerinae, and Limosininae (= Leptocerinae Hendel 1928, not Leach 1815, Trichoptera) (Marshall and Richards, Ch. 93). They are almost entirely saprophagous, breeding mainly in decaying plant matter, carrion, dung, nests of vertebrates, and fungi. The wings are reduced or absent in about 120 of the known species, and at least 44 species have been introduced with people as they moved from one continent or region to another (Hackman 1969). The Neotropical fauna is probably the most diverse and also contains representatives of the oldest branches of the family, e.g. *Archiborborus* Duda (Copromyzinae). The Australian fauna (Richards 1973) is restricted to the Limosininae, the most abundant and highly evolved subfamily. The Nearctic region supports an estimated 350 species, many of which are undescribed (Marshall and Richards, Ch. 93). About 170 species are recorded from the Palearctic region (Hackman 1969), 174 from the Neotropical region (Richards 1967), 85 from the Australian region (Richards 1973), nearly 300 species from the Ethiopian region (Richards 1980), and 67 from the Oriental region (Hackman 1977), but all these numbers are believed to be low. No fossils are known.

Mormotomyiidae. This family is known from a single, spider-like, bat-cave species, *Mormotomyia hirsuta* Austen (Austen 1936) from Kenya. Austen considered it as perhaps distantly related to the Sphaeroceridae. Its morphology, family status, and systematic relationships were discussed in detail by van Emden (1950), Hennig (1971*a*), and Griffiths (1972). Van Emden concluded that "the genus represents a well-founded family intermediate between Cordiluridae (Scathophagidae) and the Acalyptrata," which led some to believe that it belongs to the Calyptratae. Hennig discussed the pros and cons for placing it in the Calyptratae and concluded that the conspicuous cleft in the pedicel is really the only plausible indication that *Mormotomyia* might belong there. However, as was pointed out by himself and others, a similar antennal cleft also occurs in a number of families of the Acalyptratae. The absence of prestomal teeth, hyoid sclerite, and a greater ampulla, coupled with the presence of many characters of the male terminalia that agree with the Acalyptratae rather than with the Calyptratae, indicate that *Mormotomyia* does not belong to the Calyptratae. Griffiths concluded that it is an aberrant member of his prefamily Tephritoinea. He aligned it with two heleomyzid genera, which he also treated as separate families, i.e. *Neossos* Malloch (= *Chiropteromyza* Frey) (Chiropteromyzidae) and *Prosopantrum* Enderlein (= *Cnemospathis* Enderlein) (Cnemospathidae). Both these genera are now assigned to the Heleomyzidae (D. K. McAlpine 1985*b*). I agree with Griffiths that *Mormotomyia's* closest relatives probably will be found in or near the Heleomyzidae. The large size of tergite 6 in the males is the only feature that seems to speak against its inclusion in the Sphaeroceroidea, but in a species with so many distorted features, the enlargement of tergite 6 may also be a secondary condition. However, I am more inclined to believe that its large size reflects a plesiomorphic state comparable to that found in several plesiotypic members of the Heleomyzidae, especially *Dichromyia* Robineau-Desvoidy and *Amphidysis* D. K. McAlpine (see D. K. McAlpine 1985*b*, Figs. 72–73).

Mormotomyiidae shows the following apomorphic conditions with respect to the ground plan of the Sphaeroceroidea:

- puparium without protruding respiratory horns;
- entire body, including legs and wing vestiges, covered with the long, wavy hairs (longer and denser in males than in females); most macrochaetae not, or scarcely, distinguishable;
- frons equally wide in both sexes;
- compound eye much reduced, gena exceptionally broad;
- ocelli lost;
- ocellar triangle enlarged, extended anteriorly nearly to lunule and with numerous macrochaetae;
- orbital plate extended anteriorly nearly to lunule, and with 4–6 macrochaetae;
- frontal vitta with 2–4 macrochaetae on each side just above lunule;
- pedicel lengthened and with a well-developed dorsal seam;
- first flagellomere reduced, partly concealed by enlarged pedicel;
- vibrissal area well differentiated from gena, and with 2–5 vibrissal macrochaetae;
- clypeus large and prominent;
- precoxal bridge well developed;
- wings vestigial;
- halteres reduced;
- abdominal spiracles 7 lost in both sexes;
- tergite 6 of the male unusually large (possibly a plesiomorphic condition);
- sternite 6 of the male reduced to a narrow band, fused with sternites 6 and 8 in left side;
- tergite 7 of the male reduced to a narrow band in right side, joining with sternite 7 ventrally, and with a vestige of tergite 8 in right side;

  Griffiths (1972) apparently failed to distinguish between the vestiges of sternites 6 and 7, which are present and quite separable in a left lateroventral position.
- surstylus mostly enclosed within lateral extensions of epandrium, reduced to a small, anteroventrally directed process, the apex of which lies laterally to both gonopod and paramere;

  Van Emden (1950), Hennig (1971*a*), and Griffiths (1972) failed to recognize the surstylus in *Mormotomyia*. Griffiths showed what appears to be part of it in his Fig. 126, lying laterally to POG (paramere) and X (gonopod), but he omitted it from his Fig. 124. At its base the surstylus articulates with the inner surface of the epandrium; at this point it is expanded and shows what may be the remains of a reduced posterior lobe. At its extreme base it joins a well-developed, plate-like sternite 10. Each surstylus bears a few setulae on its outer and apical surfaces. The paramere and gonopod are fairly accurately depicted by Hennig (1971*a*) in his Fig. 94; however, he did not show that the gonopod also bears, on its median ventral surface, a bristle that is somewhat smaller than the one on the paramere.
- hypandrium with a prominent trough-like aedeagal guide (fulcrum of van Emden 1950);
- cerci of male somewhat reduced;

  Hennig (1971*a*) investigated and discounted the possibilities that the reduced cerci of *Mormotomyia* might be homologous with the adanalia of the Glossinidae and, thereby, a possible indicator of close relationship.
- tergite 7 of female membranized along midline anteriorly;
- sternite 7 of female weakly sclerotized in middle;
- tergite and sternite 8 of female divided along midline;
- sternite 8 of female with a prominent, probably sensory, lobe at inner apex of each half; and
- two small, sclerotized spermathecae present.

The nature of the fronto-orbital plates and the bristling of the head in *Mormotomyia* may provide useful clues to its systematic relationships in the Acalyptratae. Almost certainly the orbital plates have been secondarily extended anteriorly in concert with a similar extension of the ocellar triangle, and, as on the ocellar triangle, the number of macrochaetae is increased. It appears that the frontal plate has been virtually obliterated as in the Heleomyzidae.

On the vertex the medioclinate inner vertical bristle is usually easily distinguishable, especially in the female, at the uppermost (posteriormost) limits of the orbital plate. Medially to the inner vertical bristle, a series of six to eight more or less lateroclinate bristles arises on the vertex; also, near the midline and slightly behind this row of bristles, there is usually a pair of fine convergent bristles or hairs. These may be true postocellar bristles, and, if so, their convergent condition (as in Heleomyzidae and its relatives) would be a highly significant feature.

The vibrissal angles are well differentiated and strongly bristled in *Mormotomyia*, which indicates clearly that it descended from an ancestor with vibrissae as found in all members of the Sphaeroceroidea.

The large, prominent clypeus and mouthparts exhibited by *Mormotomyia* are reminiscent of the type found in the Sphaeroceridae. The structure of the prosternum and precoxal bridge is the "type V" form (Speight 1969) that occurs so commonly throughout the Heleomyzidae.

Perhaps as more information becomes known about the diversity within the Sphaeroceroidea, especially within the Heleomyzidae and its subgroups, a more precise placement of *Mormotomyia* will become evident. Provisionally, however, I would regard it as a separate family of the Sphaeroceroidea (Fig. 116.7).

Superfamily Ephydroidea

The concept adopted for this superfamily (Fig. 116.7) follows that proposed by Hennig (1958, 1971*a*, 1973) as Drosophiloidea, which was also followed by Griffiths (1972) and Colless and D. K. McAlpine (1975). Accordingly, it includes only five families, i.e. the Curtonotidae, Camillidae, Drosophilidae, Diastatidae, and Ephydridae. This delimitation is much more restricted than the provisional arrangement outlined in the table of classification in Volume 1 of the Manual (J. F. McAlpine *et al.*, Ch. 1). There we included the Chloropidae, Cryptochetidae, Canacidae, and Tethinidae, following Colless and D. K. McAlpine (1970), but as shown earlier (see under Carnoidea) these latter families are not closely related to the Ephydroidea.

Previous workers usually called this cluster of families Drosophiloidea, but the family-group name, Ephydridae (proposed by Zetterstedt 1837, as Ephydrinae), predates Drosophilidae (proposed by Rondani 1856, as Drosophilina) and all the other included family-group names. Therefore, in accordance with Article 36 of the rules (Ride and Sabrosky 1985), Ephydroidea is the correct name to use.

The Ephydroidea is one of the most surely grounded monophyletic subgroups of the Acalyptratae (Hennig 1958, 1971*a*; Griffiths 1972). The following ground-plan conditions are apomorphic with respect to the ground plan of the Acalyptratae; those that are considered to be unequivocally autapomorphic are marked (AA), and those that are synapomorphic with the ground plan of the Sphaeroceroidea are marked (AS):

— larvae probably without a filter apparatus for particle feeding;
— puparium without protruding respiratory horns;
— flagellum decumbent, meeting pedicel at a sharp angle;
— base of flagellum with a dorsolateral process inserted into pedicel (AA);
— pedicel with an elongate dorsolateral seam (AA);
— fronto-orbital plates not extending far forward (AS);
— fronto-orbital bristles reduced to three;
— one fronto-orbital bristle proclinate (AA);
— postocellar bristles convergent (AS);
— vibrissa present (AS);
— precoxal bridge present (AA);
— proepimeral bristle weak or absent (AA);
— metasternal area bare (AS);
— disc of proanepisternum bare (AA);
— C spinose (AS);
— C with a subcostal break (AS);
— pterostigma absent;
— Sc lying close to R_1;
— R_1 bare (AS);
— A_1 short, not attaining wing margin (AS);
— A_2 absent;
— closing vein of cell cu*p* (CuA_2) recurved;
— posterior boundary of cell cu*p* (A_1) with a wide interruption (AA);
— preapical dorsal tibial bristles present (AS);
— sternite 8 of the male reduced or absent (AA);
— aedeagus rather short, rigid, and in a relatively fixed position (AA); and
— two spermathecae present.

Two sets of head bristles that occur in the Ephydroidea require special discussion, i.e. the fronto-orbital bristles and postocellar bristles. With regard to the fronto-orbital bristles, I concur more or less with Hennig's (1958, 1965*b*) hypothesis that three pairs are present in the ground plan of the Ephydroidea, i.e. two reclinate ones and one proclinate one (an autapomorphic condition with respect to the ground plan of the Acalyptratae). He concluded that the proclinate bristle originated on the orbital plate between the reclinate bristles and the eye margin, by enlargement of one of the orbital setulae which are often found in that position. Later, it became shifted anteriorly (below) or posteriorly (above) in relation to the level of the reclinate orbital bristles and, ultimately, medially toward the frontal vitta. As a result of this shifting, its position varies considerably in different subgroups of the Ephydroidea. In the Curtonotidae and Camillidae it always arises below the reclinate orbital bristle and vertically almost in line with it; in the Diastatidae it arises laterally to the reclinate orbital bristle, sometimes a little below, and sometimes a little above the level of the latter; in the Ephydridae, where it is sometimes absent and sometimes duplicated, its position when present is as in the Diastatidae; in the Drosophilidae it may arise above or below, and inside or outside, the reclinate orbital bristle (see Wheeler, Ch. 95, Figs. 2–4).

With regard to postocellar bristles, they are present and convergent in the ground plan of the Ephydroidea. They are located on the vertex where the upper limits of the occipital plate merge with the posterior limits of the ocellar triangle. They arise behind and laterally to the posterior ocelli. They are present in the ground plans of all the families of the Ephydroidea except the Ephydridae. In the Ephydridae, true postocellar bristles are absent, but they are usually replaced by one or more pairs of weakly to strongly differentiated, divergent, pseudopostocellar bristles. In some Drosophilidae, e.g. *Amiota* Loew, *Sinophthalmus* Coquillett, *Rhinoleucophenga* Hendel, divergent pseudopostocellar bristles are present in addition to the usual convergent postocellar bristles (see Máca 1980, Fig. 1). Pseudopostocellar bristles originate from divergent ocellar setulae that are normally present on the posterior portion of the ocellar triangle. In the Ephydridae they sometimes arise in front of, or between, the posterior ocelli, but usually they arise behind the latter, i.e. in a "postocellar" position. Hence, in the Ephydridae where true postocellar bristles are absent, the pseudopostocellar bristles have usually been incorrectly called postocellar bristles (see, for example, Clausen and Cook 1971, Fig. 3). Unlike true postocellars in other Ephydroidea, they arise farther forward on the vertex, the distance between them is usually less than the distance between the posterior ocelli, and they are divergent. A similar loss of true postocellar bristles and their replacement by pseudopostocellar bristles occurs in the Tethinidae and Canacidae (see discussion under Tethinidae).

Griffiths' (1972) review of the characters and literature relating to the delimitation and cladistics of the Ephydroidea is especially useful. I agree with most of his interpretations, but I disagree on several points. First, Griffiths emphasized the fact that he found no trace of the asymmetric reduction of the sclerites of segment 7 of the male so characteristic of most Schizophora, and he assumed that the relatively symmetric conditions of these sclerites in the Ephydroidea is a plesiomorphic state with respect to the ground plan of Schizophora. I believe that the relatively symmetric form evident in these families is probably a secondarily derived (apomorphic) condition. Secondary symmetry in these sclerites occurs sporadically throughout the Muscomorpha, and it is not particularly surprising to find it in this group. The fact that the remnants of sternite 8, when present in the Drosophilidae, are in the form of two isolated, more or less symmetric sclerites (a highly apomorphic condition) indicates that secondary symmetry might also be expected in the preceding segment. Second, I do not agree with Griffiths that the aedeagus necessarily "remains posteriorly or posteroventrally directed in the rest position" in all members of the Ephydroidea. It is strongly anteroventrally directed in the Curtonotidae, for example. Third, I do not agree with Griffiths' placement of the Ephydroidea between his Lauxanioidea and Nothyboidea. The seven autapomorphic characters (marked AA) of the Ephydroidea leave no doubt about its monophyly. I draw special attention to the pecularities in the venation in the anal angle of the wing in the ground plan of the superfamily. Cell *cup* is strongly, convexly closed, i.e. CuA_2 is strongly recurved on A_1; there is a wide interruption in A_1 midway along the posterior boundary of cell *cup* (see discussion under Cryptochetidae); A_1 is short (secondarily elongated by a fold in Curtonotidae, see Hennig 1958, p. 671); and A_2 is absent, except, perhaps, at the extreme base (see especially, Hendel 1928, Fig. 109).

The ground plan of the Ephydroidea compares more favorably with that of the Sphaeroceroidea than with any other group. At least nine characters (marked AS) can be considered synapomorphic with those of the Sphaeroceroidea; and several others (precoxal bridge present, tendency for two breaks in C, two spermatheca present) are foreshadowed in the Sphaeroceroidea, especially in the Heleomyzidae. For these reasons, it seems probable that the Ephydroidea is the more recent sister group of the Sphaeroceroidea (Fig. 116.7).

Hennig (1971*a*, 1973) divided the Ephydroidea into two suprafamilies, the Drosophilidea consisting of the Curtonotidae, Camillidae, and Drosophilidae, and the Ephydridea consisting of the Diastatidae and Ephydridae. Griffiths (1972) tabulated the plesiomorphic and apomorphic characteristics of all five families, but left open the question of how the superfamily should be divided into suprafamilies. Hennig (1971*a*) based the supposed monophyly of his Drosophilidea mainly on the fact that the basal process of the first flagellomere is relatively long (apomorphic condition), whereas it is relatively short (plesiomorphic condition) in the second group. To me this difference is simply a matter of degree and is scarcely sufficient to prove the monophyly of Hennig's Drosophilidea. In my opinion the primary evolutionary dichotomy in the Ephydroidea is between the Curtonotidae and the remaining four families. This opinion is based mainly on the fact that Sc is complete and free from R_1 in the Curtonotidae (plesiomorphic condition), as opposed to being apically reduced and fused with R_1 in the remaining four families (synapomorphic condition). The peculiar set of plesiomorphic and autapomorphic characters, which sets the Curtonotidae apart from the remainder of the Ephydroidea, is more fully discussed under the Curtonotidae.

Two main evolutionary lines are evident in the remaining four families, i.e. the Camillidae + Drosophilidae in which the proximal process of the first flagellomere penetrates deeply into the pedicel and a strong humeral break is expressed (synapomorphies), and the Diastatidae + Ephydridae in which a humeral break is weak or absent in the ground plan (plesiomorphic condition) and cell *cup* is reduced or absent (synapomorphy). This last pair of families is equivalent to Hennig's Ephydridea, but in my scheme of classification (Fig. 116.7) they should not be ranked as a suprafamily. Under this scheme, Hennig's Drosophilidea is also inapplicable.

Curtonotidae. Members of this little family (J. F. McAlpine, Ch. 94) of hump-backed flies have been assigned to various families including the Heleomyzidae, Opomyzidae, Drosophilidae, Diastatidae, and Ephydridae (see J. F. McAlpine, Ch. 94). Duda (1934*a*) treated them as a separate family, which has generally been followed since that time. The family name is based on *Curtonotum* Macquart (Macquart 1843), which Agassiz (1847) unjustifiably emended to *Cyrtonotum*. Unfortunately, Agassiz's emendation was accepted by many early workers, including Enderlein (1914) who first proposed a family-group name for it (subfamily Cyrtonotinae in the family Ephydridae). The priority of *Curtonotum* over *Cyrtonotum* was recognized by Curran (1933) and Duda (1934*a*). The latter author correctly changed the family name to Curtonotidae.

Almost all the ground-plan characters of the Ephydroidea are expressed in the ground plan of the Curtonotidae. The following are particularly significant; those that are symplesiomorphic with respect to the ground plan of the Ephydroidea are marked (P), those that are apomorphic are marked (A), and those that are clearly autapomorphic are marked (AA):

- — arista with long dorsal and ventral rays (A);
- — convergent postocellar bristles present (P);
- — proclinate orbital bristle arising lateroventrally to reclinate orbital bristle (P);
- — anterior reclinate frontal bristle weak or absent (A);
- — vibrissae strongly differentiated (P);
- — postpronotum with supplementary bristles (AA);
- — scutum unusually strongly humped (AA);
- — anepisternum setulose (P);
- — anepisternum with strong bristles (A);
- — prescutellar acrostichal bristles strong (A?);
- — disc of scutellum setulose (P);
- — C spinose (P);
- — humeral costal break present (A);
- — Sc complete, free, from R_1 (P);
- — cells bm and dm confluent, i.e. crossvein bm-cu absent (AA);
- — cell cu*p* complete (P);
- — A_1 secondarily lengthened (AA);
- — all tibiae with a preapical dorsal bristle (P);
- — spiracles 6 and 7 present in both sexes (P);
- — tergites 6 and 7 of the male much reduced (A);
- — sternites 6 and 7 of the male asymmetric (P);
- — sternite 8 of the male absent (AA);
- — aedeagus enlarged, C-shaped, distiphallus anteroventrally directed (AA);
- — aedeagal apodeme closely and broadly joined to basiphallus (AA); and
- — two spermathecae present (a third, imperfectly formed one present in *Axinota* Wulp) (P).

The seven autapomorphic characters (AA), together with at least six other apomorphic conditions (A), attest to the monophyly of the family. It is more generalized than any other family of the Ephydroidea with respect to the completeness and separateness of Sc, and perhaps also with respect to the asymmetric conditions of sternites 6 and 7 in the male. These plesiomorphic features, combined with the autapomorphies mentioned, indicate that the Curtonotidae is probably a sister group of the remainder of the Ephydroidea. It would be interesting to know whether a filter apparatus is present or absent in the fulcrum of the Curtonotidae.

The family has a worldwide distribution. It consists of three genera, *Curtonotum* (about 25 described spp. in Nearctic, Neotropical, Palearctic, Ethiopian, and Oriental regions), *Axinota* (9 described spp. in Ethiopian and Oriental regions), and *Cyrtona* Séguy (2 described spp. in Ethiopian region). *Axinota* is more specialized than the other genera in that the vibrissae and proclinate orbital bristles are much reduced or absent. Also, *Axinota* is unusual in that the males have a narrowed frons and the females have a third, peculiarly formed spermatheca reminiscent of the abortive, mutant-type, third spermatheca that occurs in some species of *Drosophila* (see Sturtevant 1925–1926). Probably both these latter characters in *Axinota* are secondarily derived conditions, i.e. reversions to earlier ancestral conditions. No fossils are known.

Camillidae. For a long time most authors included *Camilla* (J. F. McAlpine, Ch. 97) in the Drosophilidae. Frey (1921) accorded it separate family rank, which has been generally accepted (Duda 1934*c*; Wheeler 1952; Collin 1956; Hennig 1958, 1965*b*, 1971*a*, 1973; J. F. McAlpine 1960; Hackman 1960; Griffiths 1972; Papp 1972, 1974, 1978*a,b*). Its ground-plan characters are similar to those of the Curtonotidae except as follows; those that are clearly autapomorphic are marked (AA):

- — fulcrum without a filter apparatus (condition unknown in Curtonotidae);
- — postpronotum with one bristle;
- — scutum moderately humped;
- — prescutellar acrostichal bristles undifferentiated;
- — Sc apically reduced and fused with R_1;
- — veins CuA_2 and A_1 reduced, hence cell cu*p* incomplete;
- — spiracles 5 and 6 of the male enclosed in margins of tergites (AA);
- — spiracle 7 absent in both sexes (AA);
- — tergite 5 of the male greatly reduced (except in fossil species; *Protocamilla succini* Hennig) (AA);
- — sternites 6 and 7 of the male reduced or lost (AA);
- — terminalia of female extensively membranized (AA); and
- — spermathecae extremely small (AA).

The six autapomorphic conditions (AA) prove the monophyly of the family. These combined with the fact that it has retained a setulose anepisternum, a plesiomorphic condition that was lost in the Drosophilidae, supports its classification as a separate family. At first, Hennig (1958, 1965*b*) considered it to be the sister group of the Ephydridae, but later (1971*a*, 1973) he allied it with the Curtonotidae and Drosophilidae mainly because the first flagellomere bears a long basal process that penetrates deeply into the scape. Its closer relationship with the Drosophilidae is confirmed by the apical reduction and fusion of Sc with R_1, the position of the proclinate orbital bristle slightly medial to the reclinate orbital bristle, and the presence of two sclerotized spermathecae.

The family consists of eleven described species in three genera (keyed by Papp 1978*a*): *Camilla* with nine species in Holarctic and Ethiopian regions, *Katacamilla* Papp (Papp 1978*a*) with one species in West Africa, and *Protocamilla* Hennig (Hennig 1965*b*) with one fossil species in Baltic amber.

Drosophilidae. Like the Ephydridae, this large family has a worldwide distribution (Wheeler, Ch. 95). More than 2500 species in 57 genera (excluding numerous subgenera) are now recognized (cataloged by Wheeler 1981*a*). Rondani (1856) was the first to propose a family-group name (Drosophilida) for the group. Thanks to its role in the development of modern genetics, it is one of the most studied families of the Diptera. However, much of the voluminous literature relating to it pertains to the genetics, evolution, and classification of the core genus, *Drosophila*, and a modern monograph treating the whole family from the standpoint of phylogenetic systematics is still greatly needed (Hennig 1965*b*; Griffiths 1972; Bock 1982). The closest approach to such a study is Throckmorton's (1975) analysis of the phylogeny, ecology, and geography of *Drosophila*, in which he briefly outlined his ideas on certain cladistic relationships of important generic and suprageneric groups within the family. A recent series of world and regionally oriented overviews, by Wheeler (1981*a,b*), Val *et al.* (1981), Bächli and Rocha Pité (1981), Tsacas *et al.* (1981), Okada (1981), Bock and Parsons (1981), Hardy and Kaneshiro (1981), Parsons and Stanley (1981), and Ashburner (1981) and published under a single cover (Ashburner *et al.* 1981), is the best compendium of systematic information yet produced on the family. This work collates information from all the important papers published over the years and is certainly a major step forward. But, as was later stated by one of those authors (Bock 1982), "the present arrangement is not entirely satisfactory" and "the classification of the family clearly merits revision." Almost certainly, some of the subgenera, genera, and higher categories are heterogeneous units.

On the other hand, almost all modern workers generally agree that the Drosophilidae, as a family group, is a monophyletic unit. All the ground-plan characters of the Ephydroidea, except a spinose C and a setulose anepisternum, also apply to it. The following ground-plan conditions are particularly significant; those that are apomorphic with respect to the ground plan of the Ephydroidea are marked (A), and those that are clearly autapomorphic are marked (AA):

- fulcrum probably with a filter apparatus;
- dorsolateral process of first flagellomere elongate, penetrating deeply into scape (A);
- arista shortly haired;
- proclinate orbital bristle arising medioventrally to lower reclinate orbital bristle (AA);
- facial carina present (AA);
- compound eyes haired (AA);
- anepisternum bare (AA);
- katepisternum with two or three bristles (AA);
- C with humeral break as well as subcostal break (A);
- Sc apically reduced and fused with R_1 (A);
- cells bm and dm separate, i.e. crossvein dm-cu present;
- cell cu*p* complete;
- spiracles 6 and 7 present in both sexes;
- tergites 6 and 7 of the male fused (A);
- sternite 7 of the male reduced or absent (A); and
- remnants of sternite 8 of the male divided into a pair of relatively symmetric dorsolateral plates (AA).

Elaboration on some of these points is required. It is uncertain whether a filter apparatus was present in the ground plan of the Drosophilidae. Certainly it is present in some genera (Frey 1921), but its occurrence and comparative structure in the more primitive genera and generally throughout the family needs further investigation.

A weakly developed facial carina, such as occurs in most *Amiota* spp., is probably closest to the ground-plan condition in the family. Similarly, the relatively sparsely hairy condition of the compound eyes, as occurs in *Stegana* Meigen and *Amiota* probably approaches the ground-plan condition for the family. Peculiar, tiny, closely spaced, tooth-like, black spinules are present ventrally near the apex of the wing in certain Steganinae, but it seems doubtful whether these are homologous with the usual widely spaced costal spines present all along the anterior margin of the wing in the Curtonotidae, Camillidae, Diastatidae, and some Ephydridae.

The seven autapomorphic characters listed (AA) firmly attest the monophyly of Drosophilidae. Perhaps the tendency for the larvae to feed on organisms causing fermentation of organic matter, i.e. specialization on yeasts, should be added to the list of specializations.

The peculiar elongation of the articulation mechanism between the first flagellomere and the pedicel, and the anteromedial position of the proclinate orbital bristle in relation to the reclinate orbitals are two synapomorphies shared with the Curtonotidae and Camillidae. These two similarities indicate that Drosophilidae is more closely related to these two families than to the Diastatidae and Ephydridae (Fig. 116.7). It is perhaps noteworthy, also, that the Drosophilidae is more generalized than either the Curtonotidae or the Camillidae in having retained crossvein bm-cu in its ground plan (as in the Diastatidae). This crossvein was repeatedly lost within the Drosophilidae, and presumably it was also lost independently several times within the superfamily.

Since Duda's (1934*b*) treatment of the family, two subfamilies, the Steganinae containing 17 genera and the Drosophilinae containing 35 genera, are usually recognized (Wheeler 1981*a*; Bock 1982). This classification is complicated by exceptions, and as yet no universally applicable diagnostic differences between these two groups has been found. As a result, five genera (see Wheeler 1981*a*) are not placed in either subfamily. This anomaly and many other similar ones at generic and subgeneric levels are still unresolved. Bock (1982) provided the most succinct analysis of these problems.

It is generally agreed that the most generalized members of the family belong to the Steganinae (Sturtevant 1942, Wheeler 1952, Hardy 1965). Throckmorton (1975) considered that *Amiota sensu lato* is the most generalized part group of the Steganinae. Contradictorily, however, he considered *Lissocephala* Malloch, which is a relatively specialized member of the Drosophilinae, as the most primitive genus of the family. Unfortunately he gave no reasons to support either assumption. Two fossil species, *Electrophortica succini* Hennig (Hennig 1965*b*) and a species of *Neotanygastrella* Duda (Wheeler 1963) are known from Baltic and Mexican amber, respectively. The first appears to be most closely related to *Amiota* or *Stegana* Meigen (Steganinae) and the second is close to *Chymomyza* Czerny (Drosophilinae).

Diastatidae. Frey (1921) was the first to treat this little group (about 40 spp.) as a separate family (J. F. McAlpine, Ch. 96), and his ranking has been followed by almost all subsequent workers. It is best represented in the Holarctic region, but it also occurs in the Ethiopian and Neotropical regions. It consists of four genera, *Pareuthychaeta* Hennig containing two fossil species in Baltic amber, *Euthychaeta* Loew with one Palearctic species, *Campichoeta* Macquart with seven species in the Holarctic region, and *Diastata* Meigen containing about 30 species in the Holarctic, Ethiopian, and Neotropical regions.

Hennig (1958, 1965*b*, 1971*a*) believed that the family is probably a monophyletic unit, but he could find no autapomorphic conditions to prove the point. Following is a list of important ground-plan characters of the family that differ more or less with respect to the ground plan of the Ephydroidea. Those that I consider autapomorphic are marked (AA):

- fulcrum probably with a filter apparatus;
- lower reclinate orbital bristle weak (AA);
- orbital plate with an additional, fine, lateroreclinate setula above strong reclinate orbital bristle (AA);
- proclinate orbital bristle arising laterally to reclinate orbital bristles;
- anepisternal bristles present;
- anteroventral costal spines fine and broadly spaced;
- C with weakly developed humeral break;
- Sc reduced apically and more or less fused with R_1;
- cell cu*p* minute (AA);
- tendency for alula and anal lobe to be reduced (A);
- anterior femur with anteroventral ctenidium; and
- mid femur with posteroventral ctenidium (AA).

Some discussion of some of these characters is necessary. Although a weakly developed filter apparatus is present in the fulcrum of certain species of *Diastata* (Frey 1921), it is not known whether it occurs in *Euthychaeta* and *Campichoeta*; if not, it may be undeveloped in the ground plan of the family.

The anterior reclinate orbital bristle is present but relatively weak in the ground plan of the Diastatidae, as in the Curtonotidae and Camillidae. It is best developed in both sexes of *Campichoeta* and in females of *Euthychaeta*, but it is greatly reduced or absent in the fossil genus, *Pareuthychaeta*, and in *Diastata*. Identification of the true, anterior (lower), reclinate orbital bristle is easily confused by the presence in all Diastatidae of a small, lateroreclinate or reclinate setula above the strong, reclinate orbital bristle. This tiny, upper orbital setula is a distinctive autapomorphic character of the family. That it is an *additional* seta is proven by the presence of all three (a relatively weak lower reclinate bristle or setula, a strong middle reclinate bristle, and a weak upper lateroreclinate or reclinate setula) in both sexes of *Campichoeta* species, in females of *Euthychaeta spectabilis* Loew, and in many species of *Diastata*. I agree with Hennig (1958) that in the ground plan of the Ephydroidea only three, more or less equal-sized orbital bristles (the lowermost one proclinate) are arranged in a nearly vertical line, as in most steganine Drosophilidae.

A setulose condition of the anepisternum is a ground-plan character of the Ephydroidea, but the development of anepisternal bristles is a specialization that occurs in a number of families including the Diastatidae, Curtonotidae, Camillidae, and Ephydridae. The absence of both setulae and bristles is an autapomorphic feature of the Drosophilidae.

An anteroventral ctenidium on the front femur is possibly a ground-plan character of the Ephydroidea, but a similar posteroventral ctenidium on the mid femur is an autapomorphic character of the Diastatidae. The same character occurs in certain genera of the Ephydridae, e.g. males of *Notiphila* Fallén (Mathis 1979).

The presence of costal spines is a ground-plan character of the Ephydroidea, but the reduced size and broad spacing of these spines (as in Camillidae) is probably an autapomorphic condition of the Diastatidae.

Reduction in the anal region of the wing with ultimate reduction of the cell cu*p*, anal lobe, and alula, occurred repeatedly in the Ephydroidea. The consistently small size of cell cu*p* in all Diastatidae, and the tendency for the reduction of the anal lobe and alula (least reduced in *Euthychaeta*) are autapomorphic conditions of the family.

These autapomorphies, especially in combination with the other characteristics listed, confirm that the Diastatidae is a monophyletic unit.

Griffiths (1972) transferred *Diastata* to the Ephydridae mainly on the grounds that sclerotized spermathecae are absent and the ventral receptacle is sclerotized (as in

Ephydridae). Although I agree that these conditions in *Diastata* may foreshadow similar conditions in the Ephydridae, I believe they evolved separately as parallel developments. Sclerotized spermathecae have been lost repeatedly within the Acalyptratae, and, in the light of their much reduced condition in the Camillidae, their absence in *Diastata* and the Ephydridae is not surprising. Frequently their function is supplemented and apparently replaced by the ventral receptacle, an organ that is usually well developed in all Ephydroidea. Thus, the strongly sclerotized condition of the ventral receptacle in *Diastata* and the Ephydridae is perhaps less significant phylogenetically than was believed by Griffiths. Differences in the form (more or less coiled or C-shaped in Diastatidae, mushroom-shaped in Ephydridae) and in the position (anteriorly directed in Diastatidae, posteriorly directed in Ephydridae) generally support the idea of separate development in each family. For these reasons I do not agree either that *Diastata* is more closely related to the Ephydridae than to *Campichoeta* and *Euthychaeta* (Griffiths 1972), or that *Diastata* alone is the sister group of the Ephydridae (Cogan 1975). The common ancestor of both these families is still unknown.

The ground-plan conditions of the Diastatidae would seem more plesiomorphic than any other family of the Ephydroidea in two respects. First, it has no discrete humeral break in C (Hennig 1958). However, this costal break does occur in most members of the family, and even where it is said to be absent (in fossil genus, *Pareuthychaeta*, and in *Euthychaeta* and *Campichoeta*) a distinct attenuation in the C occurs at the point at which the humeral break normally occurs. It is possible that in these genera (as in certain Ephydridae) the humeral break is secondarily partially closed. If the semiclosed condition in these genera is truly plesiomorphic, a discrete humeral break must have developed several times within the Ephydroidea. Second, the articulation of the pedicel and first flagellomere seems more generalized in the Diastatidae and Ephydridae than in the remaining families of the Ephydroidea (Hennig 1971*a*). In the Diastatidae and Ephydridae, the proximal process of the first flagellomere is less deeply inserted into the apex of the pedicel. Hennig stressed the fact that in the Curtonotidae, Camillidae, and Drosophilidae this process penetrates deeply into the interior of the pedicel, and he used this feature as a synapomorphic character to distinguish his "Drosophilidea" from his "Ephydridea." If the family Ephydridae is in fact the plesiotypic sister group of the remainder of the Ephydroidea, it would be necessary to postulate that a complete and separate subcosta was regained in the Curtonotidae, which seems unlikely. Possibly the proximal process of the first flagellomere is less extensive in the ground plan of the Curtonotidae than was assumed by Hennig. The relative development of this structure in other genera of the Diastatidae and Ephydridae needs further investigation.

In conclusion, I agree with Hennig (1958, 1971*a*) that the Diastatidae and Ephydridae are sister groups (Fig. 116.7). As noted by Hennig, the reduced condition of cell cu*p* and the lateral position of the proclinate orbital bristle in relation to the reclinate orbital bristles are two synapomorphic characters that support this opinion. Hennig was wrong, however, in stating that both families lack sclerotized spermathecae, for sclerotized spermathecae occur in both *Campichoeta* and *Euthychaeta*, and therefore in the ground plan of the Diastatidae.

Ephydridae. Shore flies (Wirth *et al.*, Ch. 98) constitute one of the largest (more than 1300 species), most widely distributed, and best substantiated monophyletic subgroups of the Acalyptratae. Morphologically, at least, it is the most highly specialized family of the Ephydroidea. The accepted family name, based on *Ephydra* Fallén (Fallén 1810) was proposed as Ephydrinae by Zetterstedt (1837). In response to a proposal by Mathis (1981), the International Commission on Zoological Nomenclature (1985, Opinion 1321) declared Ephydridae as the official name of the family. The classificatory history is well summarized by Wirth *et al.* (Ch. 98).

Hennig (1958) enumerated 12 characters that he considered to be apomorphic ground-plan conditions of the family. Griffiths (1972) listed seven that he regarded as apomorphic with respect to the ground plan of the Ephydroidea, but his concept of the family was complicated by the inclusion of *Diastata* within the Ephydridae (see discussion under Diastatidae). I consider the following 15 characters of the Ephydridae apomorphic with respect to the ground plan of the Ephydroidea; those that are autapomorphic are marked (AA):

- fulcrum with a filter apparatus;
- anterior reclinate orbital bristle weak;
- proclinate orbital bristle arising laterally to reclinate orbital bristles;
- postocellar bristles absent (AA);
- subcranial cavity enlarged (AA);
- anepisternal bristles present;
- C with humeral break as well as subcostal break;
- Sc reduced apically and more or less fused with R_1;
- crossvein bm-cu absent, cells bm and dm confluent;
- veins CuA_2 and A_1 atrophied, hence cell cu*p* absent (AA);
- abdominal spiracles 2–5 within tergites (AA);
- abdominal spiracles 7 absent in both sexes;
- sclerites of abdominal segments 6–8 atrophied in male;
- sclerotized spermathecae absent (AA); and
- ventral receptacle heavily sclerotized (AA).

Probably the microphagous feeding habits of the larvae and adults (Wirth *et al.*, Ch. 98), especially their tendency to feed primarily on atrophic algal cells (as opposed to proclivities of the Drosophilidae to feed on heterotrophic yeast cells, Foote 1981), should also be cited as a specialized character of the family.

From this complement of apomorphies, including six autapomorphies, it is clear that the Ephydridae is a monophyletic unit. As discussed under the Diastatidae, it is probably the younger sister group of Diastatidae (Fig. 116.7).

Papp (1977*b*, 1980) considered the Risidae to be closely related to the Ephydridae, but important differences in the antennal structure, frontal bristling, and mouthparts preclude its inclusion in Ephydroidea (see discussion under Carnoidea).

No general phylogenetic treatment of the Ephydridae has been produced; the closest approach to such a study is Mathis' (1979) cladistic analysis of the Ephydrinae. Unfortunately, the four subfamilies currently recognized, i.e. the Psilopinae, Notiphilinae, Parydrinae, and Ephydrinae (Wirth *et al.*, Ch. 98), are not clearly defined in that analysis, so that the subfamilial positions of some genera are still uncertain. No complete world list of genera has been published, but Hennig (1973) outlined typical genera of each of the subfamilies.

A rough count from various sources shows that the family consists of about 140 recognized genera. At present, 52 genera and 250 species are known from the Ethiopian region (Cogan 1980*b*), 43 genera and 120 species from the Oriental region (Cogan and Wirth 1977), 60 genera and 466 species from the Palearctic region (Cogan 1984), 68 genera and 425 species from the Nearctic region (Wirth *et al.*, Ch. 98), and 68 genera and 320 species from the Neotropical region (Deonier 1979, based mainly on Wirth 1968). No figures are available for the Australian region, but 11 genera and 24 species are known from New Zealand (Harrison 1959). No fossils are known.

GROUND-PLAN CHARACTERS AND MONOPHYLY OF THE CALYPTRATAE

As stated earlier, the monophyly of the Calyptratae, including the Hippoboscoidea, has been confirmed increasingly by many workers (see especially Roback 1951; Hennig 1958, 1965*a*, 1971*a*; Griffiths 1972). Of 128 ground-plan characters listed for the Calyptratae (see Table 116.2), 48 occur in an apomorphic condition relative to that found in the ground plan of the Acalyptratae (Table 116.2) and the Schizophora as a whole (see Table 116.1). This number in itself is ample verification of the monophyly of the Calyptratae.

SUBORDINATE GROUPS OF THE CALYPTRATAE

The question of how the Calyptratae should be subdivided is still unsettled. Proposals offered vary from author to author, depending on the characters they employ and the importance they attribute to them. The major steps in the classification of the group were well summarized by Hall (1948), Roback (1951), Singh and Judd (1966), and Hori (1967). From an evolutionary standpoint, perhaps the most important contributions are the works of Hennig (1955, 1958, 1965*a*, 1971*a*, 1976*a*,*b*), Herting (1957), and Griffiths (1972). These papers are referred to repeatedly in the following discussion and should be consulted by anyone interested in understanding the phylogeny of the Calyptratae.

In my opinion the Calyptratae consists of three monophyletic superfamilies, the Hippoboscoidea (= Glossinoidea of Hennig 1971*a*, 1973) containing the families Glossinidae, Hippoboscidae, Streblidae, and Nycteribiidae; the Muscoidea containing the Scathophagidae, Anthomyiidae, Fanniidae, and Muscidae (including Eginiini); and the Oestroidea containing the Calliphoridae, Mystacinobiidae, Sarcophagidae, Rhinophoridae, Tachinidae, and Oestridae *sensu lato* (Fig. 116.8).

The semiapterous cave species, *Mormotomyia hirsuta* Austen (Austen 1936) (Mormotomyiidae), was incorrectly referred to the Calyptratae by van Emden (1950) and Hennig (1958, 1971*a*, 1973). Hennig (1973) treated it as a possible sister group of all remaining Calyptratae and ranked it as a separate superfamily, the Mormotomyioidea. I agree with Griffiths (1972) that Mormotomyiidae is definitely a member of the Acalyptratae, but I have placed it in the Sphaeroceroidea rather than in the Tephritoidea (for detailed discussion, see under Mormotomyiidae).

Superfamily Hippoboscoidea

The classification of the Hippoboscoidea (Fig. 116.8) has received much attention from the beginning of dipterology, but many aspects of its morphology and systematics are still controversial. Key works, for anyone taking up the subject now, are those by Bequaert (1953, 1954–1957), Griffiths (1972, 1976), Hennig (1941, 1965*a*, 1971*a*), Newstead *et al.* (1924), Potts (1973), Schlein (1970), Schlein and Theodor (1971), Theodor (1963, 1967, 1975), Wentzel (1976), and Zaka-ur-Rab (1979).

The Hippoboscoidea, as a group, is more generalized in some respects than any other group of the Calyptratae. For example, in the ground plan (usually best exemplified in the Glossinidae), abdominal tergite 6 of the male is almost as long and as unmodified as tergite 5, spiracles 6 and 7 never occur together in the pregenital sclerite, the surstyli are not closely articulated with the cerci, sternite 10 is not divided into discrete bacilliform sclerites, and the gonopods are solidly fused with the hypandrium (possibly a pseudoplesiomorphic feature). In the female, abdominal segments 6 and 7 are relatively unmodified and are not retractile, and spiracle 7 is not shifted forward as in most Calyptratae. In both sexes vibrissae are undifferentiated (possibly a pseudoplesiomorphic condition). For these reasons this superfamily is considered to be the sister group of the remainder of the Calyptratae (Muscoidea and Oestroidea, see Fig. 116.8).

The ground plan of the Hippoboscoidea includes the following autapomorphic characters (synapomorphic characters of the families of the Hippoboscoidea) with respect to the ground plan of the Calyptratae:

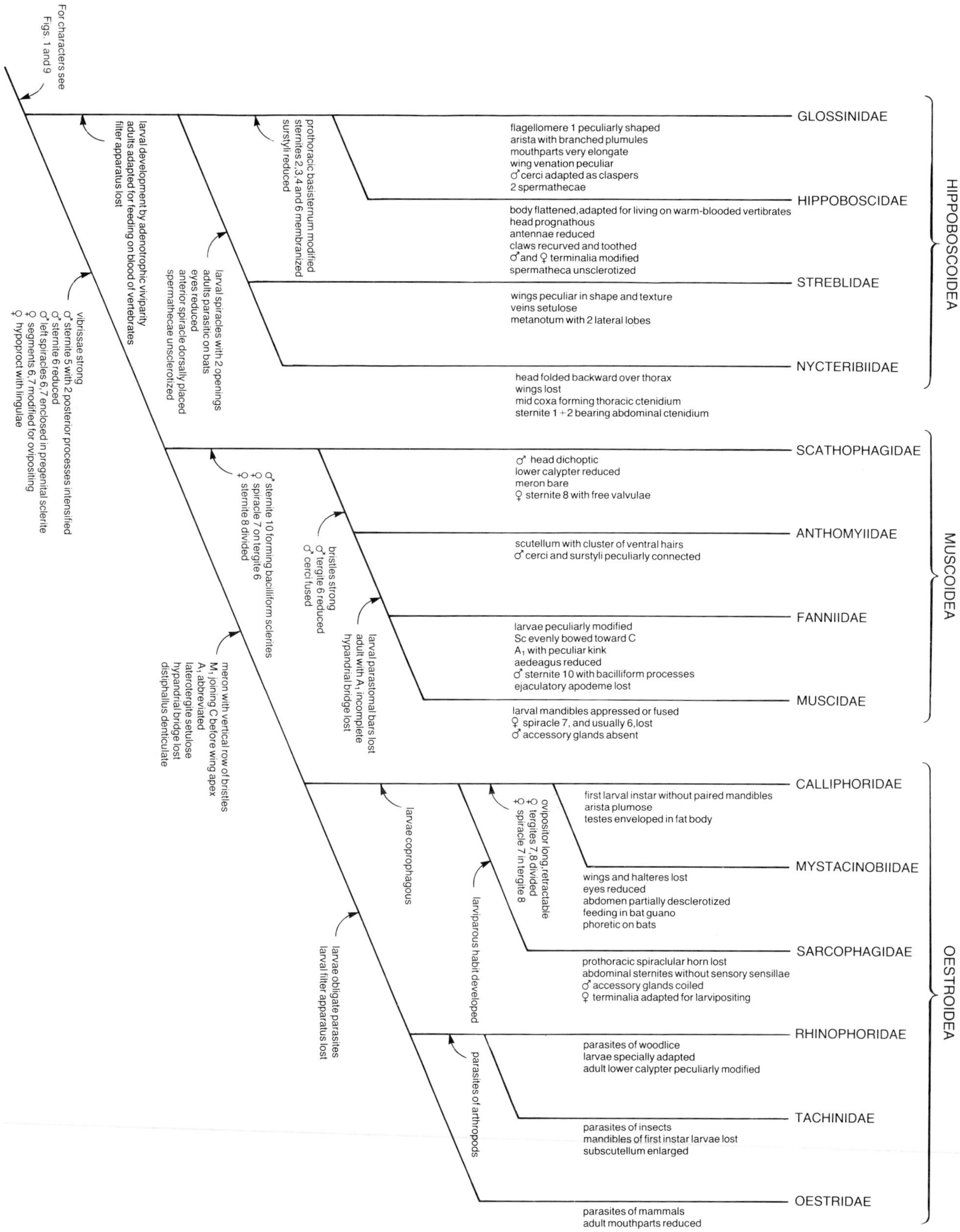

Fig. 116.8. Cladogram of the Calyptratae summarizing the relationships and apomorphies of the superfamilies, subgroups, and families recognized. (See text for additional details.)

- free larval stage suppressed;
- larvae feeding to maturity on glandular secretions in female abdomen (adenotrophic viviparity);
- larvae without pharyngeal filter;
- larvae without salivary glands;
- larvae with greatly reduced cephalopharyngeal skeleton;
- larvae metapneustic, i.e. with anterior spiracle absent;
- larvae with no direct connection between mid and hind gut;
- pupae with outer spiracular process of each anterior spiracle not penetrating puparium (Bequaert 1953);
- adults feeding exclusively on fresh, fluid blood of vertebrates by sucking blood from living hosts;
- males dichoptic, holoptic condition lost;
- females with internal abdominal secretory (milk) glands and associated structures for rearing larvae internally;
- arista with branched plumules;
- proboscis peculiarly adapted for piercing and blood-sucking, i.e. with bulb-like basal swelling and needle-like apical section;
- salivary pump absent;
- palpi modified to form sheath for proboscis;
- postocellar bristles reduced;
- outer vertical bristles absent;
- prosternum (presternum and basisternum) largely membranized (Hennig 1965*a*);
- basalar apodeme and anapleural (episternal) suture reduced (Schlein 1970);
- transverse suture of scutum complete;
- metabasisternum expanded and exposed, with double floor (Hennig 1941, Schlein 1970);
- metepimera fused ventrally behind hind coxa forming neck between thorax and abdomen (Schlein 1970);
- vein A_1 reduced, not reaching wing margin;
- abdominal sternite 1 enlarged;
- abdominal spiracles 3–5 secondarily situated in membrane;
- epiphallus lost;
- ejaculatory apodeme reduced or lost;
- testes forming a pair of long, coiled tubules;
- female abdomen enlarged and extensively membranized;
- accessory glands modified to form secretory glands for nursing larvae;
- ovaries with reduced number of ovarioles; and
- ovaries dissimilar in size, alternating in production of mature ova.

Within the Hippoboscoidea are two monophyletic sister groups. One consists of the Glossinidae and Hippoboscidae, and the other consists of the Streblidae and Nycteribiidae (Fig. 116.8).

Most of the similarities between the Glossinidae and Hippoboscidae are symplesiomorphic including basic similarities in the structures of the head, thorax, and abdomen. Most workers agree with Bequaert (1954, p. 54) that "the Glossinidae are most closely related to the Hippoboscidae" and "that the two families show so many striking similarities, that the homologies of many external structures of the Hippoboscidae can most easily be traced by assuming that they are dorsoventrally flattened, *Glossina*-like flies."

Synapomorphic features of the Glossinidae and Hippoboscidae include the following:

- dorsal surface of tibiae specially adorned or modified;
- tergal branch of the depressor muscle of the midtrochanter absent;
- prothoracic basisternum with a median apodeme;
- abdominal sternites 2–4 and 6 membranized;
- males with surstyli reduced, nonfunctional as claspers; and
- aedeagal apodeme closely and securely joined to hypandrium.

Glossinidae. The free-living Glossinidae have retained more of the primitive muscomorphan features than any of the ectoparasitic Hippoboscoidea. More than any other family, the Glossinidae shows the primitive conditions for most of the adult characters of the Hippoboscoidea (as listed here, and see also Bequaert 1953 pp. 55–56). Only in *Glossina* Wiedemann does the larva retain even the basal piece of the cephalopharyngeal skeleton, i.e. the tentoropharyngeal sclerites (Hennig 1952).

Much confusion exists in interpreting certain parts of the male terminalia of the Glossinidae and other families of the Hippoboscoidea, particularly the cerci and surstyli. In the Glossinidae, I agree with Hennig (1971*a*) that the cerci (called telomeres by Griffiths 1972) are enlarged and are peculiarly adapted for clasping. I also agree that the so-called adanalia are secondarily sclerotized areas on either side of the anus. In addition, I interpret the so-called edita as reduced surstyli, as suggested by Hennig; they connect, as in most Muscomorpha, posteriorly with the cerci and anteriorly with the hypandrium, through the intermediacy of sternite 10. In the Hippoboscidae, the processes called penis valves by Bequaert (1953), parameres by Theodor (1963), postgonites by Schlein and Theodor (1971) and Theodor (1975), and claspettes by Zaka-ur-Rab (1979), are the gonopods; true external parameres are present in the Glossinidae, but they are absent in the Hippoboscidae. The lateroventral processes that were called gonocoxites by Bequaert (1953), Seitenfortsatz by Theodor (1963), and praegonites by Schlein and Theodor (1971) and Theodor (1975) are, in fact, surstyli as was correctly indicated by Zaka-ur-Rab (1979). For correct interpretations of the male terminalia of the Hippoboscidae see Maa and Peterson (Ch. 111).

Autapomorphic characters of the Glossinidae include the following:

- first flagellomere elongate, dorsally concave in profile, and with a dorsoapical point;
- arista enlarged and flattened, bearing numerous, long, branched plumules on dorsal surface only;
- lunule setulose;

— reclinate orbital bristles absent;
— proclinate orbitals extending far forward, five in number;
— proboscis and palpi very elongate;
— sensillar areas of prosternum situated on a single plate;
— anapleural suture and its inner apodeme reduced (Schlein 1970);
— metabasisternum with lateral margins turned downward and forward, anterior end forming a cone that projects between mid coxae;
— vein M_{1+2} forwardly inclined from junction of crossvein dm-cu, and meeting C before wing apex;
— cell dm hatchet-shaped;
— cell cu*p* strongly concavely closed;
— tibiae and tarsi with a middorsal row of stout, closely placed setulae;
— male with abdominal sternite 7 and tergites 7 and 8 fused and forming a secondarily symmetric pregenital ring;
— cerci of male adapted as clasping processes;
— female with all abdominal sternites, including sternite 8, reduced and replaced by an elastic membrane; and
— spermathecae reduced to two.

Hippoboscidae. In a few respects, the Hippoboscidae (Maa and Peterson, Ch. 111) are more generalized than the Glossinidae. For example, the lunule is bare, the sensillar areas of the prosternum are still separate, and three spermathecae were sometimes retained, although they are not sclerotized (Bequaert 1953). Autapomorphic characters of the Hippoboscidae include the following:

— adults ectoparasitic, living continuously on host;
— adults feeding exclusively on warm-blooded vertebrates, exclusive of bats;
— body adapted for movement among hairs and feathers;
— head, thorax, and abdomen strongly dorsoventrally flattened;
— head retracted between anterior coxae;
— head prognathous;
— antennal scape strongly reduced, partially to completely fused with lunule;
— first flagellomere very short;
— face with deep antennal grooves;
— lunule bare;
— mesoscutum with a median suture;
— anterior thoracic spiracle dorsally situated (Webb 1945);
— legs laterally inserted;
— coxae broadly separated medially;
— mid coxal prong lost;
— tarsal segments flattened and shortened;
— claws recurved and toothed at base;
— wing veins concentrated in anterobasal area of wing blade;
— crossvein dm-cu absent;
— intermediate abdominal tergites and sternites extensively membranized;
— male with pregenital sclerites reduced; at most, represented by a dorsal plate (sternite 8);
— terminalia of male retracted into a deep genital pouch;
— hypandrium very elongate;
— parameres absent;
— cerci reduced;
— aedeagus elongate, remaining posteriorly directed at rest;
— male with four or more accessory glands; and
— female with two or three unsclerotized spermathecae (Bequaert 1953).

Synapomorphic characters of the Streblidae and Nycteribiidae are as follows:

— larval spiracles with two openings (posterior or third opening lost);
— adults ectoparasitic, living continuously on host;
— adults feeding exclusively on bats;
— body adapted for movement among hairs;
— compound eyes reduced;
— ocelli lost;
— antennal scape strongly reduced, fused with frons;
— first flagellomere very short;
— face with deep antennal grooves;
— lunule reduced;
— slender apical portion of proboscis consisting of elongated labella;
— first pair of cervical sclerites lost;
— sensillar areas of prothorax situated on a single plate;
— anterior spiracle dorsally situated;
— metanotum with two internal metanotal processes that project posteriorly into the abdomen (Schlein 1970);
— legs laterally inserted;
— coxae broadly separated medially;
— coxal spines commonly forming ctenidia;
— mid coxal prong lost;
— tarsal segments flattened and shortened (except basal tarsomere of some Nycteribiidae);
— claws recurved and toothed at base;
— lower calypter reduced;
— cerci of male reduced (probably replaced by adanalia);
— aedeagal apodeme rod-like, not closely associated with hypandrium; and
— spermathecae desclerotized.

A few of the ground-plan conditions of the Streblidae + Nycteribiidae are more plesiomorphic than those of the Glossinidae + Hippoboscidae. For example, the tergal branch of the depressor muscles for the mid trochanter is present, there is less membranization of the intermediate abdominal segments (segments 1 to 6 are relatively well developed in both sexes of Streblidae, e.g. species of *Nycterophilia* Ferris). The surstyli are well developed and adapted for clasping, as in Nycteribiidae, e.g. species of *Penicillidia* Kolenati, and the gonopods are distinctly separated from the hypandrium by a membranous area.

Streblidae. A number of characters in the ground plan of the Streblidae (Wenzel and Peterson, Ch. 113) are more plesiotypic than the same characters in the Nycteribiidae. For example, the head is not folded back on the dorsum of the thorax, the sclerotization of the dorsal surface of the thorax is relatively complete, the body is less flattened, (not at all flattened in *Nycterophilia* spp.), functional wings (though peculiarly specialized) are retained, and the segmental sclerites of the abdomen are more complete. For these reasons the Streblidae are considered to be the older sister group of the Nycteribiidae.

Autapomorphic characters of the Streblidae include the following:

- wing with peculiar shape, venation, and texture;
- veins setulose;
- C without humeral and subcostal breaks (pseudoplesiomorphy);
- Sc incomplete, fused with base of R_1;
- basal crossveins (r-m, bm-cu, and CuA_2) shifted far outward in wing blade;
- cell dm open (crossvein du-cu absent);
- axillary sclerite 3 with club-shaped projection (Schlein 1970);
- metanotum with two lateral lobes, externally; and
- spermathecae absent.

Nycteribiidae. Autapomorphic characters of the Nycteribiidae (Peterson and Wenzel, Ch. 112) include the following:

- head folded backward; in resting position with dorsal surface adjacent to mesoscutum;
- thorax membranized in middle of dorsum;
- sclerotization and musculature of thorax drastically modified (Schlein 1970);
- legs inserted dorsally;
- wings absent;
- mid coxae with a grasping organ, the thoracic ctenidium; and
- abdominal sternites 1 and 2 fused, bearing abdominal ctenidium.

Certain peculiar similarities in the Hippoboscidae, Streblidae, and Nycteribiidae, e.g. ectoparasitic habits of adults, flattened bodies, drastically modified sclerites of head and thorax, spider-like orientation of legs, and toothed tarsal claws, are sometimes suggested as synapomorphies that indicate monophyly. It now seems probable that such similarities between the Hippoboscidae, on the one hand, and the Streblidae + Nycteribiidae, on the other hand, are convergent adaptations to similar ways of life.

The sister group of the Hippoboscoidea is the Muscoidea + Oestroidea (Fig. 116.8). The monophyly of these two superfamilies is demonstrated by the following characters, which are autapomorphic (= synapomorphies of the Muscoidea and Oestroidea) with respect to the ground plan of the Calyptratae:

- vibrissae well developed;
- male with abdominal sternite 6 reduced, retracted anteriorly to lie above sternite 5, asymmetric, shifted to left side, and partially fused with syntergosternite 7 + 8;
- left abdominal spiracles 6 and 7 enclosed in pregenital sclerite on left side;
- surstyli closely connected with cerci (Hennig 1976*a*);
- female with abdominal segments 6 and 7 modified for oviposition, fully retractible within preceeding segments; and
- hypoproct with lingulae (Herting 1957).

Superfamily Muscoidea

The name Muscoidea has probably been used in a wider variety of senses than any other suprageneric name in Diptera. The most important of these usages were reviewed by Griffiths (1972) and need not be repeated here. I restrict the name to a superfamily, in the usual sense, within the Calyptratae as proposed by Roback (1951) and as adopted by Hennig (1973). As indicated previously, it contains the families Scathophagidae, Anthomyiidae, Fanniidae, and Muscidae (including Eginiini).

The superfamily Muscoidea (Fig. 116.8), by virtue of primitive conditions found in the Scathophagidae and Anthomyiidae, is more generalized in its ground plan than is the superfamily Oestroidea (Roback 1951). Most important, the meron although sometimes bearing hairs and bristles, does not have a vertical row of bristles as in the ground plan of the Oestroidea. In addition, the apical section of vein M is relatively straight and ends well behind the wing tip, vein A_1 continues to the wing margin, and the cerci are not rigidly fused.

Autapomophic characters in the ground plan of the Muscoidea (= synapomorphic characters of the Scathophagidae, Anthomyiidae, Fanniidae, and Muscidae, include the following:

- male with anus situated above bases of cerci (Hennig 1976*a*;
- sternite 10 divided along midline, forming discrete bacilliform sclerites;
- female with abdominal segments 6 and 7 relatively strongly differentiated from preceding segments;
- abdominal spiracle 7 shifted forward onto tergite 6; and
- abdominal sternite 8 divided.

Scathophagidae. Within the Muscoidea, the family Scathophagidae (Vockeroth, Ch. 103) has retained more plesiomorphic conditions in certain ground-plan characters than any other subordinate group. For example, the bristling of the head, thorax, and abdomen is weaker and sparser (occiput with pale hairs only, cruciate interfrontal absent, and katepisternum with one bristle); abdominal tergite 6 of the male is but little shorter than tergite 5; the cerci are separate and reciprocally movable; the sclerotized hypandrial bridge, behind the basiphallus and between the hypandrial arms, is still complete; and each bacilliform sclerite is discrete, i.e. distinctly articulated with both the surstylus and the hypandrial arm. Autapomorphic characters of the Scathophagidae are as follows.

- head equally dichoptic in both sexes;
- lower calypter greatly reduced;
- meron bare;
- ventral surface of scutellum bare;
- male without accessory glands (Hori 1960);
- sternite 8 of female with free valvulae (Herting 1957);
- adult predaceous on insects and other invertebrates (Vockeroth, Ch. 103); and
- mating taking place while both sexes are grounded.

The monophyly of the group, Anthomyiidae + Fanniidae + Muscidae, is demontrated by the following autapomorphic ground-plan characters (synapomorphies of included taxa):

- bristles of head, thorax, and abdomen numerous and strong (occipital bristles strongly differentiated, cruciate interfrontals present, and katepisternum with two or three bristles);
- male with abdominal tergite 6 much shorter than tergite 6;
- cerci fused, operating as a single unit; and
- acrophallus reduced (Hennig 1976*a*).

Plesiomorphically, the holoptic condition in males was retained in the ground plan of this group, but it was lost independently in various branches in many subordinate groups.

Anthomyiidae. The family Anthomyiidae (Huckett, Ch. 104) is the sister group of the remainder of the Muscoidea. The following plesiomorphic conditions were retained in its ground plan, but were lost in the Fanniidae + Muscidae:

- larvae with parastomal bars present (Roback 1951);
- dorsal cornu flared and with window present (Roback 1951);
- adults with vein A_1 continuing to wing margin; and
- ventral surface of scutellum with hairs.

Autapomorphic characters of the Anthomyiidae are as follows:

- scutellum with a cluster of fine, erect hairs on ventral surface near apex;
  The presence of fine, scattered, more of less decumbent hairs on the ventral margins of the scutellum is judged to be a plesiomorphic condition of the Muscomorpha, but the relatively constant position, erect condition, and clustered disposition of these hairs in the Anthomyiidae is the clearest autapomorphic ground-plan feature of the family; these hairs were lost several times within the family.
- base of tarsomere 1 of hind tarsus with an outstanding ventral bristle (A. Pont, personal commun.); and
- cerci and surstyli connected by a complex membrane-fold articulation (Hennig 1976*a*).

The monophyly of the Fanniidae + Muscidae is shown by the following autapomorphies (synapomorphies of the Fanniidae and Muscidae):

- larvae with parastomal bars lost;
- dorsal cornu unflared and without windows;
- vein A_1 abbreviated, not reaching hind margin of wing (except in some Eginiini);
- male with each surstylus linked directly to hypandrial arm, i.e. without discrete bacilliform sclerites; and
- hypandrial bridge absent, hypandrial arms widely separated behind basiphallus.

Fanniidae. The Fanniidae (Huckett and Vockeroth, Ch. 105) are more generalized than the Muscidae in at least three respects: the larval mandibles have remained separate; in the adults, the female has retained abdominal spiracles 6 and 7; and the male has retained a pair of accessory glands (Hori 1960).

The monophyly of the Fanniidae is abundantly established. The following list of ground-plan autapomorphies is based mainly on evidence presented by Herting (1957), Chillcott (1960), and Hennig (1965*a*):

- larva with body flattened and with peculiar, usually branched processes;
- posterior spiracles on raised processes;
- adult with proclinate orbital bristle absent in female;
- frontal plate of female broadened;
- apical portion of vein Sc evenly bowed toward vein C;
- anepimeron and meron bare;
- ventral surface of scutellum bare;
- vein A_1 peculiarly arched posteriorly, thence anteriorly;
- mid tibia of male thickened on distal half and densely pubescent on ventral surface;
- hind tibia with a mid dorsal bristle, the calcar;
- gonopods reduced, fused into hypandrium;
- aedeagus reduced, posteriorly directed;
- sternite 10 peculiarly modified, bearing "bacilliform processes," not forming elongate bacilliform sclerites, and not directly connected to hypandrium;
- surstylus connected to hypandrial arm by extension from anteromedial base of surstylus (*vice versa* in Muscidae);
- epiphallus lost; and
- ejaculatory apodeme absent (Griffiths 1972).

Fanniidae is frequently considered as a subfamily of the Muscidae (Chillcott 1960; Fonseca 1968; Hennig 1955–1964, 1965*a*, 1973); indeed, it is treated thus by Huckett and Vockeroth in Chapter 105 of this Manual. However, I agree with Roback (1951), Griffiths (1972), and Pont (1977) that ranking the group as a full family alongside the Scathophagidae, Anthomyiidae, and Muscidae is warranted.

Muscidae. The family Muscidae (Huckett and Vockeroth, Ch. 105), including the Eginiini, retained plesiomorphic counterparts of all the autapomorphic characters listed for the Fanniidae (Hennig 1965*a*). At the same time, the family has a number of autapomorphic conditions as follows:

- larva with mandibles closely appressed or fused (Roback 1951);

— dental sclerites fused ventrally (Hennig 1965*a*);
— adult female with abdominal spiracles 7, and usually 6, absent (Herting 1957);
— male with each surstylus connected to hypandrium by an extension from posterior extremity of hypandrial arm (*vice versa* in Fanniidae);
— accessory glands absent (Hori 1960); and
— the primarily predaceous life habits of the larvae is also an autapotypic ground-plan character of the Muscidae (Roback 1951).

The Eginiini have been considered to represent a distinct family (Eginiidae) (Stackelberg 1969, 1970), but their systematic position has not been satisfactorily resolved (Griffiths 1972; Hennig 1971*a*, 1973, 1976*a*). The group has frequently been referred to the Oestroidea because the meron (= hypopleuron) is either hairy or setose, or both, in most of its members. A hairy meron is a plesiomorphic condition present in the ground plan of the Schizophora. When present in the Eginiini these hairs or setae are relatively weak and are arranged in a more or less horizontal cluster near the upper margin of the meron. Similar vestiture occurs in some Anthomyiidae, e.g. species of *Eremomyioides* Malloch, and in some Muscidae, e.g. species of *Dichaetomyia* Malloch and *Phaonia* Robineau-Desvoidy, and other forms, and it should not be confused with the vertically aligned row of bristles found in almost all Oestroidea. All Eginiini also agree with *Dichaetomyia* in having bristles on the rim of the posterior thoracic spiracle.

I believe that the Eginiini belong to the Muscidae as indicated by Herting (1957), Griffiths (1972), and Pont (personal commun.), and as accepted by Huckett and Vockeroth (Ch. 105). The most compelling reason for placing the Eginiini in the Muscidae is the absence of abdominal spiracles 6 and 7 in the females, a virtually unique autapomorphy of the Muscidae (Herting 1957). In addition, the male terminalia are relatively small, with the basiphallus and distiphallus relatively weakly sclerotized and differentiated, the gonopods are reduced and fused with the hypandrium, and the arms of the hypandrium are connected directly to the bases of the surstyli without intervening bacilliform sclerites. These characters also agree with conditions in the Muscidae, but not with those in the Anthomyiidae or Oestroidea. Other important, but symplesiomorphic points of agreement between the Eginiini and Muscidae are as follows: vein M relatively straight and ending behind wing apex (not sharply bent forward and ending before wing apex as in Oestroidea), and infrasquamal setulae absent (present in Calliphoridae and related families). In addition, the base of tarsomere 1 of the hind tarsus is without an outstanding ventral bristle (present in Anthomyiidae).

The weakly notched to completely undivided abdominal sternite 8 in females of the Eginiini may be an autapomorphic adaptation (Hennig 1976*a*) associated with the peculiar eggs (large and flattened) and ovipositing habits of the group (on millipeds). The position of the anus of the male is between the bases of the cerci, rather than above them. However, this character shows considerable variation within the Muscoidea and Oestroidea, and its significance in the Eginiini is uncertain.

A complete vein A_1 in some Eginiini (Herting 1957) might suggest that the group belongs to the Anthomyiidae, or at least might exclude it from the Muscidae (Hennig 1973). However, this condition is plesiomorphic in both the Scathophagidae and Anthomyiidae, and its presence in the Eginiini cannot be taken as evidence of a close relationship between that group and the Anthomyiidae. It is uncertain whether a complete vein A_1 is a ground-plan character of the Eginiini. In specimens of *Eginia ocypterata* (Meigen) examined, it is incomplete in the male and complete in the female; it is incomplete in both sexes of *Syngamoptera armurensis* Schnable, but it is complete in both sexes of an unnamed species of *Xenotachina* Malloch. The possibility that it is incomplete in the ground plan of the Eginiini, and that it is secondarily lengthened (pseudoplesiomorphy) in some members, should be considered.

It has been stated that abdominal spiracles are absent in the Eginiini (Hennig 1971*a*). According to my observations seven pairs are present in the male, and five pairs in the female, of *E. ocypterata*. At least five pairs are present in the male, and five in the female also, of *S. armurensis*. The male and female of *Xenotachina* also appear to have five pairs. It is quite possible, however, that some of these spiracles are nonfunctional (see Griffiths 1972).

In summary, there is no convincing evidence for a close relationship between the Eginiini and the Oestroidea or Anthomyiidae, but there are several strong indications that the group is related to certain Muscidae, particularly to those of the subfamily Phaoniinae. Skidmore (1985) treated it as one of ten separate subfamilies that he recognized within the Muscidae.

Superfamily Oestroidea

According to the *International Code of Zoological Nomenclature*, (Ride, Sabrosky, *et al.* 1985), the oldest family-group name included in a superfamily should be used for the name of the superfamily. Accordingly, the name Oestroidea is used for the group of families that have a vertically aligned row of bristles on the meron, i.e. the Calliphoridae, Mystacinobiidae, Sarcophagidae, Rhinophoridae, Tachinidae, and Oestridae (Fig. 116.8). Use of other names such as Calliphoroidea (Hennig 1958, 1973) and Tachinoidea (Rohdendorf 1977) for the same group contravenes the code.

Autapomorphic characters in the ground plan of the Oestroidea (synapomorphic characters of the included families) are as follows:

— meron with a vertical row of bristles;
— anepimeron with bristles;
— laterotergite with hairs or setulae;

— vein M_1 forwardly deflected, joining C before wing apex;
— vein A_1 not attaining wing margin; and
— male terminalia without a hypandrial bridge, i.e. hypandrium open behind basiphallus.

Some characters in the ground plan of the Oestroidea are more plesiomorphic than they are in the ground plan of the Muscoidea. In the male, sternite 10 is still incompletely divided, i.e. discrete bacilliform sclerites are absent, e.g. in some Calliphoridae (especially in the Angioneurini) and in the Mystacinobiidae (but sternite 10 is divided into discrete bacilliform sclerites within the Calliphoridae, and in the ground plan of most of the remaining families of the Oestroidea). In addition, the anus of the male is still situated between (not above) the cerci (Hennig 1976*a*). In the female, abdominal spiracle 7 is not shifted forward to tergite 6 (but this migration has occurred repeatedly within the Oestroidea), and sternite 8 is undivided.

The Oestroidea (Fig. 116.8) appears to consist of two monophyletic subgroups. The more generalized group contains the Calliphoridae, Mystacinobiidae, and Sarcophagidae; the more specialized group contains the Rhinophoridae, Tachinidae, and Oestridae. In the ground plan of the first subgroup, the larvae retained a filter apparatus, but they became primarily coprophagous. This habit probably involved changes in digestive enzymes and physiology and is considered apotypic in relation to the plesiotypic saprophagous habits in the ground plan of the Calyptratae (Roback 1951). Various more highly evolved clades within this subgroup have become sarcophagous and parasitic (facultative and obligate), and in these cases the filter apparatus was lost (Dowding 1967). In the ground plan of the second group, the larvae are obligate parasites of animals and, in all cases, the filter apparatus was lost. This development is clearly apotypic with respect to the ground plans of the first subgroup and of the Calyptratae. Important changes must have occurred in the digestive systems of both subgroups. For instance, the larval crop in the Calliphoridae is large and stalked (Dowding 1967; Hennig 1973).

Calliphoridae. Almost certainly, Calliphoridae (Shewell, Ch. 106) is the most generalized family of the Oestroidea (Roback 1951), and, consequently, it is a key family for reaching an understanding of the evolution and phylogeny of the remainder of the Calyptratae. It is a large and varied group, and many details required for a synthesis of this kind are still unknown. A few excellent analyses have been made, such as Hall's (1948) systematic treatment of Nearctic blowflies and Salzer's (1968) study of the male terminalia of *Calliphora erythrocephala* (Meigen), but much comparative work throughout the family is still needed. Consequently, evolutionary proposals made here should be regarded as provisional.

Plesiomorphic characters in the ground plan of the Calliphoridae include most of those features listed for the ground plan of the Calyptratae, except for the changes considered as autapomorphies in the ground plan of the Muscoidea + Oestroidea. The family appears to have stemmed from an anthomyiid-like ancestor that acquired a vertical row of setae on the meron, a cluster of infrasquamal hairs on the laterotergite, a sharply bent vein M_1 and a reduced vein A_1. Like the Anthomyiidae, it had hairs on the under surface of the scutellum, as in *Opsodexia* Townsend and related genera, but these hairs were not clustered near the apex of the scutellum and erect as in the Anthomyiidae. Also, like the Anthomyiidae, abdominal spiracle 7 of the female was shifted forward onto tergite 6; this adaptation appears to have occurred independently several times in the Oestroidea. Unlike the Anthomyiidae, sternite 10 of the male remained incompletely divided along the midline, as in most if not all Angioneurini. In the female, abdominal sternite 8 remained undivided as in the ground plan of the Calyptratae.

Although subordinate groups within the Calliphoridae are relatively well marked, the problem of establishing the monophyly of the entire family presents the same difficulties as do other taxa throughout the order that are characterized by relatively generalized conditions. A paucity of autapomorphic characters is characteristic in such groups, and the Calliphoridae is no exception. Hence, the possibility that the group is paraphyletic (Hennig 1973) should be recognized.

The following characters are tentatively proposed as autapomorphic ground-plan features of the Calliphoridae:

— larval food habits coprophagous (and probably sarcophagous) involving changes in digestive enzymes and physiology (Roback 1951);
— first instar larvae without paired mandibles (Hennig 1973);
— dorsal cornu of larva without a window or cleft (Roback 1951);
— arista plumose;
— intrapostocular setulae present (lost in Rhiniini, and occasionally in Angioneurini and *Onesia* Robineau-Desvoidy (Hall 1948);
— prosternum setulose (lost in Polleniini and Angioneurini);
— thoracic spiracles large;
— posterior thoracic spiracle occluded by one or a pair of complex shutters (Crosskey 1977);
— female with abdominal spiracle 7 situated in tergite 6;
— female with abdominal tergites 7 and 8 divided;
— male with anal lobes present in perianal membrane (Salzer 1968); and
— testes enveloped in fat body (Hori 1960).

Most authors divide the family into about five subfamilies (Hall 1948; Crosskey 1965; Hennig 1973; Shewell, Ch. 106), i.e. the Mesembrinellinae, Rhiniinae, Chrysomyiinae, Ameniinae, and Calliphorinae (including the Polleniini). Guimãraes (1977) raised the Neotropical subfamily Mesembrinellinae to family status, but this move has not been generally accepted. Rognes (1986) transferred the genus *Helicobosca* Bezzi from the Sarcophagidae to the Calliphoridae where he assigned it to a new subfamily, the Helicoboscinae.

Mystacinobiidae. The family Mystacinobiidae was erected for a wingless New Zealand fly, *Mystacinobia zelanica* Holloway (Holloway 1976*a*). Both adults and larvae of this peculiar species live in the guano of an endemic, short-tailed bat, *Mystacina tuberculata* Gray, which lives in colonies in hollow trunks of giant kauri trees (*Agathis australis*). Holloway placed *M. zelanica* in the Acalyptratae, mainly on the basis of its large ptilinum and the shape and dehiscence of the puparial cap; she referred it to the Ephydroidea (as Drosophiloidea), based on the following characters of the adult and immature stages:

— labella with seven pseudotracheae in each labellar lobe;
— tarsal segments elongate;
— acropod truncate at apex, with well-developed unguitractor plates;
— all tibiae with a preapical dorsal bristle;
— fore tibia with cleaning brush;
— vibrissae present;
— proclinate fronto-orbital bristles present;
— aedeagus short;
— base of first flagellomere broadly inserted into apex of pedicel;
— pedicel with a longitudinal cleft;
— egg with respiratory horns;
— mature larva with elongate, multiporous, anterior spiracles and tubular, contiguous, posterior spiracles; and
— larvae with five pairs of anal papillae.

Although these characters agree with conditions found in the Ephydroidea, none of them can be accepted as true synapomorphies. Closer analysis of the characters of *M. zelanica* show that its placement in the Ephydroidea, and consequently in the Acalyptratae, cannot be sustained. The following combination of characters indicate that it is, in fact, a member of the Calyptratae:

— dorsal seam of pedicel complete;
— medioclinate frontal bristles arising medially to origins of the orbital bristles;
— prestomal teeth present;
— abdominal spiracles situated in margins of tergites;
— male with abdominal sternite 5 bilobate, and with sternite 6 lying above it;
— hypandrial bridge absent, i.e. hypandrial arms not fused behind aedeagus; and
— surstyli closely linked with cerci.

All these characters are autapomorphic ground-plan characters of the Calyptratae; none of them occur in the ground plan of the Acalyptratae.

Despite its highly modified adaptation, *M. zelanica* shows all the autapomorphic characters of the Oestroidea (listed earlier), with the exception of those relating to the wings. The presence of a vertical row of bristles on the meron is particularly significant, especially when considered in conjunction with all the other oestroid characters.

Within the Oestroidea, *M. zelanica* shows more affinities with the Calliphoridae than with any other family from the standpoints of both biology and morphology. It is unlikely to be closely related to the Sarcophagidae because the female is oviparous, and it is unlikely to be related to any of the parasitic families because the larvae are coprophilous. The elongate ovipositor of the female, with tergites 7 and 8 divided, and spiracle 7 situated in tergite 6 is the same as in the Calliphoridae. In the male terminalia, the relatively elongate aedeagus with a spiculose distiphallus is similar to that in most Calliphoridae. Also, sternite 10 is undivided as in primitive Calliphoridae.

Because *Mystacinobia* Holloway has many autapomorphies in all stages (see Holloway 1976*a*) that set it apart from all Calliphoridae and from all other families of the Oestroidea, it seems best to treat it as a separate family within the Oestroidea. It probably arose from the line leading to Calliphoridae, rather than from the line leading to the remainder of the Oestroidea. The possibility that it is a sister group or subgroup of the Calliphoridae and, therefore, a peculiar member of that family as recently proposed by Griffiths (Griffiths 1982), cannot be definitely excluded.

Sarcophagidae. Compared with the Calliphoridae, the Sarcophagidae comprise a relatively homogeneous family. It consists of two subfamilies, the more generalized Miltogramminae and its more specialized sister group, the Sarcophaginae (W. L. Downes 1955, 1958; Shewell, Ch. 108). As in the Calliphoridae, the groundplan nutritive habit for the family is coprophagous, with a strong tendency toward sarcophagous habits. Besides having entered the coprophagous and insect parasitism zones, both these families have entered three new zones; the sarcophagous, the facultative parasitism of animals, and the obligate parasitism of animals (Roback 1951). In a few aspects the Sarcophagidae are more generalized in their ground plan than are the Calliphoridae. For example, first-instar larvae have paired mandibles (single in Calliphoridae), and mature larvae have retained a window (dorsal cleft) in the dorsal cornu (absent in Calliphoridae). Hennig (1973) indicated that the presence of paired mandibles in the first-instar larvae is probably a derived condition, but I agree with Downes (1955) that it is a plesiomorphic condition retained in the Sarcophagidae (and in the Rhinophoridae and Oestridae). In the female, abdominal tergites 7 and 8 are undivided (divided in Calliphoridae, and abdominal spiracle 7 is situated in tergite 7 (Herting 1957) (moved forward to tergite 6 in all Calliphoridae). It is true that, in parallel with most Calyptratae, abdominal spiracle 7 of the female is situated in tergite 6 in most Sarcophagidae. But the well-developed condition of tergite 7 containing spiracle 7, as exists for example in species of *Macronychia* Rondani, provides convincing evidence that this condition occurs in the ground plan of the family (W. L. Downes 1958).

Autapomorphic characters in the ground plan of the Sarcophagidae, include the following:

— all species larviparous;
— labral sclerite probably absent;
— prothoracic spiracular horn not protruding through puparium (Downes 1955);
— posterior spiracles of mature larva and puparium deeply recessed;
— arista bare on at least apical half;
— abdominal sternites without sensory sensilla (Downes 1955);
— male accessory glands coiled (Hori 1967);
— females with a series of adaptations for larvipositing;
— ovipositor short, genital opening large (Hori 1961);
— uterus expanded ventrally and posteriorly to form a bilobed incubatory diverticulum;
— uterovaginal tube unusually wide (Hori 1961);
— dorsal wall of genital chamber with sclerotized sigma and accessory plates (Hori 1961); and
— ovarian ducts and common oviduct peculiarly looped (Hori 1961).

These factors together indicate that the Sarcophagidae may be the sister group of the Calliphoridae + Mystacinobiidae (Fig. 116.8). As stated by Roback (1951), "the evidence seems to point to a development of the Sarcophagidae from a primitive calliphorid-like stock."

As indicated previously, the second more specialized subgroup of the Oestroidea consists of the Rhinophoridae, Tachninidae, and Oestridae. As far as is known, all species of the group are obligate parasites of other animals. This habit is presumed to be a primary, autapotypic adaption of the subgroup (synapotypic character of the component families). Other autapomorphic characters in its ground plan include the following:

— larvae without a pharyngeal filter;
— larval cephalopharyngeal skeleton without a parastomal bar;
— dorsal cornu without a window;
— puparium fully inflated, with both ends more or less equally hemispherical; and
— subscutellum, more or less developed.

The following plesiomorphic characters, also present in the ground plan of the calliphorid + sarcophagid subgroup were retained in the ground plan of the rhinophorid + oestrid subgroup:

— first larval instar with labrum (median labral tooth) present;
— first larval instar with paired mandibles present;
— pupal respiratory horns penetrating puparium;
— oviparous reproduction;
— arista plumose;
— pleural sclerites (proepisternum, anepisternum, anepimeron, laterotergite and meron) setulose;
— abdominal sternites with sensory setulae present;
— male with abdominal tergite 6 well developed;
— female with abdominal tergites 6, 7, and 8 well developed;
— abdominal spiracle 7 situated in tergite 7; and
— abdominal sternite 8 undivided.

The rhinophorid + oestrid subgroup appears to consist of two monophyletic sister groups, namely the Rhinophoridae + Tachinidae, and the Oestridae. All members of the rhinophorid tachinid subgroup are obligate parasites of Arthropoda; all members of the oestrid subgroup are obligate parasites of Mammalia. On this basis alone it is obvious that the rhinophorid tachinid line is more generalized than its counterpart.

Rhinophoridae. Crosskey's (1977) excellent review of the characteristics and classification of the Rhinophoridae has greatly enhanced the knowledge of this family. On the basis of his synthesis, the following ground-plan autapomorphies can be accepted:

— obligate parasites of woodlice (terrestrial isopods);
— first-instar larvae with specialized external adaptations (integument warty, and with scales or tubercles or pseudopod-like processes);
— labrum (median labral tooth) probably absent;
— mandibles of first-instar larvae short, deep and with two or more teeth;
— pharyngeal sclerite of larvae with anterior end extraordinarily long and slender;
— adults with prosternum and laterotergites bare;
— lower calypter widely removed from scutellum; and
— abdominal spiracle 7 of female situated in membrane between tergite 6 and 7.

Important plesiomorphic characters in the ground plan of the Rhinophoridae include the following:

— eggs deposited away from host;
— larvae emerge from eggs that incubate outside of the mother flies;
— first-instar larvae with two mandibles;
— adult mouthparts fully developed;
— arista plumose;
— prosternum bare;
— subscutellum not, or very slightly, convexly swollen;
— posterior thoracic spiracle subcircular, without distinct operculum, and margined with erect fringe of hairs;
— coxopleural stripe well developed;
— margins of abdominal tergite 2 overlapping margins of sternite 2;
— male with sternite 5 bifid, with large, lateral lobes;
— sternite 6 well developed, free;
— female with long, retractile ovipositor; and
— abdominal tergites 7 and 8 present and undivided.

As noted by Crosskey (1977), it is particularly significant that the rhinophorids appear to have a uniquely distinctive biology, combining the obligatory parasitic habit on an unusual host group with highly modified first-instar larvae emerging from eggs that incubate outside the mother flies. Nothing quite comparable is known to occur in any other group of flies. Because the first-instar larvae have two mandibles (absent in Tachinidae) and because the females lay unincubated eggs (usually incubated in Tachinidae), both of which are plesiotypic conditions, it seems reasonable to assume that Rhinophoridae is the older sister group of the Tachinidae.

Tachinidae. Tachinidae is probably the largest, most heterogeneous, and, therefore, most complex family of Diptera (Crosskey 1980). Paradoxically, the group is biologically discrete. Larvae of all members are obligate parasites of other insects (Hexapoda) and related arthropods (Myriopoda and Scorpionida). Moreover, they always kill their hosts. The invasion of the insect parasite zone by the entire family is clearly an apotypic adaptation. Other autapomorphic characters in the ground plan of Tachinidae are as follows:

— eggs without dorsal "hatching seams," larvae leaving eggs through softer ventral part of chorion (Hennig 1973);
— mandibles absent in first-instar larvae (Hennig 1973);
— labrum (median labral tooth) of first-instar larvae strongly developed for penetrating cuticle of host (Hennig 1973); and
— subscutellum of adult strongly and convexly developed.

The females of Tachinidae are basically oviparous, as in the ground plan of the Oestroidea, and even though the vast majority retain their eggs in a distensible oviduct or "ovisac" and deposit them openly in a fully incubated condition, a few species, presumably the most generalized ones, deposit unincubated eggs on the integument of the host (Wood 1979). Thus, deposition of unincubated eggs is the ground-plan condition of the family, as in the Rhinophoridae. Other plesiomorphic conditions for Tachinidae and related families were discussed by Richter (1980).

The taxonomy of the family is difficult and confused. Various systems of classification proposed depend largely on adult morphology and chaetotaxy, and on the points of view of the classifiers. At present, there is no agreed arrangement of subfamilies and tribes, but most specialists recognize from 4 to 6 subfamilies and 50 or more tribes. Wood (Ch. 110) provided references to the most useful taxonomic works; he adopted a system of four subfamilies, namely the Tachininae, Goniinae, Phasiinae, and Dexiinae, but the limits and components of each subfamily are not yet fully settled.

The family Stackelbergomyiidae was erected by Rohdendorf (1948) for a single eremic species, *Stackelbergomyia arenaria* Rohdendorf. He placed it near the Rhinophoridae, but contrary to the implication of the title of his paper, it is not known that the species is, in fact, parasitic (Crosskey 1977). A recent study of type material by Herting (Herting 1981) showed that it belongs to the Tachinidae (Phasiinae, tribe Catherosiini).

Oestridae. The family Oestridae, as defined in this Manual (Wood, Ch. 107), is comprised of four distinctive subfamilies, the Cuterebrinae, Gasterophilinae, Hypodermatinae, and Oestrinae. Some or all of these subfamilies are frequently treated as separate families (Grunin 1964–1969; Zumpt 1965; Hennig 1973; Papavero 1977). As stated by Wood (Ch. 107), "Phylogenetic studies have led some recent workers to regard them as a single family (Hennig 1952; Herting 1957; Downes 1958), or as three families that are more closely related to one another than to any other (Hennig 1973), hence monophyletic."

The most outstanding specialization of the family relates to the larval food habits, i.e. the larvae of all Oestridae are obligatory parasites of mammals. Other autapomorphic characters in the ground plan of the family include the following:

— first-instar larvae with one or two bands of spines on each segment;
— mature larvae with integument heavily armed with spines or platelets;
— adults with very stout bodies;
— adults with bristles undifferentiated, bodies pilose to densely long-haired;
— antenna small;
— mouthparts reduced;
— subcranial cavity reduced;
— crop and alimentary tract reduced (Singh and Judd 1966);
— rectal valve far removed from rectal sac (Singh and Judd 1966);
— male with abdominal sternite 5 small, simple, without deep apical cleft;
— tergite 6 combined with syntergosternite 7 + 8;
— aedeagus short, distiphallus scarcely differentiated;
— female with abdominal spiracle 7 situated in or near tergite 6; and
— sternite 8 partially to completely divided.

The phylogeny of the four subfamilies has not been resolved (Downes 1958), but Wood (Ch. 107) has provided a sound basis for understanding the evolutionary patterns involved.

SUMMARY AND CONCLUSIONS

The cladistic relationships of the 15 superfamilies and higher categories of the infraorder Muscomorpha, as discussed in detail in the foregoing text, are summarized in Fig. 116.9. The classification adopted for 82 families is presented in Table 116.10.

Two main sections, the Aschiza and Schizophora, are easily recognized. The more recent section, the Schizophora, is amply supported by autapomorphies (synapomorphies of the component subgroups), i.e. ptilinum, lunule, and ptilinal fissure well developed. The older section, the Aschiza, although readily defined on the basis of diagnostic (plesiomorphic) characters, is less strongly supported by autapomorphies. The most important of these

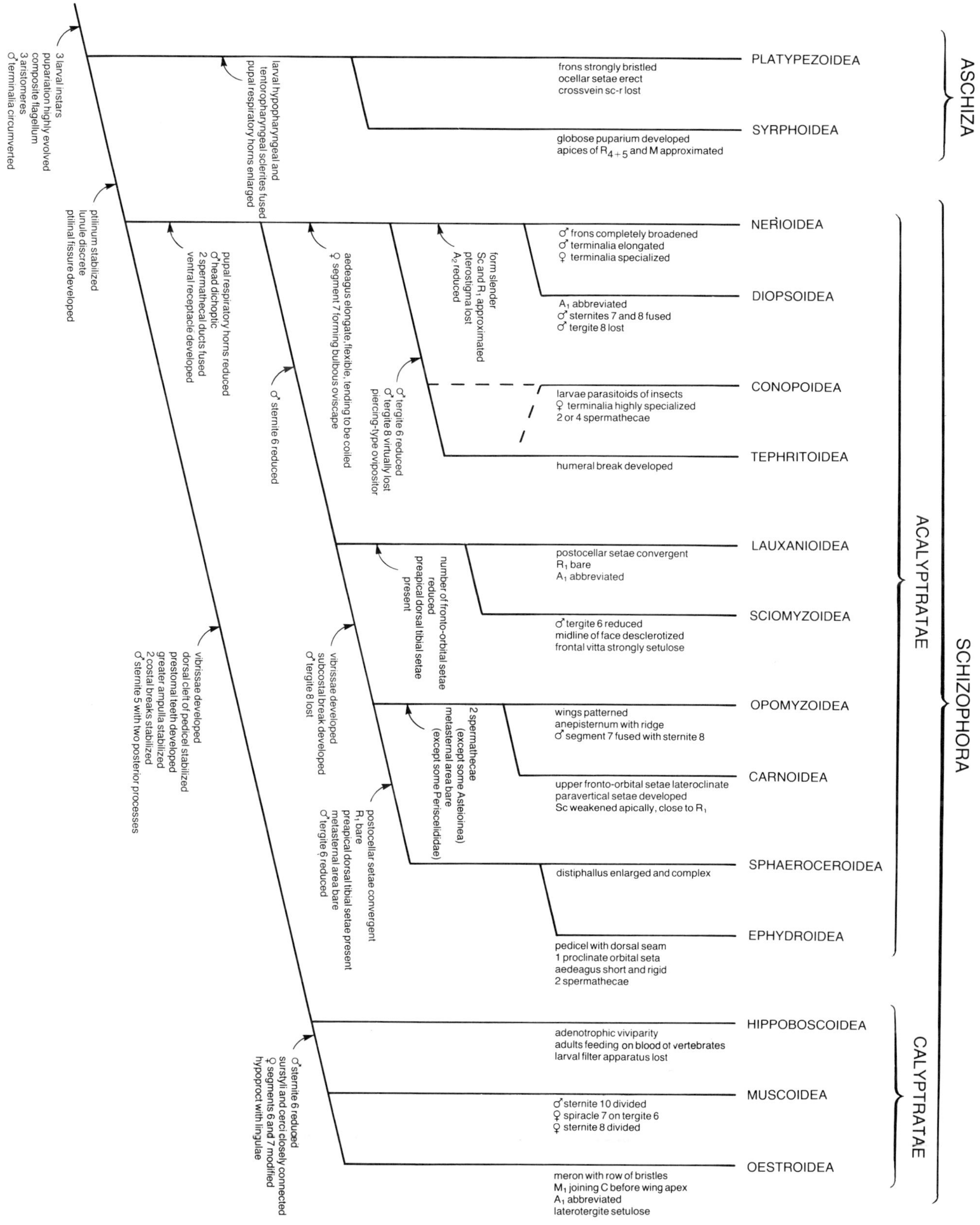

Fig. 116.9. Cladogram of the Muscomorpha summarizing the relationships and apomorphies of the sections, subsections, and superfamilies recognized. (See text for additional details.)

Table 116.10 Classification of the Muscomorpha

- Suborder Brachycera
  - Infraorder Muscomorpha (= Cyclorrhapha)
    - Section Aschiza
      - Superfamily Platypezoidea
        - Family Platypezidae (incl. Opetiidae)
        - Family Lonchopteridae
        - Family Ironomyiidae
        - Family Sciadoceridae
        - Family Phoridae (incl. Termitoxeniidae)
      - Superfamily Syrphoidea
        - Family Syrphidae (incl. Microdontidae)
        - Family Pipunculidae
    - Section Schizophora
      - Subsection Acalyptratae
        - Subgroup 1
          - Superfamily Nerioidea
            - Family Micropezidae (incl. Calobatidae, Taeniapteridae, and Tylidae)
            - Family Neriidae
            - Family Cypselosomatidae (incl. Pseudopomyzidae)
          - Superfamily Diopsoidea
            - Family Tanypezidae
            - Family Strongylophthalmyiidae
            - Family Psilidae
            - Family Somatiidae
            - Family Nothybidae
            - Family Megamerinidae
            - Family Syringogastridae
            - Family Diopsidae (incl. Centriocidae)
          - Superfamily Conopoidea
            - Family Conopidae (incl. Stylogasteridae)
          - Superfamily Tephritoidea
            - Family Lonchaeidae
            - Family Otitidae (incl. Euxestidae, Pterocallidae, and Ulidiidae)
            - Family Platystomatidae
            - Family Tephritidae
            - Family Pyrgotidae
            - Family Tachiniscidae
            - Family Richardiidae
            - Family Pallopteridae (incl. Eurygnathomyiidae)
            - Family Piophilidae (incl. Neottiophilidae and Thyreophoridae)
        - Subgroup 2
        - Subgroup 2.1
          - Superfamily Lauxanioidea
            - Family Lauxaniidae
            - Family Eurychoromyiidae
            - Family Celyphidae
            - Family Chamaemyiidae (incl. Cremifaniidae)
          - Superfamily Sciomyzoidea
            - Family Coelopidae
            - Family Dryomyzidae (incl. Helcomyzidae)
            - Family Helosciomyzidae
            - Family Sciomyzidae (incl. Phaeomyiidae)
            - Family Ropalomeridae
            - Family Sepsidae
        - Subgroup 2.2
        - Subgroup 2.2.1
          - Superfamily Opomyzoidea
            - Suprafamily Clusioinea
              - Family Clusiidae
              - Family Acartophthalmidae

(concluded)

            - Suprafamily Agromyzoinea
              - Family Odiniidae
              - Family Agromyzidae
              - Family Fergusoninidae
            - Suprafamily Opomyzoinea
              - Family Opomyzidae
              - Family Anthomyzidae
            - Suprafamily Asteioinea
              - Family Aulacigastridae (incl. Stenomicridae)
              - Family Periscelididae
              - Family Neurochaetidae
              - Family Teratomyzidae
              - Family Xenasteiidae (= Tunisimyiidae)
              - Family Asteiidae
          - Superfamily Carnoidea
            - Family Australimyzidae
            - Family Braulidae
            - Family Carnidae
            - Family Tethinidae
            - Family Canacidae
            - Family Milichiidae
            - Family Risidae
            - Family Cryptochetidae
            - Family Chloropidae (incl. Mindidae and Siphonellopsidae)
        - Subgroup 2.2.2
          - Superfamily Sphaeroceroidea
            - Family Heleomyzidae (incl. Borboropsidae, Chiropteromyzidae, Cnemospathidae, Heteromyzidae, Notomyzidae, Rhinotoridae, and Trixoscelididae)
            - Family Mormotomyiidae
            - Family Chyromyidae
            - Family Sphaeroceridae
          - Superfamily Ephydroidea
            - Family Curtonotidae
            - Family Camillidae
            - Family Drosophilidae
            - Family Diastatidae (incl. Campichoetidae)
            - Family Ephydridae
      - Subsection Calyptratae
        - Superfamily Hoppoboscoidea
          - Family Glossinidae
          - Family Hippoboscidae
          - Family Streblidae
          - Family Nycteribiidae
        - Superfamily Muscoidea
          - Family Scathophagidae
          - Family Anthomyiidae
          - Family Fanniidae
          - Family Muscidae (incl. Eginiidae)
        - Superfamily Oestroidea
          - Family Calliphoridae (incl. Mesembrinellidae)
          - Family Mystacinobiidae
          - Family Sarcophagidae
          - Family Rhinophoridae
          - Family Tachinidae (incl. Stackelbergomviidae)
          - Family Oestridae (incl. Cuterebridae, Gasterophilidae, and Hypodermatidae)

(Total 82 families)

are the fusion of the hypopharyngeal and tentoropharyngeal sclerites in the larva, and the enlarged pupal respiratory horns. Paradoxically, the pupal spiracular horns were apparently lost in one of the oldest members, the Platypezidae.

A good case can be made for recognizing only two monophyletic superfamilies within the Aschiza, namely the Platypezoidea containing five families and the Syrphoidea with two families (Fig. 116.1, Table 116.9).

Two main subsections are separable in the Schizophora, the Acalyptratae and Calyptratae. The monophyly of the Acalyptratae is supported by relatively few autapomorphies, i.e. reduced pupal respiratory horns, dichoptic condition of the male head, fusion of two of three spermathecal ducts so that only two ducts enter directly into the oviduct, and development of a ventral receptacle. The monophyly of the Calyptratae is well supported by many autapomorphic conditions including vibrissae developed, dorsal cleft of pedicel stabilized, prestomal teeth developed, greater ampulla stabilized, humeral and subcostal breaks stabilized, male sternite 5 with two posterior lobes, and so on (Figs. 116.1, 9).

Within the Acalyptratae, at least 10 superfamilies comprising two main subgroups seem justifiable (Fig. 116.9). In the ground plan of subgroup 1 (Figs. 116.2–3), which consists of the Nerioidea (three families), Diopsoidea (eight families), Conopoidea (one family), and Tephritoidea (nine families), abdominal segment 7 of the female is specialized to form a more or less bulbous oviscape, and in the male, the aedeagus is rather elongate, flexible, and tending to be looped or coiled. In subgroup 2 (Figs. 116.4–7), which consists of the Sciomyzoidea (six families), Lauxanioidea (four families), Opomyzoidea (13 families), Carnoidea (nine families), Sphaeroceroidea (four families), and Ephydroidea (five families), sternite 6 of the male is reduced. Further subgroupings are evident within subgroups 1 and 2.

Autapomorphies of subgroup 1.1 (Nerioidea + Diopsoidea) include a slenderized form, the approximated condition of Sc and R_1, loss of the pterostigma, and a reduced A_2. Autapomorphies in the ground plan of subgroup 1.2 (Conopidae + Tephritoidea) include the development of a piercing-type ovipositor in the female and the reduction and loss of tergites 6 and 8 in the male.

Autapomorphies in the ground plan of subgroup 2.1 (Sciomyzoidea + Lauxanioidea) include a reduced number of fronto-orbital setae and the development of preapical dorsal tibial setae. Autapomorphies of subgroup 2.2 (Opomyzoidea and Carnoidea + Sphaeroceroidea and Ephydroidea) include the development and relative stabilization of vibrissal setae, the stabilization of a subcostal break in C, and the loss of male tergite 8. Subgroup 2.2 appears to be composed of two clades, subgroup 2.2.1 (Opomyzoidea + Carnoidea) and subgroup 2.2.2 (Sphaeroceroidea + Ephydroidea). Autapomorphies of subgroup 2.2.1 are the reduction of spermathecae to two and the loss of setulae on the metasternal area. The status of both these conditions is somewhat equivocal, because exceptions (possible pseudoplesiomorphic conditions) occur in rare cases in one suprafamily (Asteioinea) of one superfamily (Opomyzoidea). Autapomorphies of subgroup 2.2.2 (Sphaeroceroidea + Ephydroidea) include the development of convergent postocellar setae; the appearance of preapical dorsal tibial setae (as in subgroup 2.1, Sciomyzoidea + Lauxanioidea); a bare R_1; a bare metasternal area (as in subgroup 2.2.1, Opomyzoidea + Carnoidea); and a reduced tergite 6 in the male.

Only three superfamilies (Hippoboscoidea, Muscoidea, and Oestroidea) seem justifiable within the Calyptratae (Fig. 116.8). The Hippoboscoidea (four families) is the oldest group, and its sister group is the Muscoidea (four families) + Oestroidea (six families). Autapomorphic or autapotypic conditions of the Hippoboscoidea include the loss of a filter apparatus in the larvae, the development of adenotrophic viviparity, and the acquiring of vertebrate blood-feeding habits and adaptations by the adults. Autapomorphic characters of the Muscoidea + Oestroidea include the reduction of sternite 6 and the development of a very close connection between the cerci and surstyli in the males, and special modifications of segments 6 and 7 (fully retractile) and the development of lingulae in the females.

This study began with an *a priori* working hypothesis that the traditional suprafamilial categories and higher groupings of the infraorder, founded by classical workers and accepted by most systematists of the Diptera, are in fact monophyletic clades.

I conclude that this hypothesis has been strengthened to some degree by supportive autapomorphies advanced for each clade. The ease with which many smaller less well-known taxa have fitted into this general scheme has been rather satisfying. The classification adopted (Table 116.9), which is consistent with the evolutionary pattern outlined, is suprisingly simple and appears to provide a practical systematic framework for the Muscomorpha.

However, I realize that my conclusions are really only opinions or judgements supported by a degree of evidence that renders some of them probable at best, whereas others still remain questionable. This basic limitation, common to all life sciences, is a humbling factor, but it should not deter us from continuing our endeavours to understand the evolutionary patterns before us.

To me, biological evolution, including the evolution of the Muscomorpha, is but one aspect of the mystery of creation as a whole. As a final note, I wish to acknowledge the Divine Author of all in the words of St. Paul:

> *How great are God's riches! How deep are his wisdom and knowledge! How impossible to explain his decisions or to understand his methods! As Scripture says, "Who could ever know the mind of the Lord"? Who could ever be his counselor? ... For all things were created by him, and all things exist through him and for him. To him be the glory forever! Amen.* [Romans, Ch. 11:33–36]

References

Acxél, M. L. 1955. Nothybidae, a new family of Diptera. Treubia 23: 1–15.

Acxél, M. L. 1956. Revision parcial de las Pyrogotidae neotropicales y antarcticas, con sinopsis de los géneros y especies (Diptera Acalyptratae). Revta bras. Ent. 4: 161–184; 5: 1–70; 6: 1–38.

Acxél, M. L. 1958. Pyrgotidae (Diptera Acalyptrata). Explor. Parc natn. Upemba Miss. G. F. de Witte, 50 (4): 35–53.

Agassiz, L. (1846) 1847. Nomenclatoris zoologici index universalis, continens nomina systematica classium, ordinum, familiarum et generum animalium omnium, tam viventium quam fossilium, secundum ordinem alphabeticum unicum disposito, adejectis homonymiis plantarum, nec non variis adnotationibus et emendationibus. Soloduri, Switzerland.

Aldrich, J. M. 1905. A catalogue of North American Diptera. Smithson. misc. Collns 46 (2 = publ. 1444): 1–680.

Andersson, H. 1971. The Swedish species of Chyromyidae (Diptera), with lectotype designations. Ent. Tidskr. Årg. 92: 95–99.

Andersson, H. 1976*a*. Revision of the *Anthomyza* species of northwest Europe (Diptera: Anthomyzidae) I. The *gracilis* group. Ent. scand. 7: 41–52.

Andersson, H. 1976*b*. Revision of the genus *Formicosepsis* de Meijere (Diptera: Cypselosomatidae). Ent. scand. 7: 175–185.

Andersson, H. 1976*c*. *Chyromya miladae* n. sp. from Czechoslovakia (Diptera: Chyromyidae). Ent. scand. 7: 187–189.

Andersson, H. 1977. Taxonomic and phylogenetic studies on Chloropidae (Diptera) with special reference to Old World genera. Ent. scand. Suppl. 8: 1–200.

Andersson, H. 1984. Revision of the *Anthomyza* species of Northwest Europe (Diptera: Anthomyzidae) II. The *pallida* group. Ent. scand. 15: 15–24.

Ashburner, M. 1981. 10. Entomophagous and other bizarre Drosophilidae. Pages 375–429 *in* M. Ashburner *et al*., eds. The genetics and biology of *Drosophila*. Vol. 3a. Academic Press, New York, N.Y.

Ashburner, M., H. L. Carson, and J. N. Thompson, Jr., eds. 1981. The genetics and biology of *Drosophila*. Vol. 3a. Academic Press, New York, N.Y.

Ashlock, P. D. 1971. Monophyly and associated terms. Syst. Zool. 20: 63–69.

Austen, E. E. 1936. 23. A remarkable semi-apterous fly (Diptera) found in a cave in East Africa, and representing a new family, genus and species. Proc. zool. Soc. Lond. 1936: 425–431.

Bächli, G., and M. T. Rocha Pité. 1981. Pages 169–196 *in* M. Ashburner *et al*., eds. The genetics and biology of *Drosophila*. Vol. 3a. Academic Press, New York, N.Y.

Barnes, J. K. 1979. The taxonomic postion of the New Zealand genus *Prosochaeta* Malloch (Diptera: Sciomyzidae). Proc. ent. Soc. Wash. 8: 285–297.

Barnes, J. K. 1981. Revision of the Helosciomyzidae (Diptera). Jl R. Soc.N.Z. 11: 45–72.

Baud, F. 1973. Biologie et cytologie de cinq espèces du genre *Lonchoptera* Meig. (Dipt.) dont l'une est parthénogénétique et les autres bissexuées, avec quelques remarques d'ordre taxonomique. Revue suisse Zool. 80: 473–515.

Bautz, A. M. 1971. Chronologie de la mise en place de l'hypoderme imaginal de l'abdomen de *Calliphora erythrocephala* Meigen (Insecte, Diptère, Brachycère). Arch. Zool. exp. gen. 112: 157–178.

Becher, E. 1882. Zur Kenntnis der Kopfbildung der Dipteren. Wien. ent. Ztg 1: 49–54.

Becker, T. 1907. Die Ergebnisse meiner dipterologischen Frühjahreise nach Algier und Tunis, 1906. Z. syst. Hymenopt. Dipterol. 7: 369–407.

Becker, T., M. Bezzi, K. Kertész, and P. Stein. 1905. Katalog der Palaarktischen Dipteren. Vol. 4. Budapest.

Bequaert, J. C. 1953. The Hippoboscidae or louse-flies (Diptera) of mammals and birds. Part I. Structure, physiology and natural history. Entomologica am. 32: 1–232; 33: 211–442.

Bequaert, J. O. 1954–1957. The Hippoboscidae or louse-flies (Diptera) of mammals and birds. Part II. Taxonomy, evolution and revision of American genera and species. Entomologica am. 34: 1–232; 35: 344–416; 36: 417–611.

Berg, C. O., and L. Knutson. 1978. Biology and systematics of the Sciomyzidae. A. Rev. Ent. 1978: 239–258.

Bickel, D. J. 1982. Diptera. Page 563 *in* S. P. Parker, ed. Synopsis and classification of living organisms Vol. 2. McGraw-Hill Book Co., New York, N.Y.

Bigot, M. 1852. Essai d'une classification générale et synoptique de l'ordre des insectes Diptères. Annls Soc. ent. Fr. 10: 471–489.

Billberg, G. J. 1820. Enumeratio insectorum in Museo Gust. Joh. Billberg. Stockholm.

Black, V. H. 1966. The pupal development of the male genital disc of *Eucalliphora lilaea*. Sacramento State College (unpublished M.A. thesis). 40 pp.

Bock, I. A. 1982. Drosophilidae of Australia. V. Remaining genera and synopsis (Insecta: Diptera). Aust. J. Zool. Suppl. Ser. 89: 1–164.

Bock, I. R., and P. A. Parsons. 1981. 7. Species of Australia and New Zealand. Pages 291–308 *in* M. Ashburner *et al*., eds. The genetics and biology of *Drosophila* Vol. 3a. Academic Press, New York, N.Y.

Bonhag, P. F. 1951. The skeleto-muscular mechanism of the head and abdomen of the adult horsefly. Trans. Am. ent. Soc. 77: 131–202.

Brauer, F. 1880. Die Zweiflügler des kaiserlicken Museums zu Wien. I, II. Bemerkungen zur Systematik der Dipteren. Denkschr. Akad. Wiss., Wien 42: 108–118.

Brauer, F. 1883. Die Zweiflügler des kaiserlicken Museums zu Wien. III. Systematische Studien auf Grundlage der Dipteren-Larven nebst einer Zusammenstellung von Beispielen aus der Literatur über dieselben und Beschreibung neuer Formen. Denkschr. Akad. Wiss., Wien 47: 1–100.

Brauer, F. 1890. Ueber die Verbindungsglieder zwischen den orthorrhaphen und cyclorrhaphen Dipteren und

solche zwischen Syrphiden und Muscarien. Verh. zool.-bot. Ges. Wien 40: 273–275.
Brauns, A. 1954. Puppen terricoler Dipterenlarven. Unters. angew. Bodenbiol. 2: 1–156.
Broadhead, E. C. 1984. Adaptations for fungal grazing in lauxaniid flies. J. nat. Hist. 18: 639–649.
Brues, C. T., and A. L. Melander. 1932. Classification of insects. Bull. Mus. comp. Zool. Harv. 73: 1–672.
Chandler, P. 1981. *Nemedina alamirabilis* sp. n., a new genus and species of Diptera Eremoneura, of uncertain affinities, from Hungary. Acta zool. Acad. Sci. Hung. 27: 103–113.
Chillcott, J. G. 1960. A revision of the Nearctic species of Fanniinae (Diptera: Muscidae). Can. Ent. supplement 14: 1–295.
Christophers, S. R., and P. J. Barraud. 1926. The development of the male and female hypopygium of *Phlebotomus*. Ind. J. med. Res. 13: 853–870.
Chvála, M. 1983. The Empidoidea (Diptera) of Fennoscandia and Denmark. II. General part. The families Hybotidae, Atelestidae and Microphoridae. Fauna ent. scand. 12: 1–279.
Clausen, P. J., and E. F. Cook. 1971. A revision of the Nearctic species of the tribe Parydrini (Diptera: Ephydridae). Mem. Am. ent. Soc. 27: 1–150.
Cogan, B. H. 1971. The Heleomyzidae of the Ethiopian region (Diptera). Ann. Natal Mus. 20: 627–696.
Cogan, B. H. 1975. New taxa in two families previously unrecorded from the Ethiopian region (Diptera: Odiniidae and Diastatidae). Ann. Natal Mus. 22: 471–488.
Cogan, B. H. 1977. New African species of *Trixoscelis* Rondani (Diptera: Trixoscelididae), with a short discussion of related genera. Stuttg. Beitr. Naturk. Ser. A (Biol.) 297: 1–15).
Cogan, B. H. 1980*a*. 41. Family Tachiniscidae. Page 555 *in* R. W. Crosskey *et al.*, eds. Catalogue of the Diptera of the Afrotropical Region. Publ. No. 821. British Museum (Natural History), London. Fletcher & Son Ltd., Norwich, England.
Cogan, B. H. 1980*b*. 71. Family Ephydridae. Pages 655–669 *in* Crosskey *et al.*, eds. Catalogue of the Diptera of the Afrotropical Region. Publ. No. 821. British Museum (Natural History), London. Fletcher & Son Ltd., Norwich, England.
Cogan, B. H. 1984. Family Ephydridae. Pages 126–176 *in* A. Soós and L. Papp, eds. Catalogue of Palaearctic Diptera, Vol. 10. Clusiidae-Chloropidae. Akadémiai, Budapest.
Cogan, B. H., and H. K. Munro. 1980. 40. Family Tephritidae. Pages 518–554 *in* R. W. Crosskey *et al.*, eds. Catalogue of the Diptera of the Afrotropical Region. Publ. No. 821. British Museum (Natural History), London. Fletcher and Son Ltd., Norwich, England.
Cogan, B. H., and W. W. Wirth. 1977. Family Ephydridae. Pages 321–339 *in* M. D. Delfinado and D. E. Hardy, eds. A Catalogue of the Diptera of the Oriental Region, Vol. 3. Suborder Cyclorrhapha excluding Division Aschiza. University Press of Hawaii, Honolulu.
Colless, D. H. 1962. A new Australian genus and family of Diptera (Nematocera: Perissommatidae). Aust. J. Zool. 10: 519–535.
Colless, D. H., and D. K. McAlpine. 1970. 34. Diptera (flies). Pages 656–740 *in* The insects of Australia. Melbourne University Press, Carlton, Victoria.
Colless, D. H., and D. K. McAlpine. (1974) 1975. 34. Diptera (flies). Pages 91–96 *in* Supplement 1974, The insects of Australia. Melbourne University Press, Carlton, Victoria.
Collin, J. E. 1943. The British species of Heleomyzidae. Entomologist's mon. Mag. 79: 234–251.
Collin, J. E. 1949. 10. The Palaearctic species of the genus *Aphaniosoma* Beck. (Diptera, Chiromyidae). Ann. Mag. nat. Hist. 12: 127–147.
Collin, J. E. 1956. On the identity of Fallén's *Drosophila glabra* (Diptera, Camillidae). Entomologist's mon. Mag. 92: 225–226.
Cook, E. F. 1949. The evolution of the head in the larvae of Diptera. Microentomology 14: 1–57.
Crampton, G. C. 1942. Guide to the insects of Connecticut. Part VI. The Diptera or true flies of Connecticut. First Fascicle. The external morphology of the Diptera. Bull. Conn. St. geol. nat. Hist. Surv. 64: 10–165 (reprinted 1966).
Crampton, G. C. 1944*a*. A comparative morphological study of the terminalia of male calypterate cyclorrhaphous Diptera and their acalypterate relatives. Bull. Brooklyn ent. Soc. 39: 1–31.
Crampton, G. C. 1944*b*. Suggestions for grouping the families of acalypterate cyclorrhaphous Diptera on the basis of the male terminalia. Proc. ent. Soc. Wash. 46: 152–154.
Crosskey, R. W. 1965. A systematic revision of the Ameniinae (Diptera: Calliphoridae). Bull. Brit. Mus. (nat. Hist.) Ent. 16: 35–140.
Crosskey, R. W. 1977. A review of the Rhinophoridae (Diptera) and a revision of the Afrotropical species. Bull. Br. Mus. (nat. Hist.) Ent. 36: 1–66.
Crosskey, R. W. 1980. 93. Family Tachinidae. Pages 822–882 *in* R. W. Crosskey *et al.*, eds. Catalogue of the Diptera of the Afrotropical region. Publ. No. 821. British Museum (Natural History), London. Fletcher and Son Ltd., Norwich, England.
Crosskey, R. W., B. R. Cogan, P. Freeman, A. C. Pont, K. G. V. Smith, and H. Oldroyd, eds. 1980. Catalogue of the Diptera of the Afrotropical region. Publ. No. 821. British Museum (Natural History), London. Fletcher and Son Ltd., Norwich, England.
Curran, C. H. 1933. The African species of *Curtonotum* Macquart (Drosophilidae: Diptera). Am. Mus. Novit. 675: 1–4.
Curran, C. H. 1934. The families and genera of North American Diptera. Ballou Press, New York, N.Y. [reprinted with additions and corrections 1965].
Czerny, L. 1927. Heleomyzidae, Trichoscelidae, und Chiromyidae. Lfg. 22, Fasc. 53a and 53b. Pages 1–56 *in* E. Lindner, ed. Die Fliegen der palaearktischen Region. Schweizerbart, Stuttgart.
Delfinado, M. D., and D. E. Hardy, eds. 1977. A catalog

of the Diptera of the Oriental region. Vol. 3. University Press of Hawaii, Honolulu.

de Meijere, J. C. H. 1900*a*. Zur Kenntnis der Metamorphose von *Platypeza* und der verwantschaftslichen Beziehungen der Platypezinen. Tijdschr. Ent. 54: 241–254.

de Meijere, J. C. H. 1900*b*. Ueber der Larve von *Lonchoptera*. Ein Beiträg zur Kenntnis der cyclorrhaphen Dipterenlarven. Zool. Jb. (Syst. Okol. Geogr. Tiere) 14: 87–132.

de Meijere, J. C. H. 1904. Beiträge zur Kenntnis der Biologie und der systematischen Verwandtschaft der Conopidae. Tijdschr. Ent. 46: 144–225.

Deonier, D. L. 1979. Introduction—a prospectus on research in Ephydridae. Pages 1–19 *in* D. L. Deonier, ed. First symposium on the systematics and ecology of Ephydridae (Diptera). North American Benthological Society and Association of Systematics Collections, Oxford, Ohio.

Dowding, V. M. 1967. The function and ecological significance of the pharyngeal ridges occurring in the larvae of some cyclorrhaphous Diptera. Parasitology 57: 371–388.

Downes, J. A. 1958. The feeding habits of biting flies and their significance in classification. A. Rev. Ent. 3: 249–266.

Downes, J. A. 1968. Notes on the organs and processes of sperm-transfer in the lower Diptera. Can. Ent. 100: 608–617.

Downes, W. L. 1955. Notes on the morphology and classification of the Sarcophagidae and other calyptrates (Diptera). Proc. Iowa Acad. Sci. 62: 514–538.

Downes, W. L. 1958. The Nearctic Miltogramminae (Diptera, Sarcophagidae) and certain allies. Ph.D. thesis, published on demand by University Microfilms International, Ann Arbor, Mich.

Duda, O. 1927. Revision der alteweltlichen Astiidae. Dt. ent. Z. 2: 113–147.

Duda, O. 1934*a*. Curtonotidae. Lfg. 81, Fasc. 58d. Pages 1–5 *in* E. Lindner, ed. Die Fliegen der palaearktischen Region. Schweizerbart, Stuttgart.

Duda, O. 1934*b*. Drosophilidae. Lfg. 86, Fasc. 58g. Pages 1–118 *in* E. Lindner, ed. Die Fliegen der palaearktischen Region. Schweizerbart, Stuttgart.

Duda, O. 1934*c*. Camillidae. Lfg. 81, Fasc. 58f. Pages 1–7 *in* E. Lindner, ed. Die Fliegen der palaearktischen Region. Schweizerbart, Stuttgart.

Edwards, F. W. 1938. British short-palped craneflies. Taxonomy of adults. Trans. Soc. Br. Ent. 5: 1–168.

Enderlein, G. 1914. Ordnung Diptera, Fliegen (Zweiflügler). Pages 272–334 *in* P. Brohmer, ed. Fauna von Deutschland. Leipzig.

Enderlein, G. 1936. 22. Ordnung: Zweiflügler, Diptera. Abt. 16 *in* P. Brohmer *et al.*, eds. Die Tierwelt Mitteleuropas 6: Insekten III Teil. Leipzig.

Enderlein, G. 1938. 60. Die Dipteren fauna der Juan-Fernandez-Inseln und der Öster-Insel. Pages 643–680 *in* C. Skottsberg, ed. Zoology. Vol. 3. The natural history of Juan Fernandez and Easter Island. Almquist & Wicksells, Uppsala.

Enderlein, G. 1942. Klassifikation der Pyrgotiden. Sber. Ges. naturf. Freunde Berl. 1941: 98–134.

Evans, J. J. T. 1967*a*. The integument of the Queensland fruit fly, *Dacus tryoni* (Frogg.) I. The tergal glands. Z. Zellforsch. mikrosk. Anat. 81: 18–33.

Evans, J. J. T. 1967*b*. The integument of the Queensland fruit fly, *Dacus tryoni* (Frogg.). II. Development and ultrastructure of the abdominal integument and bristles. Z. Zellforsch. mikrosk. Anat. 81: 34–48.

Evans, J. J. T. 1967*c*. Development and ultrastructure of the fat body cells and oenocytes of the Queensland fruit fly, *Dacus tryoni* (Frogg.). Z. Zellforsch. mikrosk. Anat. 81: 49–61.

Fallén, C. F. 1810. Specimen entomologicum novam Diptera disponendi methodum exhibens. Lund.

Fallén, C. F. 1820*a*. Oscinides Sveciae. Lundae.

Fallén, C. F. 1820*b*. Heteromyzides Sveciae. Lundae.

Fallén, C. F. 1820*c*. Opomyzides Sveciae. Lundae.

Fallén, C. F. 1820*d*. Sciomyzides Sveciae. Lundae.

Fallén, C. F. 1823. Phytomyzides et Ochtidiae Svecia. Lundae.

Faucheux, M. J. 1971. Données générales sur l'appareil trachéen des Diptères adultes. Thèse en vue de l'obtention du titre de docteur 3e cycle, présentée à l'U.E.R. des Sciences de la Nature de l'Université de Nantes.

Feijen, H. R. 1983. Systematics and phylogeny of Centrioncidae, a new Afromontane family of Diptera (Schizophora). Zool. Verh., Leiden 202: 1–137.

Ferrar, P. 1987. A guide to the breeding habits and immature stages of Diptera Cyclorrhapha. Entomonograph Vol. 8, Parts 1–2. E. J. Brill / Scand. Sci. Press, Leiden and Copenhagen.

Fonseca, E. C. M., D'Assis. 1968. Diptera Cyclorrhapha Calyptrata. Section (b) Muscidae. R. ent. Soc. Lond. Handbk Ident. Br. Insects 10 (4b): 1–119.

Foote, R. H. 1965. Family Rhinotoridae. Page 18 *in* A. Stone *et al.*, eds. A catalog of the Diptera of America north of Mexico. U.S. Dep. Agric., Agric. Res. Serv., Agric. Handb. 276.

Foote, B. A. 1981. Biology and immature stages of *Pelina truncatula*, a consumer of blue-green algae (Diptera: Ephydridae). Proc. ent. Soc. Wash. 83 (4): 607–619.

Fraser, A. 1959. The anatomy of the central nervous system and retrocerebral endocrine organs of the larvae of *Lucilia caesar* L. and certain other Diptera Cyclorrhapha. Proc. R. ent. Soc. Lond., Ser. A 34: 186–192.

Frey, R. 1921. Studien über den Bau des Mundes der niederen Diptera Schizophora nebst Bemerkungen über die Systematik dieser Dipterengruppe. Acta Soc. Fauna Flora fenn. 48 (3): 1–245.

Frey, R. 1927. Zur Systemtik der Diptera Haplostomata. III. Fam. Micropezidae. Notul. ent. 7: 65–76.

Frey, R. 1941*a*. Diptera. Enum. Insect. Fenn. Suec. 6: 1–63.

Frey, R. 1941*b*. Die Gattungen und Arten der Dipterenfamilie Celyphidae. Notul. ent. 21: 3–16.

Frey, R. 1952. Über Chiropteromyza n. gen. und *Pseudopomyza* Strobl. (Diptera Haplostomata). Notul. ent. 32: 5–8.

Frey, R. 1958*a*. Studien über osasiatische Dipteren VI.

Nothybidae, Micropezidae, Opomyzidae. Notul. ent. 38: 37–51.

Frey, R. 1958*b*. Zur Kenntnis der Diptera Brachycera p.p. der Kapverdischen Inseln. Comment. Biol. 18 (4): 61 pp.

Frey, R. 1960. Studien über indo-australische Clusiiden (Dipt.) nebst. Katalog der Clusiiden. Commentat. biol. 22: 1–31.

Frick, K. E. 1952. A generic revision of the family Agromyzidae (Diptera) with a catalogue of New World species. Univ. Calif. Publs Ent. 8: 339–452.

Gerstäcker, C. F. A. 1863. Handbuch der Zoologie. Bd 2: Wirbellose. Engelmann, Leipzig.

Gill, G. D. 1962. The heleomyzid flies of America North of Mexico (Diptera: Heleomyzidae). Proc. U.S. natn. Mus. 113: 495–603.

Gill, G. D. 1968. 85. Family Heleomyzidae (Helomyzidae), including the Trixoscelididae (Trichoscelidae) (86). Part 85 *in* N. Papavero, ed. A catalogue of the Diptera of the Americas south of the United States. Departamento de Zoologia, Secretaria da Agricultura, São Paulo, Brasil.

Girschner, E. 1893. Beiträg zur Systematik der Musciden. Berl. ent. Z. 38: 297–312.

Gouin, F. 1949. Recherches sur la morphologie de l'appareil buccal des diptères. Mém. Mus. natn. Hist. nat., Paris (n.s.) 28: 167–269.

Greathead, D. J. 1967. The genus *Trichopsidea* Westwood, with a discussion of its relation to other genera of Nemestrinidae (Diptera). J. nat. Hist. 1: 305–313.

Griffiths, G. C. D. 1964. The agromyzid fauna of Iceland and the Faroes, with appendices on the *Phytomyza milii* and *robustella* groups (Diptera, Agromyzidae). Ent. Meddr. 32: 393–450.

Griffiths, G. C. D. 1972. The phylogenetic classification of Diptera Cyclorrhapha, with special reference to the structure of the male postabdomen. Series entomologica, 8. Dr. W. Junk, N.V., The Hague.

Griffiths, G. C. D. 1976. Comments on some recent studies on tsetse-fly phylogeny and structure. Syst. Ent. 1: 15–18.

Griffiths, G. C. D. 1981. Book review. Manual of Nearctic Diptera, Volume 1. J. F. McAlpine, B. V. Peterson, G. E. Shewell, H. J. Teskey, J. R. Vockeroth, D. M. Wood, eds. Minister of Supply and Services. Bull. ent. Soc. Can. 13: 49–55.

Griffiths, G. C. D. 1982. On the systematic position of *Mystacinobia* (Diptera: Calliphoridae). Mem. ent. Soc. Wash. 10: 70–77.

Griffiths, G. C. D. 1983. Book Review. The Empidoidea (Diptera) of Fennoscandia and Denmark. II. General part. The families Hybotidae, Atelestidae and Microphoridae. Fauna Entomologicae Scandianvica. Vol. 12. Chvála, M. 1983. Scandinavian Science Press Ltd., Målǿv, Denmark. Quaest. Ent. 19: 483–485.

Griffiths, G. C. D. 1984. For discussion on "Male genitalia in the classification of Chloropidae" XVII International Congress of Entomology, Hamburg, August 1984, and Note on characterization of Eremoneura and Cyclorrhapha. Privately published, 2 pages.

Grunin, K. J. 1964–1969. 64*b*. Hypodermatidae, 64*a*[1]. Oestridae, and 64*a*. Gasterophilidae. Lfg. 252, 254, 255, and 257, pages 1–154; Lfg. 264, 265, and 266, pages 1–96; and Lfg. 279, pages 1–61, respectively. Bd. 8, *in* E. Lindner, ed. Die Fliegen der palaearktischen Region. Schweizerbart, Stuttgart.

Guimãraes, J. H. 1977. A systematic revision of the Mesembrinellidae, *stat. nov.* (Diptera, Cyclorrhapha). Archos zool. Est. S Paulo 29: 1–109.

Guimãraes, J. H., and N. Papavero. 1966. A contribution to the knowledge of the Neotropical Rhinotoridae (Diptera, Acalyptratae). Pap. Avul. Dep. Zool. S Paulo 18: 261–269.

Hackman, W. 1960. Chapter XVIII. Diptera (Brachycera): Camillidae, Curtonotidae and Drosophilidae. S. Afr. Anim. Life 7: 381–389.

Hackman, W. 1969. A review of the zoogeography and classification of the Sphaeroceridae (Borboridae, Diptera). Notul. ent. 49: 193–210.

Hackman, W. 1970. Trixoscelidae (Diptera) from southern Spain and description of a new *Trixoscelis* species from northern Europe. Ent. scand. 1: 127–134.

Hackman, W. 1977. Family Spaheroceridae (Borboridae). Pages 398–406 *in* M. D. Delfinado and D. E. Hardy, eds. A catalog of the Diptera of the Oriental region. Vol. 3. Suborder Cyclorrhapha (excluding Division Aschiza). University Press of Hawaii, Honolulu.

Hackman, W., and R. Väisänen. 1982. Different classification systems in Diptera. Ann. Zool. Fennici 19: 209–219.

Haliday, A. H. 1840. Subfamily 3. Acalypterae Mcq. Pages 143–154 *in* J. O. Westwood, ed. An introduction to the modern classification of insects. Synopsis of the genera of British Insects. London.

Hall, D. G. (1947) 1948. The blowflies of North America. Vol. 4. Thomas Say Found. Ent. Soc. Am., Baltimore, Md.

Hardy, D. E. 1962. Insects of Macquarie Island. Diptera: Coleopidae. Pacif. Insects. 4: 963–971.

Hardy, D. E. 1965. Diptera Cyclorrhapha II, Series Schizophora, Section Acalypterae I, Family Drosophilidae. Insects Hawaii 12: 1–814.

Hardy, D. E. 1973. The fruit flies (Tephritidae-Diptera) of Thailand and bordering countries. Pacif. Insects Monogr. 31: 1–353.

Hardy, D. E. 1974. The fruit flies of the Philippines (Diptera: Tephritidae) Pacif. Insects monogr. 32: 1–266.

Hardy, D. E. 1977. Family Tephritidae. Pages 44–134 *in* M. D. Delfinado and D. E. Hardy, eds. A catalog of the Diptera of the Oriental Region. Vol. 3. University Press of Hawaii, Honolulu.

Hardy, D. E. 1980. Xenasteiidae, a new family of Schizophora (Diptera) from the Pacific and Indian oceans. Proc. Hawaii. ent. Soc. 23: 205–225.

Hardy, D. E. and K. W. Kaneshiro. 1981. Drosophilidae of Pacific Oceania. Pages 309–348 *in* M. Ashburner *et al.*, eds. The genetics and biology of Drosophila. Vol. 3a. Academic Press, New York, N.Y.

Harrison, R. A. 1953. The Diptera of the Antipodes and Bounty Islands. Trans. R. Soc. N.Z. 81: 269–282.

Harrison, R. A. 1959. Acalypterate Diptera of New Zealand. New Zealand Department of Scientific and Industrial Research Bulletin 128, Whitcombe and Tombs Ltd., Wellington, N.Z.

Harrison, R. A. 1976. The Arthropoda of the southern islands of New Zealand (9) Diptera. Jl R. Soc. N.Z. 6: 107–132.

Hartley, J. C. 1963. The cephalopharyngeal apparatus of syrphid larvae and its relationship to other Diptera. Proc. zool. Soc. Lond. 141: 261–280.

Hendel, F. 1902. Ueber die systematische Stellung der Dipteren-gattungen *Pseudopomyza* Strobl u. *Rhicnoessa* Lw. Wien. ent. Ztg 21: 261–264.

Hendel, F. 1903. Kritische Bemerkungen zur Systematik der Muscidae Acalyptratae. Wien. ent. Ztg 22: 249–252.

Hendel, F. 1910. Über die Nomenklatur der Acalyptratengattungen nach Th. Becker's Katalog der paläarktischen Dipteren, Bd. 4. Wien. ent. Ztg 29: 307–313.

Hendel, F. 1913. Acalyptrate Musciden (Diptera II). Supplta. ent. 2: 77–112.

Hendel, F. 1916. Beiträge zur Systematik der Acalyptraten Musciden. Ent. Mitt. 5: 294–299.

Hendel, F. 1922–1923. Die paläarktischen Muscidae Acalyptrate Girsch. = Haplostomata Frey nach ihre Familien und Gattungen I. Die Familien. Konowia 1: 145–160 (1922); 153–265 (1923).

Hendel, F. 1928. Zweiflügler oder Diptera. II. Allgemeiner Teil. Pages 1–135 *in* F. Dahl, ed. Die Tierwelt Deutschlands. G. Fischer, Jena.

Hendel, F. 1931. Kritische und Synomymische Bemerkungen über Dipteren. Verh. zool.-bot. Ges. Wien 81: 4–19.

Hendel, F. 1933. Neue acalyptrate Musciden aus der paläarktischen Region (Dipt.). Dt. ent. Z. 1933: 39–56.

Hendel, F. 1934. Übersicht über die Gattungen der Pyrgotiden, nebst Beschreibung neuer Gattungen u. Arten. Encycl. ent. (B. II, Diptera) 7: 141–156.

Hendel, F. 1935. Bemerkungen zu "The families and genera of North-American Diptera" by C. H. Curran, New York 1934. Konowia 14: 51–57.

Hendel, F. 1936–1937. 26. Ordnung der Pterygogenea (Dreissigste Ordnung der Insecta): Diptera oder Fliegen. Pages 1729–1998 *in* Kükenthal, W. & T. Krumbach, eds., Handbk Zool. 4,2. Hälfte, 2. Teil, Insecta 3. Walter de Gruyter & Co., Berlin.

Hennig, W. 1936. Beiträge zur Kenntnis der Kopulationapparates der cyclorrhaphen Dipteren. Z. Morph. Ökol. Tiere 31: 328–370.

Hennig, W. 1937*a*. Uebersicht über die Arten der Neriiden und über die Zoogeographie dieser Acalyptraten-Gruppe. Stettin. ent. Ztg 98: 240–280.

Hennig, W. 1937*b*. Milichiidae and Carnidae. Lfg. 115, Fasc. 60a. Pages 1–91. Bd 6 *in* E. Lindner, ed. Die Fliegen der palaearktischen Region. Schweizerbart, Stuttgart.

Hennig, W. 1939*a*. Otitidae. Lfg. 126 and 128, Fasc. 46 and 47, Bd. 5. Pages 1–78 *in* E. Lindner, ed. Die Fliegen der palaearktischen Region. Schweizerbart, Stuttgart.

Hennig, W. 1939*b*. Beiträge zur Kenntnis des Kopulationsapparates und der Systematik der Acalyptraten. II. Tethinidae, Milichiidae, Anthomyzidae und Opomyzidae. Arb. morph. taxon. Ent. Berl. 6: 81–94.

Hennig, W. 1941. Die Verwantschaftsbeziehungen der Pupiparen und die Morphologie der Sternalregion des Thorax der Dipteren. Arb. morph. taxon. Ent. Berl. 8: 231–49.

Hennig, W. 1948*a*. Die Larvenformen der Dipteren. 1. Teil. Akademie-Verlag, Berlin [reprinted 1968].

Hennig, W. 1948*b*. Beiträge Zur Kenntnis der Kopulationsapparates und der systematik der Acalyptraten IV. Lonchaeidae und Lauxaniidae. Acta zool. lilloana 6: 333–429.

Hennig, W. 1952. Die Larvenformen der Dipteren. III. Teil. Akademie-Verlag, Berlin [reprinted 1968].

Hennig, W. 1954. Flügelgeäder und System der Dipteren unter Berucksichtigung der aus dem Mesozoikum beschreibenen Fossilen. Beitr. Ent. 4: 245–388.

Hennig, W. 1955–1964. 63b. Muscidae. Lfg. 182, 185, 194, 197, 199, 204, 205, 207, 209, 213, 215, 217, 223, 225, 226, 229, 233, 234, 241, 242, 248, 253, Bd. 7(2). Pages 1–1110. *in* E. Lindner, ed. Die Fliegen der palaearktischen Region. Schweizerbart, Stuttgart.

Hennig, W. 1958. Die Familien der Diptera Schizophora und ihre phylogenetischen Verwandtschaftsbeziehungen. Beitr. Ent. 8: 505–688.

Hennig, W. 1959. Pyrgotidae de Madagascar (Diptera). Mém. Inst. scient. Madagascar, Sér. E, 11: 321–353.

Hennig, W. 1965*a*. Vorarbeiten zu einem phylogenetischen System der Muscidae (Diptera: Cyclorrhapha). Stuttg. Beitr. Naturk. 141: 1–100.

Hennig, W. 1965*b*. Die Acalyptratae des Baltischer Bernsteins. Stuttg. Beitr. Naturk. 145: 1–215.

Hennig, W. 1966*a*. Phylogenetic systematics. University of Illinois Press, Urbana, Chicago and London.

Hennig, W. 1966*b*. Conopidae in Baltischen Bernstein (Diptera: Cyclorrhapha). Stuttg. Beitr. Naturk. 154: 1–27.

Hennig, W. 1967*a*. Die sogenannten "niederen Brachycera" im Baltischen Bernstein (Diptera: Fam. Xylophagidae, Xylomyidae, Rhagionidae, Tabanidae). Stuttg. Beitr. Naturk. 174: 1–51.

Hennig, W. 1967*b*. Neue Acalyptratae aus dem Baltischen Bernstein (Diptera: Cyclorrhapha). Stuttg. Beitr. Naturk. 175: 1–27.

Hennig, W. 1969*a*. Neue Gattungen und Arten der Acalyptratae. Can. Ent. 101: 589–633.

Hennig, W. 1969*b*. Neue Übersicht über die aus dem Baltischen Bernstein bekannten Acalypteratae (Diptera: Cyclorrhapha). Stuttg. Beitr. Naturk. 209: 1–42.

Hennig, W. 1970. Insektenfossilien aus der unteren Kreida II. Empididae (Diptera, Brachycera). Stuttg. Beitr. Naturk. 214: 1–12.

Hennig, W. 1971*a*. Neue Untersuchungen über die Familien der Diptera Schizophora (Diptera: Cyclorrhapha). Stuttg. Beitr. Naturk. 226: 1–76.

Hennig, W. 1971*b*. Insektenfossilien aus der unteren kreide III. Empidiformia (Microphorinae) aus der unteren Kreide und aus dem Baltischen Bernstein; ein Vertreter der Cyclorrhapha aus der unteren Kreide. Stuttg. Beitr. Naturk. 232: 1–27.

Hennig, W. 1971*c*. Die Familien Pseudopomyzidae und Milichiidae in Baltischen Bernstein. Stuttg. Beitr. Naturk. 233: 1–16.

Hennig, W. 1972*a*. Beiträge zur Kenntnis der rezenten und fossilen Carnidae, mit besonderer Berucksichtigung einer neuen Gattung aus Chile (Diptera: Cyclorrhapha). Stuttg. Beitr. Naturk. 240: 1–20.

Hennig, W. 1972*b*. Eine neue Art der Rhagionidengattung *Litoleptis* aus Chile, mit Bemerkungen über Fuhlerbildung und Verwandtschaftsbeziehungen einiger Brachycerenfamilien (Diptera: Brachycera). Stuttg. Beitr. Naturk. 242: 1–18.

Hennig, W. 1973. Ordnung Diptera (Zweiflügler). Handbk Zool. 4 (2) 2 / 31 (Lfg. 20): 1–227.

Hennig, W. 1976*a*. Anthomyiidae. Lfg. 315, Bd. 7(1). Pages 686–699 *in* E. Lindner, ed. Die Fliegen der palaearktischen Region. Schweizerbart, Stuttgart.

Hennig, W. 1976*b*. Das hypopygium von *Lonchoptera lutea* Panzer und die phylogenetischen Verwandtschaftsbeziehungen der Cyclorrhapha (Diptera). Stuttg. Beitr. Naturk., Ser. A (Biol.), 283: 1–63.

Hennig, W. 1981. Insect phylogeny. Adrian C. Pont, transl. and ed. John Wiley & Sons; The Pitman Press, Bath, England.

Hering, E. M. 1947. Neue Gattungen und Arten der Fruchtfliegen. Siruna Seva 6: 1–16.

Herting, B. 1957. Das weibliche Postabdomen der calyptraten Fliegen (Diptera) und sein Merkmalswert für die Systematik der Gruppe. Z. Morph. Ökol Tiere 45: 429–461.

Herting, B. 1981. Typenrevisionen einiger paläarktischer Raupenfliegen (Dipt. Tachnidae) und Beschreibungen neuer Arten. Stuttg. Beitr. Naturk., Ser. A, 346: 1–21.

Hinton, H. E. 1946. A new classification of insect pupae. Proc. zool. Soc. Lond. 116: 282–328.

Hinton, H. E. 1949. On the function, origin, and classification of pupae. Proc. S. London ent. nat. Hist. Soc. 1947–48: 111–154.

Hinton, H. E. 1961. The role of the epidermis in the deposition of tracheae and muscles. Sci. Prog., Lond. 49: 329–339.

Holloway, B. A. 1976*a*. A new bat-fly family from New Zealand (Diptera: Mystacinobiidae). N.Z. Jl Zool. 3: 279–301.

Holloway, B. A. 1976*b*. Pollen-feeding in hover-flies (Diptera: Syrphidae). N.Z. Jl Zool. 3: 339–350.

Holmes, E. B. 1980. Reconsideration of some systematic concepts and terms. Evolut. Theory 5: 35–87.

Hori, K. 1960. Comparative anatomy of the internal organs of the calyptrate muscoid flies. 1. Male internal sexual organs of the adult flies. Science Rep. Kanazawa Univ. 7: 23–83.

Hori, K. 1961. Comparative anatomy of the internal organs of the Calyptrate muscoid flies. 2. Female internal sexual organs of the adult fly. Science Rep. Kanazawa Univ. 7: 61–101.

Hori, K. 1967. Comparative anatomy of the internal organs of the calyptrate muscoid flies. V. Consideration on the phylogeny of the Calyptratae. Scient. Rep. Kanazawa Univ. 12: 215–254.

International Commission on Zoological Nomenclature. 1968. Opinion 847, *Chamaemyia* Meigen, 1803 (Insecta, Diptera): Designation of a type-species under the plenary powers. Bull. zool. Nomencl. 25 (1): 16–17.

International Commission on Zoological Nomenclature. 1985. Opinion 1321, Grant of nomenclatural precedence to Ephydridae Zetterstedt, 1837 over Hydrelliidae Robineau-Desvoidy, 1830 (Insecta, Diptera). Bull. zool. Nom. 42 (2): 177–179.

Irwin, M. E. 1976. Morphology of the terminalia and known ovipositing behavior of female Therevidae (Diptera: Asiloidea), with an account of correlated adaptations and comments on phylogenetic relationships. Ann. Natal Mus. 22: 913–935.

James, M. T. 1957. The larva of *Cyphomyia* (Diptera, Stratiomyidae) and its significance in classification. Ann. ent. Soc. Am. 50: 639–641.

Jervis, M. A. 1980. Studies on oviposition behaviour and larval development in species of *Chalarus* (Diptera, Pipunculidae), parasites of typhocybine leafhoppers (Homoptera, Cicadellidae). J. nat. Hist. 14: 759–768.

Jorgensen, N. M., and M. T. James. 1968. Evolution of antenna in the Stratiomyidae (Diptera). Ann. ent. Soc. Am. 61: 870–878.

Jusbaschjanz, S. 1916. Zur Kenntnis der nachembryonalen Entwicklung der Stratiomyiden. Jena. Z. Naturw. 48 (N.F. 38): 681–736.

Kanmiya, K. 1983. A systematic study of the Japanese Chloropidae (Diptera). Mem. ent. Soc. Wash. 11: 1–370.

Keilin, D. 1944. Respiratory systems and respiratory adaptations in larvae and pupae of Diptera. Parasitology 36: 1–66.

Kéler, S. von. 1963. Entomologisches Wörterbuch mit besonderer Berucksichtigung der morphologischen Terminologie. Akademie-Verlag, Berlin [third edition].

Kessel, E. L., M. E. Buegler, and P. M. Keyes. 1973. A survey of the known larvae and puparia of Platypezidae, with a key to ten genera based on immature stages (Diptera). Wasmann J. Bio. 31: 233–261.

Knuston, L. V., J. W. Stephenson, and C. D. Berg. 1970. Biosystematic studies of *Salticella fasciata* (Meigen), a snail-killing fly. Trans. R. ent. Soc. Lond. 122: 81–100.

Krivosheina, N. P. 1969. Ontogeny and evolution of the Diptera [in Russian]. Nauka, Moscow.

Krivosheina, N. P. 1979*a*. A new member of the family Pseudopomyzidae (Diptera) in the Palearctic and the position of the family in the system of Diptera [in Russian]. Ént. Obozr. 58: 179–189 [English translation in Ent. Rev., Wash. 58: 106–113 (1980)].

Krivosheina, N. P. 1979*b*. Systematics and biology of Palearctic species of the family Odiniidae (Diptera) [in

Russian]. Pages 13–157 *in* A. N. Severtsov, ed. Insects of wood and their entomophags. Nauka, Moscow.

Lamb, C. G. 1914. No. 15. Diptera: Heteroneuridae, Ortalidae, Trypetidae, Sepsidae, Micropezidae, Drosophilidae, Geomyzidae, Milichiidae. The Percy Sladen Trust Expedition to The Indian Ocean in 1905 by J. Stanley Gardiner. Trans. Linn. Soc. Lond. (Zool.) 16: 307–372.

Lameere, A. 1906. Notes pour la classification des Diptères. Mém. Soc. r. ent. Belg. 12: 105–140.

Larsson, S. G. 1978. Baltic amber—a palaeobiological study. Entomonograph Vol. I. Scand. Sci. Press, Klampenborg, Denmark.

Latreille, P. A. 1804. Tableau méthodique des Insectes. Pages 129–200 *in* Nouveau dictionnaire d'histoire naturelle. Vol. 24. Paris.

Latreille, P. A. 1829. Les Crustacés, les Arachnides et les Insectes. Tome second, Vol. 5 *in* G. L. C. D. Cuvier, ed. Le règne animal distribué d'après son organisation, pour servir de base à l'histoire naturelle des animaux, et d'introduction à l'anatomie comparée. Ed 2. Paris.

Leach, W. E. 1815. Entomology. Brewsters' Edinburgh Encycl. 9 (1): 52–172.

Leher, A. 1971. Morphologie et homologie des sclérites abdomineau femelle chez les Diptères Cyclorrhapha. 1. Fam. Syrphidae. Bull. Soc. ent. de Mulhouse 1971: 59–66.

Lewis, S. E. 1971. A new species of fossil Diptera (Diopsidae) from the Ruby River Basin (Oligocene) of Montana (USA). Ann. ent. Soc. Am. 64: 959–960.

Loew, H. 1861(1860). Über die Dipterenfauna des Bernsteins. Amtl. Ber. dtsch. Naturf. u. Aertze 35: 88–98.

Loew, H. 1862. Monographs of the Diptera of North America. Part 1. Smithson. Inst., Smithson. misc. Collns 6 (1(= pub. 141)): 1–221.

Loew, H. 1873. Monographs of the Diptera of North America. Part III. Smithson. Inst., Smithson. misc. Collns 11 (3(= pub. 233)): 1–351.

Ludwig, C. E. 1949. Embryology and morphology of the larval head of *Calliphora erythrocephala* (Meigen). Microentomology 14: 75–111.

Máca, J. 1978. Czechoslovak species of the family Odiniidae. Dipt. bohem. / Bratisl. 1: 147–153.

Máca, J. 1980. European species of the subgenus *Amiota* s. str. (Diptera, Drosophilidae). Acta ent. bohem. 77: 328–346.

Máca, J. 1983. Review of findings of the family Acartophthalmidae (Diptera) in Czechoslovakia [in Czechoslovakian]. Cas. slezsk. Mus. Opavě, Ser. A, 32: 233–237.

Mackerras, I. M. 1955. The classification and distribution of Tabanidae (Diptera). II. History: morphology: classification: subfamily Pangoiinae. Aust. J. Zool. 3: 439–511.

Macquart, J. C. H. 1835. Histoire naturelle des Insectes. Diptères, Tome deuxième. Diptera, Vol. 2. *in* N. E. Roret, ed. Collection des suites à Buffon. Paris.

Macquart, J. C. 1843. Diptères exotiques nouveaux ou peu connus. Mem. Soc. Sci. Agric. Arts Lille 1842: 162–460 (also published separately as his "Diptères exotiques nouveaux ou peu connus", Vol. 2, Pt. 3, pp. 5–304, Paris 1843).

Malloch, J. R. 1913. A synopsis of the genera of Agromyzidae, with descriptions of new genera and species. Proc. U.S. natn. Mus. 46: 127–154.

Malloch, J. R. 1914. Notes on the dipterous genus *Chyromya* R.-D. Proc. ent. Soc. Wash. 16: 179–181.

Malloch, J. R. 1919. The limits of the dipterous group Calyptrata (Cyclorrhapha). Bull. Brooklyn ent. Soc. 14: 111–112.

Malloch, J. R. 1921. Forest insects in Illinois I. The subfamily Ochthiphilini (Diptera, Family Agromyzidae). Bull. Ill. St. nat. hist. Survey 13: 345–361.

Malloch, J. R. 1923. A note on the relationships of the Pyrgotidae (Diptera). Ent. News 34: 283.

Malloch, J. R. 1924. Notes on Australian Diptera No. 3. Proc. Linn. Soc. N.S.W. 49: 329–338.

Malloch, J. R. 1925. Notes on Australian Diptera No. 6. Proc. Linn. Soc. N.S.W. 50: 90–92.

Malloch, J. R. 1926. New genera and species of Acalyptrate flies in the United States National Museum. Proc. U.S. natn. Mus. 68: 1–35.

Malloch, J. R. 1927. Notes on Australian Diptera No. 13. Proc. Linn. Soc. N.S.W. 52: 445–446.

Malloch, J. R. 1929. Notes on some Oriental sapromyzid flies (Dipt.) with particular reference to the Philippine species. Proc. U.S. natn. Mus. 2751 (74): 1–97.

Malloch, J. R. 1930*a*. New Zealand Muscidae Acalyptratae. Family Heleomyzidae (supplement). Rec. Canterbury Mus. 4: 333–334.

Malloch, J. R. 1930*b*. Notes on Australian Diptera. 26. Proc. Linn. Soc. N.S.W. 55: 488–492.

Malloch, J. R. 1931*a*. Exotic Muscaridae (Diptera). 32. Ann. Mag. nat. Hist. (10) 7: 314–340.

Malloch, J. R. 1931*b*. Notes on Australian Diptera 29. Proc. Linn. Soc. N.S.W. 56: 292–298.

Malloch, J. R. 1932*a*. Notes on Australian Diptera No. 32. Proc. Linn. Soc. N.S.W. 57: 213–216.

Malloch, J. R. 1932*b*. A new genus of diopsid-like Diptera (Periscelidae). Stylops 1: 266–268.

Malloch, J. R. 1933*a*. The genus *Coelopa* Meigen (Diptera, Coelopidae). Ann. Mag. nat. Hist. (10) 11: 330–350.

Malloch, J. R. 1933*b*. Acalyptrata. Vol. 6 (4). Pages 177–392 *in* British Museum (Natural History), Diptera of Patagonia and Southern Chile. Oxford University Press, Oxford, England.

Malloch, J. R. 1934. Acalyptrate [concl.]. Vol. 6 (5). Pages 393–489 *in* British Museum (Natural History), Diptera of Patagonia and Southern Chile. Oxford University Press, Oxford, England.

Malloch, J. R. 1936. Notes on Australian Diptera. No. 36. Proc. Linn. Soc. N.S.W. 59: 259–261.

Malloch, J. R. 1940. The North American genera of the dipterous subfamily Chamaemyiinae. Ann. Mag. nat. Hist. (11) 6: 265–274.

Malloch, J. R. 1948. Key to families of Acalyptrata, with notes of some of the families. Vol. 6 (6). Pages 491–507 *in* British Museum (Natural History), Diptera of Patagonia and Southern Chile. Oxford University Press, Oxford, England.

Malloch, J. R., and W. L. McAtee. 1924. Keys to the families Lonchaeidae, Pallopteridae, and Sapromyzidae of the eastern United States, with a list of the species of the District of Columbia region. Proc. U.S. natn. Mus. 65 (12): 1–26.

Mathis, W. N. 1979. Studies of Notiphilinae (Diptera: Ephydridae), I: Revision of the Nearctic species of *Notiphila* Fallén, excluding the *caudata* group. Smithson. Contr. Zool. 287: 1–111.

Mathis, W. N. 1981. Proposed use of the plenary powers to grant precedence to the family-group name Ephydridae over Hydrelliidae (Insecta, Diptera), Z. N. (S.) 2334. Bull. zool. Nom. 38: 201–204.

Mathis, W. N. 1982. Studies of Canacidae (Diptera), I: Suprageneric revision of the family, with revisions of new tribe Dynomiellini and new genus *Isocanace*. Smithson. Contr. Zool. 347: 1–39.

McAlpine, D. K. 1958*a*. A family of flies new to Australia (Diptera, Rhinotoridae). Proc. R. zool. Soc. N.S.W. 1956–1957: 64–65.

McAlpine, D. K. 1958*b*. A key to the families of acalyptrate Diptera (Insecta). Rec. Aust. Mus. 24 (12): 183–189.

McAlpine, D. K. 1960. A review of the Australian species of Clusiidae (Diptera, Acalyptrata). Rec. Aust. Mus. 25: 63–94.

McAlpine, D. K. 1966. Description and biology of an Australian species of Cypselosomatidae (Diptera), with a discussion of family relationships. Aust. J. Zool. 14: 673–685.

McAlpine, D. K. 1967. The Australian species of *Diplogeomyza* and allied genera (Diptera, Heleomyzidae). Proc. Linn. Soc. N.S.W, 92: 76–106.

McAlpine, D. K. 1968. The genus *Cairnsimyia* Malloch (Diptera, Heleomyzidae, Rhinotorini). Rec. Aust. Mus. 27: 263–283.

McAlpine, D. K. 1973. The Australian Platystomatidae (Diptera, Schizophora) with a revision of five genera. Mem. Aust. Mus. 15: 1–256.

McAlpine, D. K. 1974*a*. The subfamily classification of the Micropezidae and the genera of Eurbatinae (Diptera: Schizophora). J. Ent., Ser. B, 43: 231–245.

McAlpine, D. K. 1974*b*. Some field observations on *Nothybus* (Diptera, Nothybidae). Aust. ent. Mag. 1: 89–91.

McAlpine, D. K. 1976. A new genus of flies possibly referable to Cryptochetidae (Diptera, Schizophora). Aust. ent. Mag. 3: 45–56.

McAlpine, D. K. 1978. A new species of Cypselosomatidae from Lord Howe Island (Diptera, Micropezoidea). Aust. ent. Mag. 5: 61–64.

McAlpine, D. K. 1982. A new genus of Australian littoral flies (Diptera: Canacidae). Mem. ent. Soc. Wash. 10: 108–117.

McAlpine, D. K. 1983. A new subfamily of Aulacigastridae (Diptera: Schizophora), with a discussion of Aulacigastrid classification. Aust. J. Zool. 31: 55–78.

McAlpine, D. K. 1985*a*. Taxonomic notes on the genus *Zale* McAlpine (Diptera: Canacidae). Aust. ent. Mag. 11: 81–82.

McAlpine, D. K. 1985*b*. The Australian genera of Heleomyzidae (Diptera: Schizophora) and a reclassification of the family into tribes. Rec. Aust. Mus. 36: 203–251.

McAlpine, D. K. 1985*c*. Upside-down flies. Aust. nat. Hist. 21: 454–456.

McAlpine, D. K., and D. S. Kent. (1981) 1982. Systematics of *Tapeigaster* (Diptera: Heleomyzidae) with notes on biology and larval morphology. Proc. Linn. Soc. N.S.W. 106: 33–58.

McAlpine, J. F. 1960. First record of the family Camillidae in the New World (Diptera). Can. Ent. 92: 954–956.

McAlpine, J. F. 1962. The evolution of the Lonchaeidae. Ph.D. thesis, University of Illinois. Published on demand by University Microfilms Inc., Ann Arbor, Mich.

McAlpine, J. F. 1963. Relationships of *Cremifania* Czerny (Diptera: Chamaemyiidae) and description of a new species. Can. Ent. 95: 239–253.

McAlpine, J. F. 1967. A detailed study of Ironomyiidae (Diptera: Phoroidea). Can. Ent. 99: 225–236.

McAlpine, J. F. 1968. Taxonomic notes on *Eurychoromyia mallea* (Diptera: Eurychoromyiidae). Can. Ent. 100: 819–823.

McAlpine, J. F. 1970. First record of calypterate flies in the Mesozoic Era (Diptera, Calliphoridae). Can Ent. 102: 342–346.

McAlpine J. F. 1973. A fossil ironomyiid fly from Canadian amber (Diptera, Ironomyiidae). Can. Ent. 105: 105–111.

McAlpine, J. F. 1976. Systematic position of the genus *Omomyia* Coquillett and transference to the Richardiidae (Diptera). Can. Ent. 198: 849–853.

McAlpine, J. F. 1977. A revised classification of the Piophilidae including, 'Neottiophilidae,' and 'Thyreophoridae' (Diptera: Schizophora). Mem. Can. Ent. 103: 1–66.

McAlpine, J. F. 1981*a*. Chapter 2. Morphology and terminology: adults. Pages 9–63 *in* J. F. McAlpine *et al.*, eds. Manual of Nearctic Diptera. Vol. 1. Research Branch, Agriculture Canada, Ottawa. Agric. Can. Monog. 27.

McAlpine, J. F. 1981*b*. Chapter 4. Key to families: adults. Pages 89–124 *in* J. F. McAlpine *et al.*, eds. Manual of Nearctic Diptera. Vol. 1. Research Branch, Agriculture Canada, Ottawa. Agric. Can. Monogr. 27.

McAlpine, J. F. 1981*c*. *Morgea freidbergi* new species, a living sister-species of the fossil species *M. mcalpinei* and a key to world genera of Pallopteridae (Diptera). Can. Ent. 113: 81-91.

McAlpine, J. F. 1982. A review of the genus *Chaetolonchaea* Czerny, with description of a new species from North America (Diptera: Lonchaeidae). Mem. ent. Soc. Wash. 10: 118–124.

McAlpine, J. F. 1983. A revision of the genus *Protearomyia* McAlpine. Can. Ent. 115: 885–903.

McAlpine, J. F., and J. E. H. Martin. 1966. Systematics of Sciadoceridae and relatives, with descriptions of two new genera and species from Canadian amber, and erection of family Ironomyiidae (Diptera: Phoroidea). Can. Ent. 98: 527–544.

McAlpine, J. F., and J. E. H. Martin. 1969. Canadian amber: a paleontological treasure-chest. Can. Ent. 101: 819–838.

McAlpine, J. F., and D. D. Munroe. 1968. Swarming of lonchaeid flies and other insects, with description of four species of Lonchaeidae (Diptera). Can. Ent. 100: 1154–1178.

McAlpine, J. F., and C. W. Sabrosky. 1965. *Chamaemyia* Meigen, 180 (Insecta, Diptera, Chamaemyiidae); proposed designation of a type species under the plenary powers. Z. N. (S.) 1716. Bull. zool. Nom. 22: 267–268.

McAlpine, J. F., and G. C. Steyskal. 1982. A revision of *Neosilba* McAlpine with a key to the world genera of Lonchaeidae (Diptera). Can. Ent. 114: 105–137.

McAlpine, J. F., B. V. Peterson, G. E. Shewell, H. J. Teskey, J. R. Vockeroth, and D. M. Wood, coordinators. 1981. Manual of Nearctic Diptera. Vol. 1. Research Branch, Agriculture Canada, Ottawa. Agric. Can. Monogr. 27.

Meigen, J. W. 1803. Versuch einer neuen Gattungseintheilung der europaischen Zweiflügeligen Insekten. Magazin Insekten. (Illiger) 2: 259–281.

Melander, A. L. 1949. A report on some Miocene Diptera from Florissant, Colorado. Am. Mus. Novit. 1407: 1–63.

Melander, A. L. 1952. The American species of Trixoscelidae. Jl N.Y. ent. Soc. 60: 37–52.

Miller, D. 1950. Catalogue of the Diptera of the New Zealand sub-region. Bull. N.Z. Dep. scient. ind. Res. 100: 1–194.

Miller, R. M. 1977*a*. Taxonomy and biology of the Nearctic species of *Homoneura* (Diptera: Lauxaniidae) 1. Subgenera *Mallochomyza* and *Tarsohomoneura*. Iowa St. J. Res. 52: 147–176.

Miller, R. M. 1977*b*. Ecology of Lauxaniidae (Diptera: Acalyptratae) 1. Old and new rearing records with biological notes and discussion. Ann. Natal Mus. 23: 215–238.

Miller, R. M. 1980*a*. 53 Lauxaniidae. Pages 603–608 *in* R. W. Crosskey *et al.*, eds. Catalogue of the Diptera of the Afrotropical region. Publ. No. 821. British Musem (Natural History), London. Fletcher and Son Ltd., Norwich, England.

Miller, R. M. 1980*b*. 54 Family Celyphidae. Page 609 *in* R. W. Crosskey *et al.*, eds. Catalogue of the Diptera of the Afrotropical Region. Publ. No. 821. British Museum (Natural History), London. Fletcher and Son Ltd., Norwich, England.

Morge, G. 1963. Die Lonchaeidae und Pallopteridae Osterreichs und der angrenzenden Gebeite. 1. Teil: Die Lonchaeidae. Naturk. Jb. Stadt Linz 1963: 123–314.

Nagatomi, A. 1982. Ninth sternum and genital fork in female Nematocera and orthorrhaphous Brachycera (Diptera). Kontyû 50: 95–96.

Nagatomi, A., and K. Iwata. 1976. Female terminalia of lower Brachycera (Diptera). 1. Beitr. Ent. 26: 5–47.

Nagatomi, A., and K. Iwata. 1978. Female terminalia of lower Brachycera (Diptera) 2. Beitr. Ent. 28: 263–293.

Narchuk, E. P. 1983. Classification of the superfamily Chloropoidea (Diptera, Cyclorrhapha) [in Russian]. Ént. Obozr. 62: 638–648 [English translation in Ent. Rev., Wash. 62: 180–193 (1984)].

Newman, E. 1834. Art. 37. Attempted division of British insects into natural orders. Ent. Mag. 2: 379–431.

Newstead, R., A. M. Evans, and W. H. Potts. 1924. Guide to the study of tsetse-flies. Mem. Lpool Sch. trop. Med. (new series) 1: 1–382.

Nitzsch, C. L. 1818. Die Familien und Gattungen der Thierinsekten (Insecta epizoica) als Prodromus einer Naturgeschichte derselben. Magazin Ent. Halle 3: 261–316.

Nowakowski, J. T. 1962. Introduction to a systematic revision of the family Agromyzidae (Diptera) with some remarks on host selection by these flies. Annls zool., Warsz. 20: 67–183.

Okada, T. 1981. 6. Oriental species, including New Guinea. Pages 261–290 *in* M. Ashburner *et al. eds*. The genetics and biology of *Drosophila*. Vol. 3a. Academic Press, New York, N.Y.

Oldroyd, H. 1964. The natural history of flies. Wiedenfeld and Nicholson, London.

Örözi Pal, Z. 1966. Die Bienenlaus-Arten. Angew. Parasit. 7: 138–171.

Osten Sacken, C. R. 1878. Catalogue of the described Diptera of North America (Ed. 2). Smithson. misc. Collns 16 (2 = publ. 270): 1–276.

Ozerov, A. L. 1984. A new Palearctic genus of the family Thyreophoridae (Diptera) from the Soviet Far East [in Russian]. Zool. Zh. 63: 464–466.

Papavero, N. 1967–. Ed. A catalogue of the Diptera of the Americas south of the United States. Fasc. 1–110. Departamento de Zoologia, Secretaria da Agricultura, São Paulo, Brasil.

Papavero, N. 1967*a*. Explanatory Note. Pages v–xiii *in* N. Papavero, ed. A catalogue of the Diptera of the Americas south of the United States. Fasc. 1. Departamento de Zoologia, Secretaria da Agricultura, São Paulo, Brasil.

Papavero, N. 1967*b*. 87. Family Rhinotoridae. Pages 87.1–4 *in* N. Papavero, ed. A catalogue of the Diptera of the Americas south of the United States. Departamento de Zoologia, Secretaria da Agricultura, São Paulo, Brasil.

Papavero, N. 1977. The world Oestridae (Diptera), mammals and continental drift. Series entomologica, 14. Dr. W. Junk., N.V., The Hague.

Papp, L. J. 1972. Camillids from Mongolia (Diptera, Acalyptratae). Acta zool. hung. 18: 327–332.

Papp, L. J. 1974. Some Diastatidae, Camillidae, Aulacigastridae and Asteiidae (Diptera) from Asia. Acta zool. hung. 20: 177–181.

Papp, L. 1977*a*. One species each of Hyperoscelididae, Pyrgotidae, Odiniidae, Acartophthalmidae, and *Risa* Becker from Mongolia (Diptera). Folia ent. hung. (Series Nova) 30: 119–122.

Papp, L. 1977*b*. Notes on some Becker's types (Diptera,

Carnidae and Risidae fam. n.). Annls hist. nat. Mus. natn. hung. 69: 185–189.
Papp, L. J. 1978*a*. Some cavernicolous Diptera of the Geneva Museum. Revue suisse Zool. 85: 99–106.
Papp, L. 1978*b*. Species of nine Acalyptrate fly families from Tunisia (Diptera). Folia ent. hung. 31: 211–217.
Papp, L. 1979. Seven new species of the Palaearctic Lauxaniidae and Asteiidae (Diptera). Reichenbachia 17: 87–97.
Papp, L. 1980. New taxa of the Acalyptrate flies (Diptera: Tunisimyiidae fam. n., Risidae, Ephydridae: Nannodastinae subfam. n.). Acta zool. hung. 26: 415–431.
Papp, L. 1984. Family Xenasteiidae. Pages 176–177 *in* A. Soós and L. Papp, eds. Catalogue of Palaearctic Diptera. Vol. 10. Elsevier Science Publishers, Amsterdam, and Akadémiai Kiadó, Budapest.
Paramanov, S. J. 1956. Notes on Australian Diptera. 21. Mindidae: a new family of Acalyptrata (Diptera). Ann. Mag. nat. Hist. (12) 9: 779–783.
Paramanov, S. J. 1958. A revision of Australian Pyrgotidae (Diptera). Aust. J. Zool. 6: 89–138.
Paramanov, S. J. 1961. Notes on Australian Diptera 32–36. 32. Echiniidae: a new family of Diptera (Acalyptrata). Ann. Mag. nat. Hist. (13) 4: 97–100.
Parsons, P. A., and S. M. Stanley. 1981. 9. Domesticated and widespread species. Pages 349–374 *in* M. Ashburner *et al.*, eds. The genetics and biology of *Drosophila*. Vol. 3*a*. Academic Press, New York, N.Y.
Peterson, A. 1951. Larvae of insects. An introduction to Nearctic species. Part II. Edwards Bros., Inc., Ann Arbor, Mich.
Pont, A. C. 1977. A revision of Australian Fanniidae (Diptera: Calyptrata). Aust. J. Zool. Suppl. Ser. 51: 1–60.
Pont, A. C. 1979. Sepsidae. Diptera Cyclorrhapha, Acalyptrata. Handbk Ident. Br. Insects 10 (5c): 1–35.
Postlethwait, J. H., and H. A. Schneiderman. 1974. Developmental genetics of *Drosophila* imaginal disc. Ann. Rev. Genetics 7: 381–433.
Potts, W. H. 1973. Glossinidae (tsetse-flies). Pages 209–249 *in* K. G. V. Smith, ed. Insects and other arthropods of medical importance. British Museum (Natural History), London.
Prado, A. P., do. 1966. Segundo contribuicão ao conhecimento da familia' Rhopalomeridae (Diptera, Acalyptratae). Studia ent. (1965) 8: 209–268.
Prado, A. P., do. 1969. Syringogastridae, uma nova familia de dipteros Acalyptratae, com a descrição de seis espécies novas do gênera *Syringogaster* Cresson. Studia ent. 12: 1–32.
Richards, O. W. 1962. 71. Diptera. Two new East African species of *Alombus* Beck. and an apterous sepsid (Chloropidae and Sepsidae). Annls Mus. r. Afr. cent. Sér. 4 + 0 107: 464–468.
Richards, O. W. 1967. 72. Family Sphaeroceridae. Pages 72.1–28 *in* N. Papavero, ed. A Catalogue of the Diptera of the Americas south of the United States. Departmento de Zoologia, Secretaria da Agricultura, São Paulo, Brasil.
Richards, O. W. 1973. The Sphaeroceridae (= Borboridae or Cypselidae; Diptera Cyclorrhapha) of the Australian region. Aust. J. Zool. Suppl. Ser. 22: 297–401.
Richards, O. W. 1980. 57. Family Sphaeroceridae. Pages 614–626 *in* R. W. Crosskey *et al.*, eds. Catalogue of the Diptera of the Afrotropical region. Publ. No. 821. British Museum (Natural History), London. Fletcher and Son Ltd., Norwich, England.
Richter, V. A. 1980. On some plesiomorphous characters of male genitalia in the family Tachinidae (Diptera) [in Russian]. Ént. Obozr. 59: 925–934.
Ride, W. D. L., C. W. Sabrosky, *et al.* 1985. International Code of Zoological Nomenclature (ed.). Int. Trust zoo. Nomenclature. London.
Roback, S. S. 1951. A classification of the muscoid calyptrate Diptera. Ann. ent. Soc. Am. 44: 327–361.
Roberts, H. 1969. Forest insects of Nigeria with notes on their biology and distribution. Inst. Pap. Commonw. For. Inst. 44: 1–206.
Roberts M. J. 1969*a*. Structure of the mouthparts of the larvae of the flies *Rhagio* and *Sargus* in relation to feeding habits. J. Zool. Lond. (1969) 159: 381–398.
Roberts, M. J. 1969*b*. Pharyngeal grinding mills in some dipteran larvae. Entomologist 102: 279–398.
Roberts, M. J. 1971. The structure of the mouthparts of some calypterate dipteran larvae in relation to their feeding habits. Acta. zool., Stockh. 52: 171–188.
Robineau-Desvoidy, J. B. 1830. Essaie sur les Myodaires. Mém. Acad. r. Sci. Inst. Fr., (Ser.2), 2: 1–813.
Roddy, L. R. 1955. A morphological study of the respiratory horns associated with the puparia of some Diptera, especially *Ophyra aenescens* (Wied.). Ann. ent. Soc. Am. 48: 407–415.
Rognes, K. 1986. The systematic position of the genus *Helicobosca* Bezzi with a discussion of the monophyly of the calyptrate families Calliphoridae, Rhinophoridae, Sarcophagidae and Tachinidae (Diptera). Ent. Scand. 17: 75–92.
Rohácěk, J. 1983. Faunistics of the Czechoslovakian species of Anthomyzidae and Stenomicridae (Diptera) [in Czechoslovakian]. Cas. slezsk. Mus. Opavě, Ser. A, 32: 125–135.
Rohdendorf, B. B. 1948. Novae semeistvo paraziticheskikh muskoidnykh dvukrylykh iz peskov Zavolzh'ya. Doklady Akad. Nauk SSSR 63: 455–458.
Rohdendorf, B. B. 1964. The historical development of two-winged insects [in Russian]. Trudy paleont. Inst. 100: 1–311.
Rohdendorf, B. B. 1974. The historical development of Diptera [transl. from Russian] B. Hocking, H. Oldroyd, and G. E. Ball, eds. University of Alberta Press, Edmonton, Alta.
Rohdendorf, B. B. 1977. The classification and phylogeny of the Diptera. Pages 81–88 *in* O. A. Scarlato and K. B. Gorodkov, eds. Systematics and evolution of the Diptera (Insecta) [in Russian]. A collection of scientific works. Acad. Sci. U.S.S.R., Zool. Inst., Leningrad.
Rondani, C. 1856. Dipterologiae Italicae prodromus Vol. 1. A. Stoschi, Parama.

Ross, H. H. 1965. A textbook of entomology [3rd ed]. John Wiley & Sons, Inc., New York.

Rozkošný, R. 1973. The Stratiomydae (Diptera) of Fennoscandia and Denmark. Fauna ent. scand. 1: 1–140.

Rozkošný, R. 1982. A biosystematic study of the European Stratiomyidae (Diptera). Volume 1. Introduction, Beridinae, Sarginae and Stratiomyinae. Series entomologica, 21. Dr. W. Junk, N.V., The Hague.

Rozkošný R. 1983. A biosystematic study of the European Stratiomyidae (Diptera). Volume 2. Clitellariinae, Hermatiinae, Pachygastrinae, and bibliography. Series entomologica, 25. Dr. W. Junk, N.V., The Hague.

Sabrosky, C. W. 1956. Additions to the knowledge of Old World Asteiidae (Diptera). Revue fr. Ent. 23: 216–243.

Sabrosky, C. W. 1962. The "new family" Echiniidae and the chloropid genus *Anatrichus* (Diptera). Ann. Mag. nat. Hist. (13) 4: 559–560.

Sabrosky, C. W. (1976) 1977. 20. Fam. Asteiidae. La faune terrestre de l'Ile de Sainte-Hélène. Troisième Partie. Mus. R. l'Afrique Centrale, Tervuren, Belg. Ann. Ser. In-8°. Sci. zool. No. 215, 1976: 96–100.

Sabrosky, C. W. 1978. The family position of the peculiar genus *Horaismoptera* (Diptera: Tethinidae). Ent. Germ. 4: 327–336.

Salzer, R. 1968. Konstruktionsanatomische Untersuchung des männlichen Postabdomens von *Calliphora erythrocephala* Meigen (Insecta, Diptera). Z. Morph. Ökol. Tiere 63: 155–238.

Samouelle, G. 1819. The entomologist's useful compendium, or an introduction to the knowledge of British insects. Boys, London.

Schiner, J. R. 1862(1864). Fauna Austriaca. Pages 1–80 *in* Die Fliegen (Diptera). Vol. 2 (Heft. 8). Gerold, Wien.

Schiner, J. R. 1864. Ein neues System der Dipteren. Verh. zool.-bot. Ges. Wien 14: 201–212.

Schlein, Y. 1970. A comparative study of the thoracic skeleton and musculature of the Pupipara and the Glossinidae (Diptera). Parasitology 60: 327–373.

Schlein, Y., and O. Theodor. 1971. On the genitalia of the Pupipara and their homologies with those of *Calliphora* and *Glossina*. Parasitology 63: 331–342.

Schräder, T. 1927. Das Hypopygium 'circumversum' von *Calliphora erythrocephala*. Ein Beitrag Zur Kenntnis des Kopulationsapparates der Dipteren. Z. Morph. Ökol. Tiere 8: 1–44.

Scudder, S. H. 1877. The insects of the Tertiary beds at Quesnel (British Columbia). Rep. geol. Surv. Can. 1875–1876: 12.

Scudder, S. H. 1890. The Tertiary insects of North America. Rep. U.S. geol. Surv. 13: 1–734.

Séguy, E. 1938. Mission scientifique de l'Omo. Diptera I. Nematocera et Brachycera. Mém. Mus. nat. Paris (N.S.) 8: 319–380.

Séguy, E. 1950. La biologie des diptères. Encycl. ent. 26: 213, 249, 514–515.

Shewell, G. E. 1977. Family Lauxaniidae. Pages 183–214 *in* M. D. Delphinado, and D. E. Hardy, eds. A catalogue of the Diptera of the Oriental region. Vol. 3. University Press of Hawaii, Honolulu.

Shillito, J. F. 1971. The genera of Diopsidae (Insecta: Diptera). J. Linn. Soc. (Zool.) 50: 287–295.

Singh, S. B., and W. W. Judd. 1966. A comparative study of the alimentary canal of adult calyptrate Diptera. Proc. ent. Soc. Ont. 96: 29–80.

Skidmore, P. 1985. The biology of the Muscidae of the world. Series entomologica, 29. Dr. W. Junk, Dordrecht, The Netherlands.

Snodgrass, R. E. 1957. A revised interpretation of the external reproductive organs of male insects. Smithson. misc. Collns 135 (6): 1–60.

Soós, Á. 1977. Taxonomische und faunistiche Untersuchungen über die Mongolischen Trixoscelididen (Diptera: Acalyptratae). Acta zool. hung. 23: 395–413.

Soós, Á., and L. Papp, eds. 1984*a*. Catalogue of palaearctic Diptera. Vol. 9. Elsevier Science Publishers, Amsterdam, and Akadémiai Kiadó, Budapest.

Soós, Á., and L. Papp, eds. 1984*b*. Catalogue of palaearctic Diptera. Vol. 10. Elsevier Science Publishers, Amsterdam, and Akadémiai Kiadó, Budapest.

Speight, M. C. D. 1969. The prothoracic morphology of acalyptrates (Diptera) and its use in systematics. Trans. R. ent. Soc. Lond. 121: 325–421.

Speiser, P. 1903. Eine neue Dipterengattung mit rudimentaren Flügeln, und andere dipterologische Bemerkungen. Berl. ent. Z. 48: 65–72.

Spencer, A. K. 1969. The Agromyzidae of Canada and Alaska. Mem. ent. Soc. Can. 64: 1–311.

Stackelberg, A. A. 1969. Key to families. Pages 35–55 *in* G. Ya. Bie-Bienko, ed. Opredelitel nasekomykh Europeskoi chasti SSSR 5(2) (in Russian).

Stackelberg, A. A. 1970. 103. Family Eginiidae. Page 595 *in* G. Ya. Bie-Bienko, ed. Opredelitel nasekomykh Europeskoi chasti SSSR 5(2) (in Russian).

Stenhammar, C. (1853) 1854. Skandinaviens Copromyzinae. K. svenska Vetensk-Akad. Handl. (3) 1853: 257–442. [Also published separately under the title "Copromyzinae Scandinaviae," 184 pp., Holmiae (= Stockholm), 1855].

Stevens, P. F. 1980. Evolutionary polarity of character states. A. Rev. ecol. Syst. 11: 333–358.

Steyskal, G. C. 1957*a*. The postabdomen of male acalyptrate Diptera. Ann. ent. Soc. Am. 50: 66–73.

Steyskal, G. C. 1957*b*. A revision of the family Dryomyzidae (Diptera, Acalyptratae). Pap. Mich. Acad. Sci. 42: 55–68.

Steyskal, G. C. 1958. Notes on Nearctic Helcomyzidae and Dryomyzidae (Diptera, Acalyptratae). Pap. Mich. Acad. Sci. 43: 133–143.

Steyskal, G. C. 1961. The genera of Platystomatidae and Otitidae known to occur in America north of Mexico (Diptera, Acalyptratae). Ann. ent. Soc. Am. 54: 401–410.

Steyskal, G. C. 1965. The subfamilies of Sciomyzidae of the world (Diptera: Acalyptratae). Ann. ent. Soc. Am. 58: 593–594.

Steyskal, G. C. 1967. 60. Family Ropalomeridae. 60.1–7

in N. Papavero, ed. A catalogue of the Diptera of the Americas south of the United States. Departmento de Zoologia, Secretaria de Agricultura, São Paulo, Brasil.

Steyskal, G. C. 1968*a*. 49. Family Neriidae. Pages 49.1–7 *in* N. Papavero, ed. A catalogue of the Diptera of the Americas south of the United States. Departmento de Zoologia, Secretaria da Agricultura, São Paulo, Brasil.

Steyskal, G. C. 1968*b*. Notes and descriptions of Egyptian acalyptrate Diptera. Bull. Soc. ent. Égypte 50: 109–125.

Steyskal, G. C. 1968*c*. 53. Superfamily Tephritoidea. Pages 53.1–26 *in* Papavero, ed. A catalogue of the Diptera of the Americas south of the United States. Departmento de Zoologia, Secretaria da Agricultura, São Paulo, Brasil.

Steyskal, G. C. 1968*d*. The genus *Somatia* Schiner, II. (Diptera, Somatiidae). Proc. ent. Soc. Wash. 70: 255–257.

Steyskal, G. C. 1971. The genus *Paraleucopis* Malloch (Diptera: Chamaemyiidae), with one new species. Ent. News 82: 1–4.

Steyskal, G. C. 1972*a*. A catalogue of species and key to genera of the family Diopsidae. Stuttg. Beitr. Naturk. 234: 1–20.

Steyskal, G. C. 1972*b*. African Pyrgotidae (Diptera). Stuttg. Beitr. Naturk. 238: 1–11.

Steyskal, G. C. 1974. Recent advances in the primary classification of the Diptera. Ann. ent. Soc. Am. 67: 513–517.

Steyskal, G. C. 1976. The terminology of bristles on the upper back of the head in the higher Diptera. J. Kans. ent. Soc. 49: 155–159.

Steyskal, G. C. 1977. Family Neriidae. Pages 8–11 *in* M. C. Delphinado and D. C. Hardy, eds. A catalog of the Diptera of the Oriental region. Vol. 3. Suborder Cyclorrhapha (excluding Division Aschiza). University Press of Hawaii, Honolulu.

Steyskal, G. C. 1978. Synopsis of the North American Pyrgotidae (Diptera). Proc. ent. Soc. Wash. 80: 149–155.

Steyskal, G. C. 1980. 45. Family Neriidae. Page 578 *in* R. W. Crosskey *et al.*, eds. Catalogue of the Diptera of the Afrotropical region. Publ. No. 821. British Museum (Natural History), London. Fletcher and Son Ltd., Norwich, England.

Steyskal, G. C. 1981. A new "bobo" fly from the Gulf of California (Diptera: Chamaemyiidae: *Paraleucopis mexicana*). Proc. ent. Soc. Wash. 83: 403–405.

Stone, A. C., C. W. Sabrosky, W. W. Wirth, R. H. Foote, and J. R. Colson, eds. 1965. A catalog of the Diptera of America north of Mexico. U.S. Dept. Agric., Agric. Res. Serv., Agric. Handbk 276.

Strickland, E. H. 1953. The ptilinal armature of flies (Diptera, Schizophora). Can. J. Zool. 31: 263–299.

Stuckenberg, B. R. 1971. A review of the Old World genera of Lauxaniidae (Diptera). Ann. Natal Mus. 20: 449–610.

Sturtevant, A. H. 1925–1926. The seminal receptacles and accessory glands of the Diptera, with special reference to the Acalyptratae. Jl N.Y. ent. Soc. 33: 195–215; 34: 1–24.

Sturtevant, A. H. 1942. The classification of the genus *Drosophila* with descriptions of nine new species. Univ. Tex. Publs 4213: 5–51.

Sturtevant, A. H. 1954. Nearctic flies of the family Periscelidae (Diptera) and certain Anthomyzidae referred to the family. Proc. U.S. natn. Mus. 103: 551–561.

Sturtevant, A. H. 1963. A fossil periscelid (Diptera) from the amber of Chiapas, Mexico. J. Paleont. 37: 121–122.

Süss, H., and W. R. Müller-Stoll. 1975. Durch *Palaeophytobia platani* n.g., n.sp. (Agromyzidae, Diptera) verursachte Markflecken im Holz fossiler Platanen aus dem ungarischen Miozän. Wiss. Z. Humboldt-Univ. Berl. 24: 515–519.

Tanasijtschuk, V. N. 1984. Families Chamaemyiidae and Cremifaniidae. Pages 220–232 *in* A. Soós and L. Papp, eds. Catalogue of palaearctic Diptera. Vol. 9. Elsevier Scientific Publishers, Amsterdam, and Akadémiai Kiadó, Budapest.

Tenorio, J. M. 1969. A revision of the Celyphidae (Diptera) from the Philippine Islands. Pacif. Insects 11: 579–611.

Tenorio, J. M. 1972. A revision of the Celyphidae (Diptera) of the Oriental region. Trans. R. ent. Soc. Lond. 123: 359–453.

Tenorio, J. M. 1979. Family Celyphidae. Pages 214–222 *in* M. D. Delfinado and D. E. Hardy, eds. A catalog of the Diptera of the Oriental region. Vol. 3. University Press of Hawaii, Honolulu.

Teskey, H. J. 1981*a*. Chapter 3. Morphology and terminology: Larvae. Pages 65–88 *in* J. F. McAlpine *et al.*, eds. Manual of Nearctic Diptera. Vol. 1. Research Branch, Agriculture Canada, Ottawa. Agric. Can. Monogr. 27.

Teskey, H. J. 1981*b*. Chapter 5. Key to families: Larvae. Pages 125–147 *in* J. F. McAlpine *et al.*, eds. Manual of Nearctic Diptera. Vol. 1. Research Branch, Agriculture Canada, Ottawa. Agric. Can. Monogr. 27.

Theodor, O. 1963. Über den Bau der Genitalien bei den Hippobosciden (Dipt.). Stuttg. Beitr. Naturk. 108: 1–15.

Theodor, O. 1967. An illustrated catalogue of the Rothschild collection of Nycteribiidae (Diptera) in the British Museum (Natural History). Whitefriars Press Ltd., London.

Theodor, O. 1975. Fauna Palaestina. Insecta 1: Diptera Pupipara. Israel Academy of Sciences and Humanities. Jerusalem Post Press, Jerusalem.

Theodor, O. 1976. On the structure of the spermathecae and the aedeagus in the Asilidae and their importance in the systematics of the family. Israel Academy of Science and Humanities, Jerusalem.

Theodor, O. 1983. The genitalia of Bombyliidae (Diptera). Israel Academy of Sciences and Humanities, Jerusalem.

Thorpe, W. H. 1941. A description of six new species of the genus *Cryptochaetum* (Diptera-Agromyzidae) from East Africa and East Indies, together with a key to adults and larvae of all known species. Parasitology 3: 131–148.

Throckmorton, L. H. 1975. The phylogeny, ecology and geography of *Drosophila*. Ch. 17. Pages 421–469 *in* R. C. King, ed. Handbook of Genetics. Vol. 3. Plenum Pub. Corp., New York, N.Y.

Tonnoir, A. L. 1916. A new and primitive sub-family of the Phoridae (Dipt.). Rec. Canterbury Mus. 3: 32–38.

Tonnoir, A. L. 1937. Revision of the genus *Fergusonina* Mall. (Diptera, Agromyzidae). Proc. Linn. Soc. N.S.W. 62: 126–146.

Townsend, C. H. T. 1935. Manual of myiology in twelve parts. Part 2: Muscoid classification and habits. São Paulo, Brasil.

Tsacas, L., D. Lachaise, and J. R. David. 1981. 5. Compostion and biogeography of the Afrotropical drosophilid fauna. Pages 197–260 *in* M. Ashburner *et al.*, eds. The genetics and biology of *Drosophila*. Vol. 3a. Academic Press, New York, N.Y.

Val, F. C., C. R. Vilela, and M. D. Marques. 1981. Pages 123–168 *in* M. Ashburner *et al.*, eds. The genetics and biology of *Drosophila*. Vol. 3a. Academic Press, New York, N.Y.

van Emden, F. I. 1950. *Mormotomyia hirsuta* Austen (Diptera) and its systematic position. Proc. R. ent. Soc. Lond., Ser. B 19: 121–128.

van Emden, F. I. 1965. Muscidae. Fauna India, Diptera 7: 1–647.

Verbeke, J. 1950. Sciomyzidae (Diptera Cyclorrhapha). Explor. Parc. natn. Albert Miss. G. F. de Witte (1933–1935) 66: 1–97.

Verbeke, J. 1961. Sciomyzidae (Diptera Brachycera Malacophaga). Explor. Parc. natn. Upemba Miss. G. F. de Witte (1946–1949) 61: 1–37.

Verhoeff, C. 1893. Vergleichende Untersuchungen über die Abdominalsegmente und die Copulationsorgane der männlichen Coleoptera. Dt. ent. Z. 1893: 113–170.

Watrous, L. E., and Q. D. Wheeler. 1981. The out-going comparison method of character analysis. Syst. Zool. 30: 1–11.

Webb, J. E. 1945. On the respiratory mechanism of *Melophagus ovinus* L. (Diptera). Proc. zool. Soc. Lond. 115: 218–250.

Wentzel, R. L. 1976. The streblid batflies of Venezuela (Diptera: Streblidae). Sci. Bull. Biol. Ser. Brigham Young Univ. 20: 1–77.

Westwood, J. O. 1840. Order 13. Diptera Aristotle (Antliata Fabricius. Halteriptera Clairv.). Pages 125–158 *in* Synopsis of the genera of British insects: *in* An introduction to the modern classification of insects. Vol. 2. A. Spottiswoode, London.

Wheeler, M. 1952. 11. The Drosophilidae of the Nearctic region, exclusive of the genus *Drosophila*. Univ. Tex. Publs 5204: 162–218.

Wheeler, M. R. 1954. *Rhinotora diversa* Giglio Tos from the southwestern United States (Diptera: Rhinotoridae). Wasmann J. Biol. 12: 35–39.

Wheeler, M. R. 1963. A note on some fossil Drosophilidae (Diptera) from the amber of Chiapas, Mexico. J. Paleont. 37: 123–124.

Wheeler, M. R. 1981*a*. 1. The Drosophilidae: A taxonomic overview. Pages 1–97 *in* M. Ashburner *et al.*, eds. The genetics and biology of *Drosophila*. Vol. 3a. Academic Press, New York, N.Y.

Wheeler, M. R. 1981*b*. 2. Geographical survey of Drosophilidae: Nearctic species. Pages 99–121 *in* M. Ashburner *et al.*, eds. The genetics and biology of *Drosophila*. Vol. 3a. Academic Press, New York, N.Y.

Whitten, J. M. 1957. The supposed pre-pupa in Cyclorrhaphous Diptera. Quart. J. micro. Sci., London 98: 241–263.

Wiedemann, C. R. W. 1824. Munus rectoris in Academia Christiana Albertina aditurus, Analecta entomologica ex Museo Regio Havniensi, etc., Kiliae (= Kiel).

Wiley, E. D. 1981. Phylogenetics. The theory and practice of phylogenetics systematics. John Wiley & Sons, New York, N.Y.

Williston, S. W. 1896. Manual of North American Diptera [2nd ed]. J. T. Hathaway, New Haven, Conn.

Wirth, W. W. 1955. Los insectos de les islas Juan Fernandez. 20. Ephydridae (Diptera). Revta. chil. Ent. 4: 51–72.

Wirth, W. W. 1968. 77. Family Ephydridae. Pages 77.1–43 *in* N. Papavero, ed. A catalogue of the Diptera of the Americas south of the United States. Departamento de Zoologica, Secretaria da Agricultura, São Paulo, Brasil.

Wood, D. M. 1979. Tachinidae and Rhinophoridae. Pages 418–19 *in* J. F. McAlpine *et al.* Chapt. 42. Diptera, pp. 389–424, *in* H. V. Danks ed. Canada and it's insect fauna. Mem. ent. Soc. Can. 108: 1–573.

Woodley, N. E. 1982. Two new species of *Neurochaeta* McAlpine (Diptera: Neurochaetidae), with notes on cladistic relationships within the genus. Mem. ent. Soc. Wash. 10: 211–218.

Zaka-ur-Rab, M. 1979. Morphology of the male genitalia of *Hippobosca maculata* L. (Diptera: Hippoboscidae). Zool. Anz. 203: 182–188.

Zetterstedt, J. W. 1837. Conspectus familiarum, generum et specierum, in fauna insectorum Lapponica descriptorum. Isis (Oken s) 1837: 28–67.

Zumpt, F. 1965. Myiasis in man and animals in the Old World. Butterworth & Co., London.

CORRECTIONS AND ADDENDA TO VOLUMES 1 AND 2

Volume 1

Chapter 4. Key to Families—Adults
p. 101, couplet 53*b* After "prong" insert "often"
p. 109, line 6 Insert "(continued on p. 114)"
p. 115, couplet 99*b* After "fewer" insert "more"

Volume 2

Authors. p. i The following two authors were inadvertedly omitted:
Kenneth G. V. Smith, M.I.Biol., F.Z.S., F.R.E.S.
Department of Entomology, British Museum (Natural History), Cromwell Road, London SW7 5BD, England
K. A. Spencer, B.A., DSc.
Exwell Farm, Bray Shop, Callington PL17 8QJ, Cornwall, England

Chapter 50. Platypezidae
p. 683, couplet 7*a* Insert parentheses around "Loew"

Chapter 52. Syrphidae
p. 737, couplet 101*a* Insert parentheses around "Fallén"

Chapter 66. Tephritidae
p. 828, couplet 45*b* Change "52" to "50"
p. 837, references, col. 2, line 8 Change "E. P. Vanzolini" to "P. E. Vanzolini"

Chapter 67. Richardiidae
p. 834, couplet 1*a* Epiplateinae is a new subfamily

Chapter 68. Pallopteridae
p. 842, couplet 2*a* Insert parentheses around "Johnson"

Chapter 73. Agromyzidae
p. 877, couplet 12*a* Insert parentheses around "Meigen"
p. 877, couplet 14*a* After "*galliivora*" insert parentheses around "Spencer"
p. 877, couplet 14*b* Change "*obscura* Spencer and Stagmaier" to "*obscura* Spencer"
p. 877, couplet 16*b* Change "*pleuralis* Melander" to "*pleuralis* (Malloch)"
p. 877, couplet 17*a* Insert parentheses around "Meigen"
p. 877, couplet 17*b* Insert parentheses around "Zetterstedt"

Chapter 77. Periscelididae
p. 898, col. 2, line 3 Change "Periscelidea" to "Periscelididea"

Chapter 80. Carnidae
p. 910, couplet 1*a* Change "Fig. 2" to "Fig. 3"

Chapter 84. Sciomyzidae
p. 934, couplet 20*a* Insert parentheses around "Walker"

Chapter 85. Ropalomeridae
p. 941, caption to Fig. 85.1 Change "*Ropolomera*" to "*Ropalomera*"

Chapter 87. Lauxaniidae
p. 953, col. 2, lines 1–2 Subfamily, Homoneurini was first used by Frey (1941, Ennumeratio Insectorum Fenniae pt. 6 (Diptera), p. 23), not Stuckenberg (1971)
p. 964, couplet 28*b* Change "(Figs. 32–38)" to "(as in Figs. 32–38)"

Chapter 90. Trixoscelididae
p. 982, col. 1, paragraph 2, line 15 Change "*Psiloplogia*" to "*Psiloplagia*"

Chapter 99. Chloropidae
p. 1065, couplet 48*b* Change "(Fig. 30)" to "(as in Fig. 30)"

Chapter 101. Tethinidae
p. 1075, couplet 2*a* The binomen *Masoniella richardsi* Vockeroth is a *nomen nudum*

Chapter 105. Muscidae
p. 1126, couplet 41*a* Change "*Neodexiopsis* Malloch" to "*Coenosia* (*Neodexiopsis* Malloch)"

Chapter 106. Calliphoridae
p. 1136, couplet 1*a* Change "Chrysomyiinae" to "Chrysomyinae"

Chapter 107. Oestridae
p. 1150, col. 1, paragraph 2, lines 7–8 Insert parentheses around "Linnaeus Jr."
p. 1152, col. 2, line 22 Change "hve" to "have"

Chapter 108. Sarcophagidae
p. 1168, couplet 15*a* *Neosarcophaga* Shewell is a *nomen nudum*
p. 1181, couplet 69*a* Change "*Erythandra*" to "*Erythrandra*"

Chapter 110. Tachinidae
p. 1220, footnote 36 Change "Polidiini" to "Polideini"

Chapter 113. Streblidae

p. 1297, col. 2, line 2 After "Wenzel" insert "in Wenzel"

Corrections and Addenda to Volume 1

p. 1305, col. 1, line 27 Change "Pelechorhynchidae" to "Pelecorhynchidae"

Index

p. 1308, col. 1 Change "*abominalis*" to "*abdominalis*"

p. 1311, col. 1 Change "Braula Nitzch" to "Braula Nitzsch"

p. 1318, col. 1 Change "*Hinei*" to "hinei"

p. 1322, col. 1 Change "*myopaeformis* Roder" to "myopaeformis Röder"

p. 1326, col. 1 Change "Polidiini" to "Polideini"

p. 1327, col. 1 Change "*Psiloplogia*" to "Psiloplagia"

p. 1330, col. 1 After "*sylvosus* (Williston)" insert parentheses around "*Gyroconops*"

p. 1330, col. 2 Change "Rohacek" to "Roháček"

p. 1330, col. 2 After *tertiariae* Hennig change "*Acartophthelmites*" to "*Acartophthalmites*"

INDEX

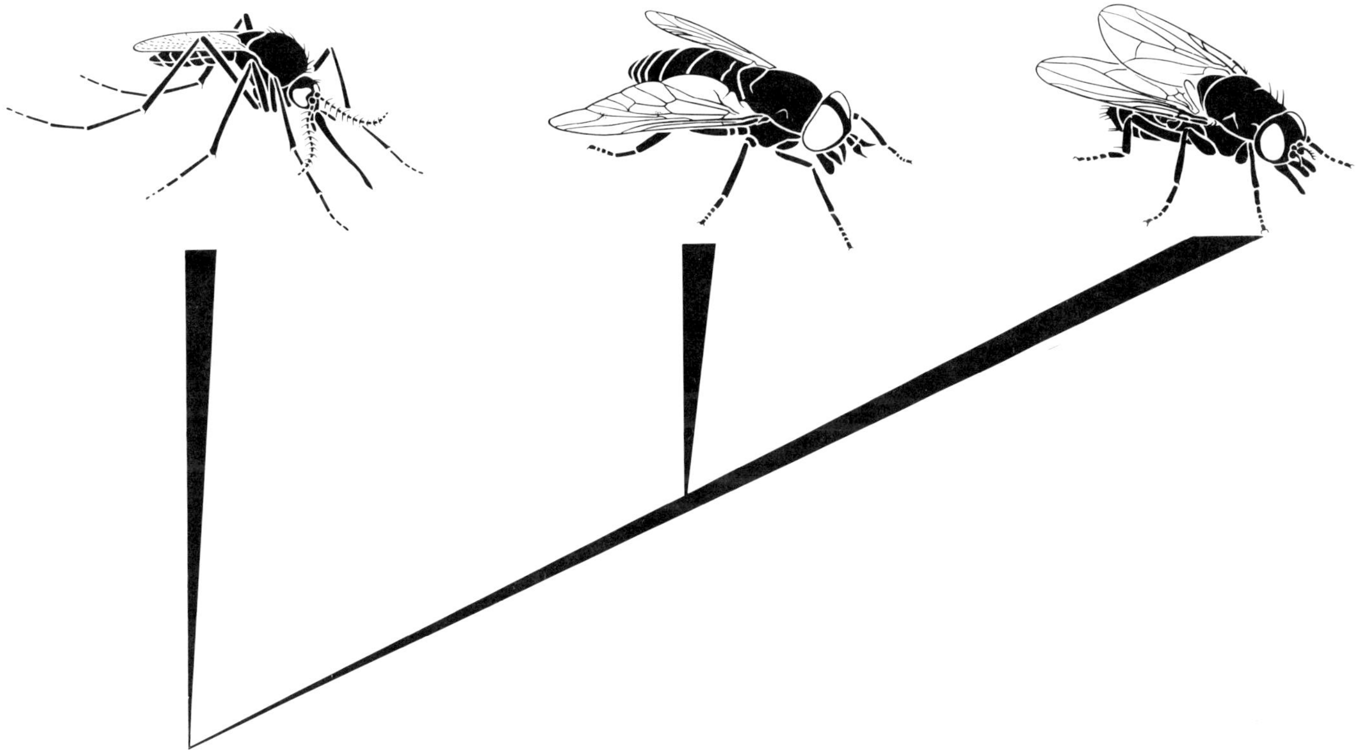

This composite index is restricted to the taxonomic names and morphological terms associated with Diptera that appear in Volumes 1, 2, and 3 of the Manual. Similar names and terms for plants and animals other than Diptera are excluded. Bibliographic references are not indexed.

Accepted taxonomic names for all categories (suborders, infraorders, superfamilies, families, subfamilies, tribes, genera, subgenera, species, and subspecies) are given in Roman type; synonyms are in italics. Every such citation throughout the Manual is listed. Author's names are provided for genera and species. Species are followed in parentheses by the names of the genera to which they are assigned. Subgeneric names are indexed in the same way as generic names; subspecies are indexed in the same way as species. Boldface page numbers indicate chapter subjects or designate where taxa appear in keys to genera; italicized numbers indicate the location of illustrations of these taxa.

Preferred morphological terms for adults and larvae are given in Roman type; synonyms are in italics. The plural spelling follows the singular if it is formed irregularly. Terms that apply only to larvae are designated by a capital "L" in parentheses. For practical reasons, page references provided for each morphological term are restricted to those places in the text where the term is first mentioned, where an explanation of its usage is given, or where its relationship is explained. Most of these citations occur in Chapters 2 and 3. Boldface page numbers indicate principal discussions of the terms. Italicized page numbers indicate locations of illustrations of the morphological features involved. Terms enclosed in single quotes are accepted for use in certain families but are not morphologically correct.